Serono Symposia, USA
Norwell, Massachusetts

FERTILIZATION IN MAMMALS
    Edited by Barry D. Bavister, Jim Cummins,
    and Eduardo R.S. Roldan

*FOLLICLE STIMULATING HORMONE: Regulation of Secretion
and Molecular Mechanisms of Action*
    Edited by Mary Hunzicker-Dunn and Neena B. Schwartz

GAMETE PHYSIOLOGY
    Edited by Ricardo H. Asch, Jose P. Balmaceda,
    and Ian Johnston

*GLYCOPROTEIN HORMONES: Structure, Synthesis, and
Biologic Function*
    Edited by William W. Chin and Irving Boime

GROWTH FACTORS IN REPRODUCTION
    Edited by David W. Schomberg

*THE MENOPAUSE: Biological and Clinical Consequences
of Ovarian Failure: Evaluation and Management*
    Edited by Stanley G. Korenman

MODES OF ACTION OF GnRH AND GnRH ANALOGS
    Edited by William F. Crowley, Jr., and P. Michael Conn

MOLECULAR BASIS OF REPRODUCTIVE ENDOCRINOLOGY
    Edited by Peter C.K. Leung, Aaron J.W. Hsueh,
    and Henry G. Friesen

NEUROENDOCRINE REGULATION OF REPRODUCTION
    Edited by Samuel S.C. Yen and Wylie W. Vale

PREIMPLANTATION EMBRYO DEVELOPMENT
    Edited by Barry D. Bavister

*SIGNALING MECHANISMS AND GENE EXPRESSION
IN THE OVARY*
    Edited by Geula Gibori

UTERINE CONTRACTILITY: Mechanisms of Control
    Edited by Robert E. Garfield

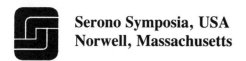

Serono Symposia, USA
Norwell, Massachusetts

Barry D. Bavister
Editor

Preimplantation Embryo Development

With 70 Figures

Springer-Verlag
New York Berlin Heidelberg London Paris
Tokyo Hong Kong Barcelona Budapest

Barry D. Bavister, Ph.D.
Department of Animal Health and Biomedical Sciences
University of Wisconsin–Madison
Madison, WI 53706
USA

Proceedings of the Symposium on Preimplantation Embryo Development, sponsored by Serono Symposia, USA, held August 15 to 18, 1991, in Newton, Massachusetts.

For information on previous volumes, please contact Serono Symposia, USA.

Library of Congress Cataloging-in-Publication Data
Preimplantation embryo development/Barry D. Bavister, editor.
    p. cm.
    "Proceedings of the Symposium on Preimplantation Embryo Development, sponsored by Serono Symposia, USA, held August 15 to 18, 1991, in Newton, Massachusetts"—T.p. verso.
    Includes bibliographical references and indexes.
    ISBN 0-387-97934-4.—ISBN 3-540-97934-4
    1. Human embryo—Physiology—Congresses. I. Bavister, Barry D.
II. Symposium on Preimplantation Embryo Development (1991: Newton, Mass.) III. Serono Symposia, USA.
    [DNLM: 1. Blastocyst—congresses. 2. Embryo—growth & development—congresses. 3. Mammals—embryology—congresses. 4. Preimplantation Phase—congresses. QS 604 P924 1991]
QP277.P74 1992
599'.0334—dc20
DNLM/DLC   92-49441

Printed on acid-free paper.

© 1993 Springer-Verlag New York, Inc.
All rights reserved. This work may not be translated or copied in whole or in part without the written permission of the publisher (Springer-Verlag New York, Inc., 175 Fifth Avenue, New York, NY 10010, USA), except for brief excerpts in connection with reviews or scholarly analysis. Use in connection with any form of information storage and retrieval, electronic adaptation, computer software, or by similar or dissimilar methodology now known or hereafter developed is forbidden.
The use of general descriptive names, trade names, trademarks, etc., in this publication, even if the former are not especially identified, is not to be taken as a sign that such names, as understood by the Trade Marks and Merchandise Marks Act, may accordingly be used freely by anyone.
While the advice and information in this book are believed to be true and accurate at the date of going to press, neither the authors, nor the editors, nor the publisher, nor Serono Symposia, USA, nor Serono Laboratories, Inc., can accept any legal responsibility for any errors or omissions that may be made. The publisher makes no warranty, expressed or implied, with respect to the material contained herein.
Permission to photocopy for internal or personal use, or the internal or personal use of specific clients, is granted by Springer-Verlag New York, Inc., for libraries registered with the Copyright Clearance Center (CCC), provided that the base fee of $5.00 per copy, plus $0.20 per page is paid directly to CCC, 21 Congress Street, Salem, MA 01970, USA. Special requests should be addressed directly to Springer-Verlag New York, Inc., 175 Fifth Avenue, New York, NY 10010, USA.

Production coordinated by Marilyn Morrison and managed by Francine McNeill; manufacturing supervised by Jacqui Ashri.
Typeset by Best-set Typesetter Ltd., Hong Kong.
Printed and bound by Edwards Brothers, Inc., Ann Arbor, MI.
Printed in the United States of America.

9 8 7 6 5 4 3 2 1

ISBN 0-387-97934-4 Springer-Verlag New York Berlin Heidelberg
ISBN 3-540-97934-4 Springer-Verlag Berlin Heidelberg New York

SYMPOSIUM ON PREIMPLANTATION EMBRYO DEVELOPMENT

Scientific Committee

    Barry D. Bavister, Ph.D., Chairman
    University of Wisconsin–Madison

    John Eppig, Ph.D.
    Jackson Laboratory

    Henry Leese, Ph.D.
    University of York

    Michael Roberts, Ph.D.
    University of Missouri

    Janet Rossant, Ph.D.
    Samuel Lunenfeld Research Institute

Organizing Secretaries

    Bruce K. Burnett, Ph.D.
    L. Lisa Kern, Ph.D.
    Serono Symposia, USA
    100 Longwater Circle
    Norwell, Massachusetts

Preface

This volume contains the Proceedings of the Serono Symposium on Preimplantation Embryo Development, held in Newton, Massachusetts, in 1991. The idea for the symposium grew out of the 1989 Serono Symposium on Fertilization in Mammals* at which preimplantation development was the predominant suggestion for a follow-up topic. This was indeed a timely subject in view of the recent resurgence of interest in this fundamental phase of embryogenesis and its relevance to basic research and applied fertility studies in humans, food-producing animals, and endangered species. The symposium brought together speakers from a broad range of disciplines in order to focus on key regulatory mechanisms in embryo development, using a wide variety of animal models, and on representative topics in human preimplantation embryogenesis.

The culmination of preimplantation development is a blastocyst containing the first differentiated embryonic tissues and capable of initiating and sustaining pregnancy. The central objective of the symposium was to throw light on the regulation of cellular and molecular events underlying blastocyst formation. It was particularly appropriate that the date of the symposium marked the 20th anniversary of the publication of the classic volume *Biology of the Blastocyst*, the proceedings of an international workshop held in 1970. This book, which summarized most of the information then available on this topic in mammals, was edited by the pioneer in blastocyst research, Dr. Richard Blandau, who was the guest speaker at the symposium. Using the symposium presentations as a retrospective overview of progress during the past two decades, it is clear that considerable advances in understanding regulatory mechanisms have been made in some key areas, such as oocyte maturation and the genetic control of embryo development, while much remains to be revealed in other areas, including relationships between the developing embryo and its external environment.

The presentation of the Symposium on Preimplantation Embryo Development was the culmination of considerable effort by many

---

*Bavister BD, Cummins J, Roldan ERS, eds. Fertilization in mammals. Norwell, MA: Serono Symposia, USA, 1990.

individuals. I am very grateful to the Scientific Committee, Drs. John Eppig, Henry Leese, Michael Roberts, and Janet Rossant, whose help was invaluable in selecting appropriate topics and speakers. I appreciate the assistance of other colleagues for their helpful suggestions. I want to thank all of the speakers for their thought-provoking contributions. I also thank the many individuals, including the members of the Scientific Committee, who reviewed the speakers' manuscripts and the abstracts of poster presentations. My appreciation is also extended to Helen Noeldner for her invaluable assistance with the editorial process. I am especially grateful to Dr. Bruce Burnett, Joan Jack, Dianne Ferreira, and Lois Svedeman of Serono Symposia, USA for their meticulous planning of every aspect of the program functions that ensured the smooth running of the symposium. Finally, I want again to express my thanks to Serono Symposia, USA, whose generous underwriting of the Symposium on Preimplantation Embryo Development made this event possible.

BARRY D. BAVISTER

Contents

Preface vii
Contributors xiii

Part I. Development of the Oocyte

1. Nuclear and Cytoplasmic Changes During Oocyte
   Maturation 3
   DAVID F. ALBERTINI, DINELI WICKRAMASINGHE,
   SUSAN MESSINGER, BRITTA A. MATTSON, AND
   CARLOS E. PLANCHA

2. Follicular Control of Meiotic Maturation in Mammalian
   Oocytes 22
   CATHERINE RACOWSKY

3. Translation Control in Oocytes: A Critical Role for the
   Poly(A) Tail of Maternal mRNAs 38
   MARCIA L. O'CONNELL, JOAQUIN HUARTE, DOMINIQUE BELIN,
   JEAN-DOMINIQUE VASSALLI, AND SIDNEY STRICKLAND

4. Establishment of Competence for Preimplantation
   Embryogenesis During Oogenesis in Mice 43
   JOHN J. EPPIG

Part II. Epigenetic Control of Embryo Development

5. Regulation of Hamster Embryo Development In Vitro by
   Amino Acids 57
   BARRY D. BAVISTER AND SUSAN H. MCKIERNAN

6. Energy Metabolism in Preimplantation Development 73
   HENRY J. LEESE

7. Growth Factors as Regulators of Mammalian
   Preimplantation Embryo Development 83
   BERND FISCHER

8. Intracellular pH Regulation by the Preimplantation Embryo 97
JAY M. BALTZ, JOHN D. BIGGERS, AND CLAUDE LECHENE

Part III. Genetics of Embryo Development

9. Activation of the Embryonic Genome: Comparisons Between Mouse and Bovine Development 115
ANDREW J. WATSON, AILEEN HOGAN, ANN HAHNEL, AND GILBERT A. SCHULTZ

10. Mutations Affecting Early Development in the Mouse 131
TERRY MAGNUSON, SHYAM K. SHARAN, AND BERNADETTE HOLDENER-KENNY

11. Parental Imprinting in Mammalian Development 144
ANNE C. FERGUSON-SMITH AND M. AZIM SURANI

12. Developmental Potential of Mouse Embryonic Stem Cells ... 157
JANET ROSSANT, ELIZABETH MERENTES-DIAZ, ELEN GOCZA, ESZTER IVANYI, AND ANDRAS NAGY

Part IV. Differentiation of the Embryo

13. Blastocyst Development and Growth: Role of Inositol and Citrate 169
M.T. KANE AND M.M. FAHY

14. Development of Human Blastocysts In Vitro 184
KATE HARDY

15. Development of Na/K ATPase Activity and Blastocoel Formation 200
CATHERINE S. GARDINER AND ALFRED R. MENINO, JR.

16. Effects of Imprinting on Early Development of Mouse Embryos 212
R.A. PEDERSEN, K.S. STURM, D.A. RAPPOLEE, AND Z. WERB

Part V. Embryo-Maternal Interactions

17. Uterine Secretory Activity and Embryo Development 229
R. MICHAEL ROBERTS, WILLIAM E. TROUT, NAGAPPAN MATHIALAGAN, MELODY STALLINGS-MANN, AND PING LING

18. In Vitro Models for Implantation of the Mammalian Embryo 244
S.J. KIMBER, R. WATERHOUSE, AND S. LINDENBERG

19. Chemical Signals in Embryo-Maternal Dialogue: Role of
    Growth Factors.................................... 264
    S.K. DEY AND B.C. PARIA

20. Regulation of Chorionic Gonadotropin Secretion by Cultured
    Human Blastocysts................................. 276
    ALEXANDER LOPATA AND KAREN OLIVA

Appendix: Poster Presentations 297
Author Index 339
Subject Index 341

Contributors

DAVID F. ALBERTINI, Tufts University Health Science Schools, Boston, Massachusetts, USA.

JAY M. BALTZ, Laboratory of Human Reproduction and Reproductive Biology and The Department of Cellular and Molecular Physiology, Harvard Medical School, Boston, Massachusetts, USA.

BARRY D. BAVISTER, Department of Animal Health and Biomedical Sciences, University of Wisconsin–Madison, Madison, Wisconsin, USA.

DOMINIQUE BELIN, Department of Morphology, University of Geneva Medical School, Geneva, Switzerland.

JOHN D. BIGGERS, Laboratory of Human Reproduction and Reproductive Biology and The Department of Cellular and Molecular Physiology, Harvard Medical School, Boston, Massachusetts, USA.

S.K. DEY, Department of Obstetrics–Gynecology and Physiology, University of Kansas Medical Center, Kansas City, Kansas, USA.

JOHN J. EPPIG, The Jackson Laboratory, Bar Harbor, Maine, USA.

M.M. FAHY, Division of Neuroscience, Oregon Regional Primate Research Center, Beaverton, Oregon, USA.

ANNE C. FERGUSON-SMITH, Department of Molecular Embryology, AFRC–Institute of Animal Physiology and Genetics Research, Babraham, Cambridge, England.

BERND FISCHER, Department of Anatomy and Reproductive Biology, Medical Faculty, RWTH Aachen, Klinikum Pauwelsstrasse, Aachen, Germany.

CATHERINE S. GARDINER, Animal Science Department, Southern Illinois University, Carbondale, Illinois, USA.

ELEN GOCZA, Department of Biochemistry, Lorand Eötvös University, H-2131 Göd, Javorka S. u. 14., Hungary.

ANN HAHNEL, Department of Medical Biochemistry, Health Sciences Centre, University of Calgary, Calgary, Alberta, Canada.

KATE HARDY, Institute of Obstetrics and Gynaecology, Royal Postgraduate Medical School, Hammersmith Hospital, London, United Kingdom.

AILEEN HOGAN, Department of Medical Biochemistry, Health Sciences Centre, University of Calgary, Calgary, Alberta, Canada.

BERNADETTE HOLDENER-KENNY, Department of Genetics, Case Western Reserve University, Cleveland, Ohio, USA.

JOAQUIN HUARTE, Department of Morphology, University of Geneva Medical School, Geneva, Switzerland.

ESZTER IVANYI, Department of Biochemistry, Lorand Eötvös University, H-2131 Göd, Javorka S. u. 14., Hungary.

M.T. KANE, Department of Physiology, University College, Galway, Ireland.

S.J. KIMBER, Department of Cell and Structural Biology, University of Manchester, Manchester, United Kingdom.

CLAUDE LECHENE, Laboratory of Cellular Physiology and Department of Medicine, Harvard Medical School and Brigham and Women's Hospital, Boston, Massachusetts, USA.

HENRY J. LEESE, Department of Biology, University of York, York, England.

S. LINDENBERG, Chromosome Laboratory, Rigshospitalet, Copenhagen, Denmark.

PING LING, Department of Animal Sciences, Animal Science Research Center, University of Missouri, Columbia, Missouri, USA.

ALEXANDER LOPATA, Department of Obstetrics and Gynaecology, The University of Melbourne, and Reproductive Biology Unit, Royal Women's Hospital, Carlton, Victoria, Australia.

TERRY MAGNUSON, Department of Genetics, Case Western Reserve University, Cleveland, Ohio, USA.

NAGAPPAN MATHIALAGAN, Department of Animal Sciences, Animal Science Research Center, University of Missouri, Columbia, Missouri, USA.

BRITTA A. MATTSON, Merck Sharp and Dohme Research Laboratories, West Point, Pennsylvania, USA.

SUSAN H. MCKIERNAN, Department of Animal Health and Biomedical Sciences, University of Wisconsin–Madison, Madison, Wisconsin, USA.

ALFRED R. MENINO, JR., Animal Science Department, Oregon State University, Corvallis, Oregon, USA.

ELIZABETH MERENTES-DIAZ, Division of Molecular and Developmental Biology, Samuel Lunenfeld Research Institute, Mount Sinai Hospital, Toronto, Canada, and Department of Biochemistry, Lorand Eötvös University, H-2131 Göd, Javorka S. u. 14., Hungary.

SUSAN MESSINGER, Tufts University Health Science Schools, Boston, Massachusetts, USA.

ANDRAS NAGY, Division of Molecular and Developmental Biology, Samuel Lunenfeld Research Institute, Mount Sinai Hospital, Toronto, Canada, and Department of Biochemistry, Lorand Eötvös University, H-2131 Göd, Javorka S. u. 14., Hungary.

MARCIA L. O'CONNELL, National Institutes of Health, Bethesda, Maryland, USA.

KAREN OLIVA, Department of Obstetrics and Gynaecology, The University of Melbourne, Parkville, Victoria, Australia.

B.C. PARIA, Department of Obstetrics–Gynecology and Physiology, University of Kansas Medical Center, Kansas City, Kansas, USA.

R.A. PEDERSEN, Laboratory of Radiobiology and Environmental Health, University of California, San Francisco, California, USA.

CARLOS E. PLANCHA, Faculdade de Medicina de Lisboa, Instituto de Histologia e Embriologia, 1699 Lisboa Codex, Portugal.

CATHERINE RACOWSKY, Department of Obstetrics/Gynecology, University of Arizona Health Sciences Center, University of Arizona, Tucson, Arizona, USA.

D.A. RAPPOLEE, Laboratory of Radiobiology and Environmental Health, University of California, San Francisco, California, USA.

R. MICHAEL ROBERTS, Departments of Animal Sciences and Biochemistry, Animal Science Research Center, University of Missouri, Columbia, Missouri, USA.

JANET ROSSANT, Division of Molecular and Developmental Biology, Samuel Lunenfeld Research Institute, Mount Sinai Hospital, Toronto, Canada.

GILBERT A. SCHULTZ, Department of Medical Biochemistry, Health Sciences Centre, University of Calgary, Calgary, Alberta, Canada.

SHYAM K. SHARAN, Department of Genetics, Case Western Reserve University, Cleveland, Ohio, USA.

MELODY STALLINGS-MANN, Department of Animal Sciences, Animal Science Research Center, University of Missouri, Columbia, Missouri, USA.

SIDNEY STRICKLAND, Department of Pharmacology, State University of New York at Stony Brook, Stony Brook, New York, USA.

K.S. STURM, Children's Medical Research Foundation, Camperdown, Australia.

M. AZIM SURANI, Department of Molecular Embryology, AFRC–Institute of Animal Physiology and Genetics Research, Babraham, Cambridge, England.

WILLIAM E. TROUT, Department of Animal Sciences, Animal Science Research Center, University of Missouri, Columbia, Missouri, USA.

JEAN-DOMINIQUE VASSALLI, Department of Morphology, University of Geneva Medical School, Geneva, Switzerland.

R. WATERHOUSE, Department of Cell and Structural Biology, University of Manchester, Manchester, United Kingdom.

ANDREW J. WATSON, Department of Medical Biochemistry, Health Sciences Centre, University of Calgary, Calgary, Alberta, Canada.

Z. WERB, Laboratory of Radiobiology and Environmental Health, University of California, San Francisco, California, USA.

DINELI WICKRAMASINGHE, Tufts University Health Science Schools, Boston, Massachusetts, USA.

Part I

Development of the Oocyte

1

Nuclear and Cytoplasmic Changes During Oocyte Maturation

DAVID F. ALBERTINI, DINELI WICKRAMASINGHE, SUSAN MESSINGER, BRITTA A. MATTSON, AND CARLOS E. PLANCHA

The success or failure of embryonic development is determined by the quality or maturational status of the gametes that participate in the formation of the conceptus. From a phylogenetic point of view, it is clear that while the genetic contribution of each gamete to the zygotic genome is a dominant goal in sexually reproducing animal forms, cytoplasmic contributions from the female gamete are vital to the early livelihood of the embryo (1). Studies of vertebrates and invertebrates on the mechanisms of maternal inheritance now illustrate with great clarity that both nutritional and informational molecules must be expressed and organized during the course of oogenesis in order for normal development to proceed upon fertilization (2). To what extent the developmental blueprint for embryogenesis is coupled with the process of meiosis itself is not at all clear, although insights into this interrelationship on a mechanistic level have derived from recent studies on the determinants of embryonic polarity evident at the terminal phases of oogenesis. One common expression of polarity in oocytes shared in diverse animal species is manifest in the asymmetric cleavage exhibited at the time of polar body extrusion, a process of dualistic function whereby nuclear reductive division occurs concomitant with the conservation of ooplasmic mass. Thus, the coupling of meiosis with unequal cytoplasmic partitioning necessarily sets the stage for the beginnings of embryogenesis. In conventional parlance, this terminal phase of oogenesis is referred to as *meiotic maturation* and must implicitly constitute a rather remarkable integration of both nuclear and cytoplasmic events in order to produce a viable and developmentally competent ovum.

This chapter seeks to define aspects of nuclear and cytoplasmic maturation that occur prior to and during the course of meiotic maturation by means of a series of studies that take advantage of advances in three areas of recent research on mammalian oocytes: (i) the developmental competence of oocytes grown and matured in culture (3), (ii) the application of fluorescence staining techniques that permit precise definition of nuclear and cytoplasmic components (4), and (iii) new findings on the mechanisms of cell cycle control (5). As a background to an analysis of meiotic maturation in the mammalian oocyte, we first review seminal findings on mitotic control mechanisms in somatic cells with respect to cytoplasmic and nuclear events.

Cell Cycle Control in Mitosis

Entry into and exit from the M-phase of the mitotic cell cycle is now known to be regulated by the formation and destruction of an M-phase kinase activity referred to as *mitosis promoting factor* (MPF) (5). MPF is a heterodimeric protein complex consisting of the p34^{cdc2} Ser/Thr kinase and cyclin, subserving, respectively, catalytic and regulatory roles in the M-phase complex that directly or indirectly alter the phosphorylation status of structural and enzymatic proteins critical to the progression of mitosis.

When somatic cells transit from G_2 of interphase to the M-phase, a characteristic series of morphological changes ensue in preparation for the karyokinetic and cytokinetic events of mitosis that include alterations in the cell surface, nucleus, and cytoplasm. These changes are summarized in Table 1.1. The most notable of these involve cellular shape, manifest as rounding and formation of numerous microvilli; chromatin condensation and nuclear envelope and lamina disassembly; and elaboration of the mitotic spindle apparatus. To elicit these changes in a timely and coordinated fashion at the onset of M-phase, the kinase activity associated with MPF is believed to act on specific protein substrates, causing their phosphorylation, which in turn results in semipermanent alteration of their assembly status and/or enzymatic activity. Examples of specific targets of MPF at M-phase entry are listed in Table 1.1 and include structural proteins such as histones, lamins, and vimentin, as well as enzymes involved in the modulation of microtubule assembly (MAP kinase) and acto-myosin-based contractility (myosin light chain kinase).

Several features of mitotic progression involving the MPF-dependent transformation in cellular architecture are important to note before comparing mitosis to the meiotic process in mammalian oocytes. First, mitotic progression is a semipermanent and reversible state, with the events leading to metaphase (e.g., chromatin condensation and nuclear envelope disassembly) being conversely deployed upon M-phase exit into the next interphase (e.g., chromatin decondensation and nuclear envelope reassembly). Mitotic events from prophase to telophase are rapid in somatic cells (<1h) and pivoted about the metaphase-anaphase transi-

TABLE 1.1. Structural modifications during mitotic progression.

Compartment/structure		Prophase	Metaphase/anaphase	Telophase	Putative M-phase kinase substrate
Nucleus	Nuclear envelope	Breakdown		Reformation	Lamin B*
	Nuclear lamina	Disassembly		Reassembly	Lamin A/C*
	Nucleolus	Breakdown		Reformation	Nucleolin*
	Chromatin	Condensation	Congression/alignment/segregation	Decondensation	Histone H1*
	Kinetochore	Activation	Centrosome splitting	Inactivation	?
Cytoplasm	Centrosome	Activation[a]	Dephosphorylation	Inactivation	?
	Microtubules	Spindle assembly	Elongation	Spindle disassembly	Tubulin, MAPs, MAP kinase
	Microfilaments (SF)	Disassembly		Reassembly	
	Intermediate filaments (IF)	Aggregation		Dispersion	Vimentin
	Golgi	Dispersion		Aggregation	?
	Mitochondria	Aggregation		Dispersion	Kinesin, dynein
Cell surface	Microvilli (#)	Increase		Decrease	?
	Shape	Spherical		Flat	?
	Endocytosis	Decrease		Increase	?
	Receptor mobility	Increase		Decrease	?
	Signal transduction	Uncoupled		Recoupled	G-proteins

Note: *Denotes known protein substrates for M-phase kinase; [a] refers to increased nucleation capacity due to centrosome phosphorylation.

tion, the point at which MPF rapidly disappears. Finally, cell division is definitively symmetric in outcome, ensuring equivalent partitioning of both cytoplasmic and nuclear components. It is now appropriate to consider aspects of meiotic progression in mammalian oocytes that bear directly on the nuclear and cytoplasmic changes required to effect and complete the maturation process.

Nuclear Events During Meiosis

The events associated with remodeling of the oocyte nucleus can be subdivided into those events involved with the cessation of transcriptional activity at the end of the growth phase of oogenesis and those involved with entry into M-phase at the onset of meiotic maturation. Much attention has been paid to changes in nucleolar function with regard to the former, and many studies have indicated that upon completion of the growth phase, ribosomal RNA production is halted or greatly diminished in various mammalian species (6). Attendant with this functional inactivation of the nucleolar compartment are a number of structural modifications in the nucleolus that occur subsequent to the formation of the antrum in the follicle of rodents.

For example, in the mouse ovarian follicle, heterochromatic foci located at the periphery of the nucleolus in oocytes of preantral follicles (Fig. 1.1A) undergo progressive spreading until in the late antral follicle, the oocyte nucleolus is completely enveloped by a heterochromatic shell (7) (Fig. 1.1B). This type of reorganization of perinucleolar chromatin has been noted to occur at later stages of folliculogenesis in other species and, interestingly, has been shown to be a useful indicator of the stage at which the developing oocyte acquires meiotic competence—that is, the ability to resume meiosis spontaneously under culture conditions (8, 9). These nucleolar transformations are readily recognized in mammalian oocytes using DNA-specific dyes, such as Hoechst 33258, as shown in Figure 1.1 (A, B, G, and H), and therefore facilitate the identification of important developmental transition states in fixed or living oocytes using fluorescence microscopy. When DNA-specific dyes are used in combination with other fluorescent probes, various structural and functional parameters of oogenesis can be monitored in an expedient and reproducible fashion relative to more traditional methods involving electron microscopy.

For example, counterstaining with the f-actin reactive probe rhodamine phalloidin permits the evaluation of changes in the cortical actin cytoskeleton during the growth phase of oogenesis. Growing mouse oocytes obtained after enzymatic dissociation of juvenile ovaries display heterogeneous patterns of cortical f-actin distribution related to the formation of

1. Nuclear and Cytoplasmic Changes During Oocyte Maturation   7

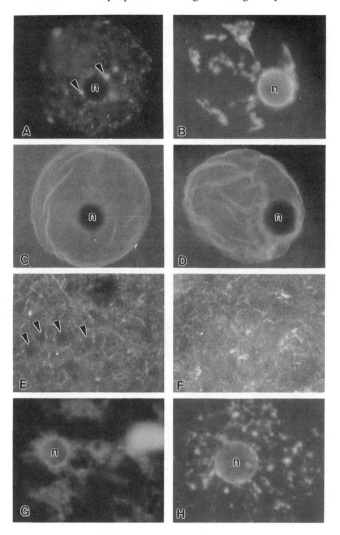

FIGURE 1.1. Fluorescence micrographs of mouse $(A-F)$, rhesus (G), and rabbit (H) germinal vesicle stage oocytes. A, B, G, and H depict DNA staining patterns in which different patterns of chromatin associated with the nucleolus (n) are depicted. Note perinucleolar chromatin granules in precompetent mouse oocyte (arrowheads, A) and envelopment of nucleoli with chromatin in B, G, and H. C and D depict lamin B disposition at the boundary of germinal vesicles for the same oocytes shown, respectively, in A and B. E and F represent cortical f-actin distribution in meiotically precompetent (E) and competent (F) mouse oocytes stained with rhodamine phalloidin. Note areas of diminished staining in E (arrowheads) $(A-F: 500\times; G$ and $H: 600\times)$.

microvillar and nonmicrovillar domains in oocytes from preantral follicles (Figs. 1.1E and 1.1F); oocytes from antral follicles develop a more uniform pattern of microvillus distribution as follicle development to the Graafian stage continues, suggesting that the final differentiation of the oocyte cortex occurs well after antrum formation during the period associated with differentiation of the cumulus.

Similar developmental studies in the mouse have addressed the question of how the relationship of the nuclear lamins with oocyte chromatin is modulated during both the growth phase of oogenesis and meiotic maturation. This work complements previous studies on changes in lamin organization in mouse oocytes prior to and following fertilization (10). *Lamins* are recognized as a subclass of intermediate filament proteins that mediate the binding of chromatin to the nuclear envelope. B-type lamins are thought to link chromatin to membrane by virtue of covalent lipid attachment, whereas A/C lamins are thought to form the filament meshwork of the nuclear lamina. Using guinea pig antibodies to rat lamins A/C or B and indirect immunofluorescence microscopy, the germinal vesicle of mouse oocytes appears to react equally at all developmental stages examined and displays homogeneous nuclear lamina staining at all portions of the nuclear periphery except where the nucleolus is positioned (Figs. 1.1C and 1.1D). This approach was also used to monitor the disposition of nuclear lamins during the course of meiotic maturation.

During diakinesis and early prometaphase of meiosis I, the onset of M-phase is highlighted by nucleolar dissolution and condensation of chromosomal bivalents (Figs. 1.2a to 1.2d). At these stages, the nuclear lamina remains intact as a complete perinuclear shell although distinct plications or infoldings are apparent (Figs. 1.2a to 1.2d). The irregular surface contour of the lamina is reminiscent of changes in the nuclear boundary observed by Calarco et al. (11), who noted complete breakdown of the nuclear envelope at this early stage of maturation. Remnants of the nuclear lamina were found to persist well into metaphase of meiosis I, and complete disappearance of this structure was not observed until anaphase (Figs. 1.2E, 1.2F, 1.2e, and 1.2f).

These results are somewhat surprising for two reasons. First, an intact nuclear lamina persists well after nuclear envelope breakdown, indicating that the regulation of the breakdown of these two structures is temporally dissociated. Second, in mitotic cells, nuclear envelope and lamina disassembly occur coordinately and abruptly in prophase at a time when MPF levels are rising sharply. It is doubtful that the latent stability of the nuclear lamina during M-phase onset in mouse oocytes is due to a deficiency in MPF levels overall at this time because so many other MPF-driven events are ongoing; perhaps the lamina is resistant to depolymerization due to unique meiotic requirements that are designed to facilitate the localization of components important to the assembly of the spindle at this critical juncture in meiotic progression.

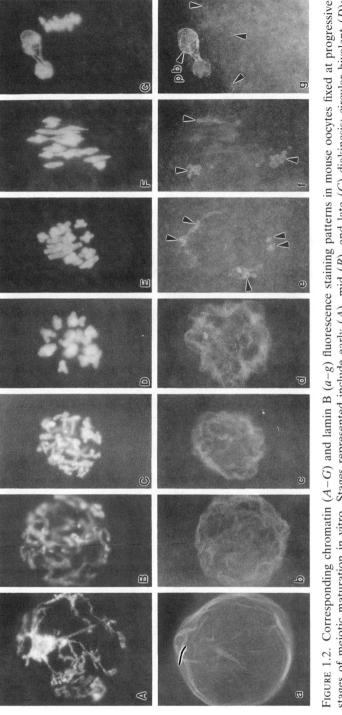

FIGURE 1.2. Corresponding chromatin (A–G) and lamin B (a–g) fluorescence staining patterns in mouse oocytes fixed at progressive stages of meiotic maturation in vitro. Stages represented include early (A), mid (B), and late (C) diakinesis; circular bivalent (D); prometaphase I (E); anaphase I (F); and metaphase II (G). Note collapse of nuclear lamina (a–d) during progressive stages of chromatin condensation (A–D) and distribution of lamin B containing vesicles and saccules at periphery of meiotic spindle (arrowheads, e and f). Reformation of nuclear lamina-like structure in first polar body (pb) is apparent in g; also, cytoplasm of metaphase II-stage oocyte contains many lamin B-bearing vesicular structures (arrowheads, g). Magnification is identical for all photographs (550×).

Meiotic Spindle Dynamics

The meiotic spindle assumes center stage in the events of maturation because of its pluralistic role in the organization and segregation of chromosomes and the positional demands it imparts on the site of polar body formation and architecture of the oocyte cytocortex. Drawing from reserves of cytoplasmic structural proteins like tubulin, the organizational capacity of spindle pole materials must regulate, in a timely and co-ordinated fashion, assembly and disassembly events for microtubules between varied interfaces, including kinetochores, the nuclear lamina, organelles, and the oocyte cortex. Since centrosomes serve these critical functions during mitotic and meiotic divisions—and are themselves targets of the M-phase kinase—it will be important to define precisely the compositional and structural aspects of this organelle's behavior in order to fully understand meiotic progression. How mammalian oocytes achieve these functions remains uncertain, especially in view of the apparent lack of centrioles during meiotic spindle formation. Ongoing studies have characterized elements of meiotic spindle behavior using a variety of immunological reagents in making comparisons between the assembly and dynamics of this structure during both meiosis I and II.

In most mammalian oocytes, the meiotic spindle exhibits two features that distinguish it from mitotic spindles, as shown in Figure 1.3. The poles of the spindle are distinctly flattened and lack astral microtubules, and one pole is intimately associated with the overlying cortex at the site of incipient polar body extrusion. While these properties are shared by oocytes from rodents, domestic animals, and primates (Figs. 1.3A to 1.3D), an additional feature of microtubule organization commonly seen in some species such as the mouse is the presence of multiple *microtubule organizing centers* (MTOCs) not associated with the spindle (Fig. 1.3E). The presence of MTOCs can be enhanced by treatment of mammalian oocytes with the tubulin assembly-promoting drug taxol and, as in untreated oocytes, MTOCs are found near the cortex displaced from the site of spindle attachment (Fig. 1.3F). Although these structures have been identified as oocyte centrosomes and are postulated to participate in spindle pole organization following fertilization (12, 13), recent studies indicate further that they undergo stage-specific oscillations in number throughout maturation and exhibit an M-phase specific phosphorylated epitope (14). Since microtubule assembly during M-phase is dynamically altered, it is of special interest to investigate posttranslational modifications of tubulin, *microtubule-associated proteins* (MAPs), and centrosomes with respect to microtubule turnover.

Tubulin is subject to a variety of posttranslational modifications, including tyrosination, acetylation, and phosphorylation, and it has been reported that spindle microtubules are posttranslationally modified in the metaphase II-arrested mouse oocyte (15). We have evaluated this

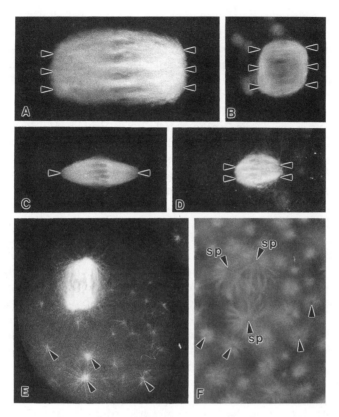

FIGURE 1.3. Antitubulin fluorescence staining patterns of mouse (*A* and *E*), goat (*B*), rat (*C* and *F*), and pig (*D*) oocytes fixed at metaphase of meiosis II (*A–D*) or during meiosis I (*E* and *F*). Figures *A–D* illustrate diversity in meiotic spindle pole organization from flat (*A* and *B*), tapered (*C*), and intermediate (*D*) forms (arrowheads). Note that in all cases the spindle pole is anastral. Multiple cytoplasmic microtubule organizing centers are shown in *E* (arrowheads) in a mouse oocyte fixed at early anaphase of meiosis I. A portion of a rat oocyte treated with 1.0-μM taxol at metaphase of meiosis I prior to fixation is shown in *F*; note that the meiotic spindle poles (sp) exhibit radiating microtubules, and numerous asters (arrowheads) appear in the cytoplasm in response to taxol (*A:* 1000×; *B*, *C*, and *D:* 600×; *E:* 450×; *F:* 800×).

problem during both meiosis I and II in mouse oocytes using antibodies that recognize specific tubulin epitopes. As shown in Figure 1.4, regional differences in the distribution of acetylated tubulin are found within what appear to be kinetochore-to-pole microtubules. Using digital imaging microscopy, we have topographically mapped such modifications along the length of the spindle relative to total tubulin patterns of staining. While confirming that heterogeneity exists with respect to acetylated and not to other posttranslational tubulin subtypes, these studies show further that a distinct polarity at one end of the spindle is present, possibly representing the site of spindle attachment (Fig. 1.4). To what extent these modifications result in differential stability properties that might distinguish metaphase in meiosis I or II remains to be determined, especially in view of the dynamic turnover of tubulin subunits noted to occur in meiotic spindles in mouse oocytes (16).

The factors regulating spindle pole organization and microtubule dynamics are likely to be integrated in order to achieve appropriate form and function of this organelle during maturation. These same factors must be expressed during oogenesis and must be subject to cell cycle cues exerted at the time of meiotic resumption. With this in mind, we evaluated physical parameters of spindle formation in growing mouse oocytes that had acquired meiotic competence.

Oocytes recovered from 15- to 30-day-old animals that resume meiosis in culture assemble spindles that reach a metaphase length (pole-to-pole distance) that is proportional to their developmental age (Fig. 1.5). Notably, spindle width was less variable over the age range studied, but the variations in length observed were comparable for metaphase of meiosis I or II.

This result suggests that a quantitative relationship exists with respect to the amount of polymer incorporated into the spindle of growing oocytes that may be modified by factors involved in the regulation of oogenesis. We therefore tested the influence of gonadotropin stimulation on juvenile mice to determine whether enhancement of growth would alter spindle size. Table 1.2 shows that in age-matched controls, a statistically significant increase in spindle size was achieved upon PMSG stimulation that presumably reflects both an enhancement of meiotic competence expression (9) and an increased pool size of the tubulin subunits that could become incorporated during assembly of the spindle. Why spindle size varies with respect to gonadotropin stimulation or animal age remains unclear; whether this response is due to differences in the size of the tubulin subunit pool or factors that organize the spindle poles is also unknown, but recent data support the latter interpretation.

Following staining for centrosomes, the actual number of centrosomes was found to fluctuate in GV stage mouse oocytes as a function of developmental state as induced by exogenous gonadotropin stimulation. Specifically, full-grown oocytes from unstimulated animals contain 4.4 ± 0.6 (SEM) centrosomes (n = 52), whereas oocytes from PMSG-primed animals contain an average of 7.5 ± 0.18 (SEM) per cell (n = 20).

Collectively, the data reviewed above indicate that the regulation of microtubule organization during both oogenesis and meiotic maturation is a complex multistep process involving critical translational and posttranslational modifications in tubulin, MAPs, and centrosomes that coordinately participate in spindle assembly. Further insight into the

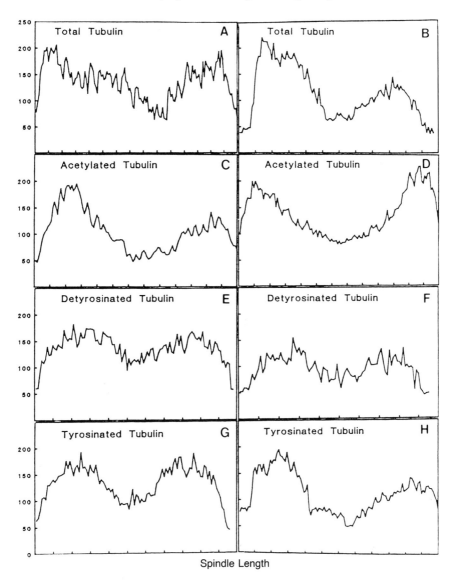

FIGURE 1.4. Densitometry scans of mouse meiotic spindles labeled with antibodies to tubulin (total B-specific, *A* and *B*) or posttranslationally modified variants (acetylated, *C* and *D*; detyrosinated, *E* and *F*; tyrosinated, *G* and *H*) using probes described in the text. Tracings for individual metaphase I (*A, C, E,* and *G*) or metaphase II (*B, D, F,* and *H*) spindles were derived from digitized images of spindles in which both poles were in the same optical plane; a single scan line was placed pole to pole (x-axis), and the gray level value (y-axis) of each pixel along the line was determined within a range of 0 (black) to 256 (white). Note distinctions in the polar density of the signal for acetylated tubulin in metaphase I (*C*) and in total (*B*) and tyrosinated (*H*) tubulin in metaphase II.

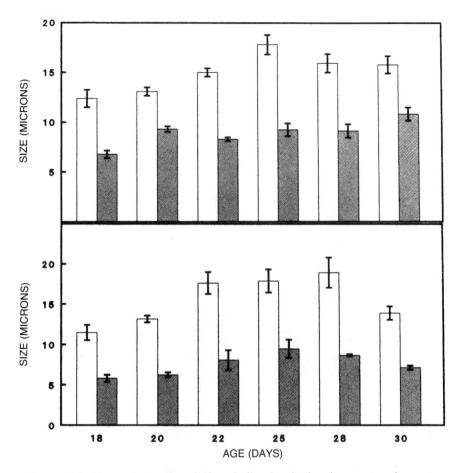

FIGURE 1.5. Comparison of meiotic spindle size in in vitro matured oocytes collected from ovaries of juvenile mice ranging in age from 18 to 30 days postpartum. The top graph represents metaphase I spindles, and the bottom graph represents metaphase II spindles; open bars show mean length, and shaded bars indicate mean width. Measurements were made from digitized images of oocytes processed with antitubulin antibodies, and indirect immunofluorescence staining was combined with Hoechst 33258 to determine precise meiotic stage.

mechanisms dictating the expression of cytoplasmic and nuclear events will require a detailed analysis of the factors controlling cell cycle transition states in order to fully understand how the oocyte achieves its normal, developmentally competent character.

TABLE 1.2. Effects of PMSG priming on meiotic spindle size.

Meiotic stage	Animal age (days)	N	Length (μm) ± SD	Width (μm)
Metaphase I				
−PMSG	22	40	15 ± 2.6	8 ± 2
+PMSG	23	10	34 ± 4.1	17 ± 2.4
Metaphase II				
−PMSG	22	7	18 ± 3.8	8 ± 3.5
+PMSG	23	10	34 ± 4.0	26 ± 4.5

Note: Oocytes were obtained by enzymatic dissociation, as described in reference 9, from 22-day-old unprimed animals (−PMSG [pregnant mare serum gonadotropin]) or by follicular puncture and retrieval of cumulus oocyte complexes from 23-day-old animals that had received 5-IU PMSG 48 h prior to oocyte collection (+PMSG). Cultures were fixed at 7 or 14 h and labeled with antitubulin and Hoechst 33258 to establish meiotic stage. Fluorescent images recorded on videotapes were digitized, and oocytes displaying parfocal spindle poles were used to make all measurements after magnification calibration with a stage micrometer. (N = number of oocytes measured; SD = standard deviation.)

Cell Cycle Mechanisms Governing Meiotic Maturation

As noted above, the systematic integration of patterns of gene expression during oogenesis with somatically derived cues controlling development of the ovarian follicle ultimately ensures the completion of nuclear and cytoplasmic maturation at the time of ovulation. While attention has been focused on specific oocyte components whose structural reorganization underlies the terminal maturative events in this long-term process, a growing body of evidence indicates that timely alterations in the expression of cell cycle control molecules, such as MPF, represent focal determinants in this process. The experimental tools are now in place for successfully integrating changes in oocyte cytoarchitecture with the developmental steps required for the formation of a viable embryo. The methodology is available for achieving maturation and fertilization of mammalian ova under defined culture conditions (3). Key cytological parameters can be monitored with the introduction of multiple labeling techniques (4, 7, 9).

In addition, the combination of these approaches will permit a thorough kinetic analysis of the translational and posttranslational requirements for the completion of meiotic progression at both the nuclear and cytoplasmic levels. From both practical and conceptual points of view, the most important gaps that need to be filled derive in part from three problematic areas: (i) How are the intrinsic mechanisms for cell cycle progression controlled within the oocyte? (ii) What pathways for interaction of somatic cells with the oocyte exist to regulate the multiple processes of competence acquisition, meiotic arrest, and meiotic resumption? and (iii) How is this interactive heterocellular system modulated by hormonal influences?

Intrinsic Mechanisms for Cell Cycle Progression

There is mounting evidence to suggest that during the course of oogenesis in mammals, both structural and enzymatic elements of cell cycle machinery are synthesized and at least partially organized in anticipation of the resumption of meiosis. The most compelling evidence favoring this idea derives from studies on the developmental expression of meiotic competence (reviewed in 9). In mouse oocytes, the ability to resume meiosis in culture has been used as an indicator of the "competence level" acquired as a function of progressive growth exhibited in oocytes obtained from juvenile animals. Coupled with the fact that GVBD and progression to the prometaphase-metaphase stage of meiosis I occurs under conditions in which protein synthesis is inhibited (17), the acquisition of meiotic competence must involve the activation of preformed stores of MPF sufficient to initiate and sustain early aspects of meiotic progression (18).

However, in the mouse there is additional evidence indicating that some processing of pre-MPF to MPF may occur even before overt meiotic resumption following oocyte removal from the follicle since clear-cut changes in centrosome phosphorylation, microtubule organization, and chromatin condensation indicative of a G_2/M-like transition occur at the time of competence acquisition in vivo (9). Since nuclear changes associated with this functional transition take place at the time of antrum formation in the mouse (7), and somewhat later in folliculogenesis in other mammals (6, 8), it seems likely that gonadotropin responsiveness and consequences of hormonal stimulation may additionally modulate the expression of cell cycle machinery in developing oocytes. Exactly what components within the oocyte are subject to regulation by MPF remain to be established, but several lines of evidence implicate the centrosome microtubule complex as a target for cell cycle control.

Centrosome phosphorylation occurs at M-phase entry in mitotic cells (19), and in vitro experiments with isolated centrosomes have demonstrated a direct correlation between their phosphorylation by cdc2 kinase and altered microtubule dynamics (20). Recently, the phosphorylation of both spindle pole and cytoplasmic centrosomes has been shown to occur during meiotic progression in mouse oocytes (14) at stages when MPF levels would be expected to be maximal (18). While the direct demonstration that centrosome activation represents an early locus of action for the M-phase kinase in triggering meiotic resumption awaits further investigation, this notion receives support from studies illustrating a protein phosphorylation and/or phosphatase requirement for GVBD in mouse oocytes (21–23). One further aspect of centrosome behavior related to the seminal role this structure plays in spindle assembly concerns the developmental expression of centrosomes during oogenesis.

As mentioned earlier, meiotically competent oocytes derived from PMSG-primed juvenile mice contain multiple cytoplasmic centrosomes

(6-8/cell), whereas competent oocytes from unprimed age-matched animals display fewer cytoplasmic centrosomes/cell. As shown in Table 1.2, smaller meiotic spindles are formed in oocytes from unprimed animals relative to oocytes obtained from gonadotropin-stimulated animals. Moreover, a correlation exists between the spindle size attained at metaphase and animal age, suggesting that larger spindles are assembled in more developmentally advanced oocytes (Fig. 1.5).

Spindle mass, as approximated by staining methods used in the present studies, may be limited by the number and/or quantity of available centrosomes that upon interaction with chromosomes establish the width and length characteristics of the anastral and acentriolar spindle poles seen in mouse oocytes (24). If this were true, then the number of centrosomes generated during oogenesis might determine the size of the meiotic spindle formed during meiotic progression, assuming that all centrosomes are equivalently activated following GVBD and that mechanisms exist for providing assembly of this material at each pole, a function presumably subserved by chromosomes. While our data are consistent with the idea that centrosomal mass influences spindle size, these results also point out that in the mouse centrosomal mass may be related to gonadotropin-dependent processes ongoing within the preovulatory follicle. How gonadotropins might control the expression and/or functional organization of oocyte centrosomes remains unknown, but is deserving of further attention.

One final perplexing aspect of centrosome behavior in mammalian oocytes concerns the origin and function of non-spindle-associated centrosomes that thus far have been observed in the mouse (12-14), hamster (Plancha, Albertini, unpublished observation), goat, and rabbit (Albertini, unpublished observation). These structures are commonly found in the oocyte cortex, are excluded from the meiotic spindle, and their numbers oscillate during meiotic progression in the mouse (14). Unlike the mouse, where GV stage oocytes contain distinct cytoplasmic centrosomes, preliminary studies on the hamster and rabbit indicate that cytoplasmic centrosomes do not appear until GVBD has occurred (Plancha, Albertini, unpublished observations).

Thus, species differences may exist with respect to the formation of cytoplasmic centrosomes during the resumption of meiosis, just as differences are known to exist with regard to the protein synthesis requirements for meiotic resumption in various mammals. It is presumptuous at this point to suggest that the metabolic determinants for resuming meiosis involve centrosomes per se in view of complex interrelationships that exist between translational and posttranslational activities in the generation of active MPF. In addition, more information is needed from mammalian model systems other than rodents to fully assess the relationship between MPF and centrosomes in the initiation of maturation. However, the persistence of centrosomes throughout maturation (14) and into the early stages of embryonic development (12, 13) suggests that these structures are maternally inherited by the embryo and may, therefore, play a vital role in establishing developmental potential of the egg (1, 2).

Somatic and Gametic Interactions

The coordinate expression of nuclear and cytoplasmic maturation occurs normally within the context of ovulation when the gametic alterations needed to establish fertilizability and developmental potential of the oocyte become manifest through the actions of hormones on the somatic elements of the ovarian follicle. It is now widely recognized that communication between the oocyte and surrounding granulosa cells is vital throughout the process of oogenesis, and much attention has been paid to the role of gap junctions in mediating the regulative interactions within the follicle (3 and Chapter 2, this volume). However, it is also becoming clear that at the interface between the oocyte and follicle cells, soluble factors originating in either of these cell types also participate in the processes of oocyte maturation and cumulus expansion through as yet unresolved mechanisms. Moreover, very little is known regarding the physicochemical properties of the zona pellucida with respect to either the transport and signaling potential of soluble factors or the physical interconnections of cumulus cells with the oocyte. Thus, in order to understand more fully how the cell cycle machinery of the oocyte is modulated by the somatic cells of the follicle, it will be necessary to define in more precise biochemical terms the signal transduction elements at the level of both cell types. Despite our ignorance in this area, data reviewed below strongly indicate that the hormonal milieu of the follicle at ovulation plays a critical role in the penultimate maturative events of oogenesis.

Hormonal Modulation of Maturation

In addition to the completion of nuclear maturation, the oocyte must acquire certain key cytoplasmic properties during its final maturation to ensure, among other things, that an effective block to polyspermy can be mounted upon fertilization and that sperm head decondensation occurs on schedule (25). It has long been recognized that the coordinate expression of nuclear and cytoplasmic maturation in most species requires a correctly timed exposure of the follicle to LH since the developmental capability of the ovum is compromised when, for example, LH (hCG) is administered too early (26). In fact, even under in vitro conditions, exogenous LH has been shown to accelerate the onset of meiotic maturation and result in improved fertilizability (27). Thus, it is not surprising that recent successes in in vitro fertilization of in vitro matured oocytes have been attributed to the inclusion of exogenous steroid and peptide hormones in media (28). However, the central questions that remain for further improvements in assisted reproductive technologies and gamete manipulation are manifold and have yet to be addressed at a mechanistic level with respect to exactly how the hormonal milieu created in vivo or in vitro influences the developmental capability of the oocyte (29).

It is our contention that the major determinants of meiotic maturation in the mammalian oocyte involve critical translational and posttranslational modifications in the M-phase kinase that are realized through a coordinated series of changes in the centrosome-microtubule complex. Furthermore, steroid and peptide hormones, through as yet unresolved mechanisms, are likely to modulate the patterns of expression of key kinases and phosphatases that may differentially influence nuclear and cytoplasmic components during the maturation process. Temporal congruity ensuring coordinate nuclear and cytoplasmic maturation is likely to be achieved when an appropriate hormonal milieu is created at the time of ovulation in order to synchronize activation of the cell cycle machinery. Such a mechanism and the experimental paradigm it involves explain, in part, distinctions in cell cycle progression expressed by different species, as well as differences noted previously with respect to the developmental potential of oocytes matured under in vivo versus in vitro conditions (30, 31).

Acknowledgments. We thank John Kilmartin, Larry Gerace, David Asai, and Chloe Bulinski for their generous gifts of antibodies used in various aspects of this work and Nancita R. Lomax of the Drug Synthesis and Chemistry Branch, Division of Cancer Treatment, National Cancer Institute, for kindly providing taxol. Mary Currier provided much-needed assistance in preparation of the manuscript. This work has been jointly supported by grants from the NIH (HD-20068), the Pharmaceutical Manufacturers Association Foundation, Inc., and the March of Dimes (15-91-68). This chapter is dedicated to the memory of Allen Schroeder for his intellectual and technical contributions to the field of mammalian oogenesis.

References

1. Wilson EB. The cell in development and heredity. New York: Macmillan, 1928.
2. Melton DA. Pattern formation during animal development. Science 1991; 252:234–40.
3. Buccione R, Schroeder AC, Eppig JJ. Interactions between somatic cells and germ cells throughout mammalian oogenesis. Biol Reprod 1990;43:543–7.

4. Albertini DF. Novel morphological approaches for the study of oocyte maturation. Biol Reprod 1984;30:13–28.
5. Lewin B. Driving the cell cycle: M phase kinase, its partners, and substrates. Cell 1990;61:743–52.
6. Daguet MC. In vivo change in the germinal vesicle of the sow oocyte during the follicular phase before the ovulatory LH surge. Reprod Nutr Dev 1980; 20:673–80.
7. Mattson BA, Albertini DF. Oogenesis: chromatin and microtubule dynamics during meiotic prophase. Mol Reprod Dev 1990;25:374–83.
8. McGaughey RW, Montgomery DH, Richter JD. Germinal vesicle configurations, and patterns of protein synthesis of porcine oocytes from antral follicles of different size as related to their competency for spontaneous maturation. J Exp Zool 1979;209:239–54.
9. Wickramasinghe D, Ebert KM, Albertini DF. Meiotic competence acquisition is associated with the appearance of M-phase characteristics in growing mouse oocytes. Dev Biol 1991;143:162–72.
10. Schatten G, Maul G, Schatten H, et al. Nuclear lamins and peripheral nuclear antigens during fertilization and embryogenesis in mice and sea urchins. Proc Natl Acad Sci USA 1985;82:4727–31.
11. Calarco PG, Donahue RP, Szollosi D. Germinal vesicle breakdown in the mouse oocyte. J Cell Sci 1972;10:369–85.
12. Maro B, Howlett SK, Webb M. Non-spindle microtubule organizing centers in metaphase-II-arrested mouse oocytes. J Cell Biol 1985;101:1665–72.
13. Schatten G, Simerly C, Schatten H. Maternal inheritance of centrosomes in mammals? Studies on parthenogenesis and polyspermy in mice. Proc Natl Acad Sci USA 1991;88:6785–9.
14. Messinger SM, Albertini DF. Centrosome and microtubule dynamics during meiotic progression in the mouse oocyte. J Cell Sci 1991;100:289–98.
15. de Pennart H, Houliston E, Maro B. Post-translational modifications of tubulin and the dynamics of microtubules in mouse oocytes and zygotes. Biol Cell 1988;64:374–8.
16. Gorbsky GJ, Simerly C, Schatten G, Borisy GG. Microtubules in the metaphase-arrested mouse oocyte turn over rapidly. Proc Natl Acad Sci USA 1990;87:6049–53.
17. Downs SM. Protein synthesis inhibitors prevent both spontaneous and hormone-dependent maturation of isolated mouse oocytes. Mol Reprod Dev 1990;27:235–43.
18. Hashimoto N, Kishimoto T. Regulation of meiotic metaphase by a cytoplasmic maturation-promoting factor during mouse oocyte maturation. Dev Biol 1988;126:242–52.
19. Centonze VE, Borisy GG. Nucleation of microtubules from mitotic centrosomes is modulated by a phosphorylated epitope. J Cell Sci 1990; 95:405–11.
20. Verde F, Labbe J, Doree M, Karsenti E. Regulation of microtubule dynamics by cdc^2 protein kinase in cell-free extracts of *Xenopus* eggs. Nature (London) 1990;343:233–8.
21. Motlik J, Rimkevicova Z. Combined effects of protein synthesis and phosphorylation inhibitors on maturation of mouse oocytes in vitro. Mol Reprod Dev 1990;27:230–4.

22. Gavin A-C, Tsukitani Y, Schorderet-Slatkine S. Induction of M-phase entry of prophase-blocked mouse oocytes through microinjection of okadaic acid, a specific phosphatase inhibitor. Exp Cell Res 1991;192:75–81.
23. Szollosi M, Debey P, Szollosi D, Rime H, Vautier D. Chromatin behavior under influence of puromycin and 6-DMAP at different stages of mouse oocyte maturation. Chromosoma 1991;100:339–54.
24. Szollosi D, Calarco P, Donahue RP. Absence of centrioles in the first and second meiotic spindles of mouse oocytes. J Cell Sci 1972;11:521–41.
25. Cran DG, Moor RM. Programming the oocyte for fertilization. In: Bavister BD, Cummins J, Roldan ERS, eds. Fertilization in mammals. Norwell, MA: Serono Symposia, USA, 1990:35–47.
26. Hunter RHF, Cook B, Baker TG. Dissociation of response to injected gonadotropin between the Graafian follicle and oocyte in pigs. Nature (London) 1976;260:156–7.
27. Kaplan R, Dekel N, Kraicer PF. Acceleration of onset of oocyte maturation in vitro by luteinizing hormone. Gamete Res 1978;1:59–63.
28. Thibault C, Szollosi D, Gerard M. Mammalian oocyte maturation. Reprod Nutr Dev 1987;27:865–96.
29. Thibault C. Are follicular maturation and oocyte maturation independent processes? J Reprod Fertil 1977;51:1–15.
30. Morgan PM, Warikoo PK, Bavister BD. In vitro maturation of ovarian oocytes from unstimulated rhesus monkeys: assessment of cytoplasmic maturity by embryonic development after in vitro fertilization. Biol Reprod 1991;45:89–93.
31. Younis AI, Zuelke KA, Harper KM, Oliveira MAL, Brackett BG. In vitro fertilization of goat oocytes. Biol Reprod 1991;44:1177–82.

2

Follicular Control of Meiotic Maturation in Mammalian Oocytes

CATHERINE RACOWSKY

In female mammals, the meiotic process is initiated during the fetal period, but in the majority of species it is interrupted in the diplotene stage of prophase I either immediately before or shortly after birth (1, 2). Following the onset of puberty, this process resumes in a species-specific number of mature selected follicles during each reproductive cycle (reviewed in 3). Thus, female germ cells spend most of their lives in the meiotically arrested, so-called dictyate or *germinal vesicle* (GV) state of development and spend a relatively short period undergoing progression through the later stages of meiosis. During this protracted period of meiotic arrest, however, oocyte growth and development ensue under stringent control by the follicle cells to which the oocyte is structurally (4–7) and metabolically (8, 9) coupled. Indeed, it is through somatic regulation, which in turn is controlled by follicle cell interactions with gonadotropins, steroids, and various intrafollicular molecules (e.g., cyclic nucleotides and purines), that oocyte meiotic competency and, ultimately, nuclear maturation are achieved.

Classically defined, the term *nuclear maturation* encompasses the events of chromatin condensation, *germinal vesicle breakdown* (GVBD), and progression of the dictyate oocyte through to metaphase of the second meiotic division (10, 11). However, it is now well established that as the mammalian oocyte is reprogrammed to resume meiosis, it undergoes a period of irreversible commitment that precedes any overt changes in chromatin configuration. Therefore, the term *meiotic maturation* must be expanded to encompass the event of irreversible commitment, and any study investigating the regulation of meiotic maturation should use onset of irreversible commitment as the reference point for meiotic resumption. In this chapter, the possible physiological mechanisms underlying such regulation are considered, with particular attention given to the roles of follicular tissue and follicular fluid and to the participation of *cyclic adenosine monophosphate* (cAMP), purines, and steroids in this process.

Follicular Control of Oocyte Meiotic Arrest

Over 50 years ago, Pincus and Enzmann (12) observed that immature rabbit oocytes would spontaneously mature in culture following removal from their follicular environment. This original observation, now confirmed in a plethora of studies with oocytes of all mammalian species so far examined (reviewed in 13), has led to general acceptance that meiotic arrest in mammalian oocytes is maintained by a follicular *arrester*. The arrester is believed to be of granulosa cell origin since cocultures of cumulus-enclosed oocytes with membrana granulosa cells (14–16), media in which granulosa cells were cultured (17), and extracts of membrana granulosa cells (18) all exhibited an arresting influence on follicle-free oocytes.

From a physiological standpoint, a membrana granulosa meiotic arrester could manifest its effect in two possible ways. First, following its secretion into follicular fluid, it may be taken up directly through the oolemma and/or indirectly via the adherent cumulus cells (reviewed in 19). Indeed, consistent with the original observation of Chang in 1955 (20) that rabbit follicular fluid suppressed spontaneous meiotic resumption in rabbit oocytes, several studies have confirmed a meiotic arresting activity in the follicular fluid of other species (21–23). However, other laboratories have failed to obtain such corroborating results (24–26), and a recent in vivo study in golden (Syrian) hamsters revealed that meiotic arrest was maintained only transiently following the transfer and incubation of oocytes in antral cavities of mature follicles (27). Taken together, these findings have given rise to considerable controversy regarding the physiological relevance of a follicular fluid meiotic arrester and indicate that meiotic arrest is unlikely to be maintained solely by a follicular fluid arrester.

The second possible way that the arrester may be transmitted to the oocyte is by transfer through the numerous gap junctions known to connect the membrana granulosa cells (4, 5, 28), the cumulus cells (6, 29), and the innermost layer of cumulus granulosa cells with the oolemma (4, 6, 7). This possibility is supported by the findings that while meiotic arrest is maintained when oocyte-cumulus complexes are grafted to experimentally opened follicles (30), meiotic resumption occurs if such graft formation is prevented. Moreover, in studies with explanted follicles in which the oocyte-cumulus complex was dislodged from the underlying membrana granulosa cells, released into the antrum, and subsequently allowed to reestablish contact during incubation, there was a significant increase in the proportion of GV-stage oocytes as the extent of recontact increased (27). Collectively, these results argue in favor of a stringent control of oocyte meiotic status by the follicle cell/oocyte syncitium.

Regulation of Meiotic Maturation: A Decrease in the Meiotic Arrester or an Induction of a Meiotic Stimulator?

While it is well established that exposure of the mature antral follicle to LH provides the physiological trigger for entry of the oocyte into the period of irreversible commitment, the mechanisms that mediate this gonadotropic action remain ill-defined. While several hypotheses have been advanced, we have invested considerable energy testing two hypotheses in particular. First, LH causes a physical disruption of the follicular gap junctional network, which thereby disrupts heterologous transfer of the meiotic arrester, cAMP, into the oocyte (31, 32). Second, the gonadotropin-induced change in follicular steroid profile is correlated with commitment of the oocyte to undergo meiotic resumption, as is the case in such nonmammalian species as the frog (33) and starfish (34). Thus, the first hypothesis proposes that LH effects a decrease in the intraoocyte level of cAMP below the threshold required for maintenance of meiotic arrest, while the second proposes that a decrease in follicular estradiol, accompanied by an increase in follicular progesterone, stimulates meiotic resumption.

As discussed below, not only are data available to support each of these hypotheses, but also it is becoming increasingly apparent that fundamental differences may exist among different species regarding the meiotic control mechanisms involved. Thus, regarding the role of cAMP, what may hold true for the mouse and rat probably does not pertain to the golden hamster and is almost certainly not applicable to such species as the pig and cow. Nevertheless, as far as the role of steroids is concerned, data exist to support the possibility that follicular steroid dynamics play a central and similar role in the regulation of meiotic maturation in at least hamster and pig oocytes.

Follicular Gap Junction Integrity and Oocyte Meiotic Status

Attention has focused particularly on the kinetics of disruption of the functional gap junctional network following the preovulatory surge of gonadotropin and on how such disruption impacts upon the intraoocyte level of cAMP. While it is quite clear from fluorescent dye (5, 35) and freeze fracture (29) studies that exposure of the mature follicle to LH ultimately leads to disruption of this network concomitant with cumulus mass expansion, inconsistent data exist regarding the timing of its disruption as it relates to release of the oocyte from meiotic arrest (reviewed in 36). The extent of intercellular communication within the oocyte-cumulus complex has been studied using metabolic coupling assays involving [^{3}H]-

uridine or [^{3}H]-choline and freeze fracture analysis. While the majority of studies using the metabolic coupling assays have failed to reveal a reduction in oocyte-cumulus cell coupling until after GVBD (37–40), one study using such a metabolic coupling assay (35) has indicated a tight temporal relationship between oocyte meiotic status and down-regulation of follicular granulosa cell gap junctions in vivo. In addition, using freeze fracture electron microscopy, a dramatic reduction was revealed in the net area of gap junction membrane, both in cumulus membrane prior to GVBD in rat oocytes (29) and in membrana granulosa membrane during irreversible commitment in hamster oocytes (36). Available data using the direct approach therefore support the hypothesis that LH causes a decrease in granulosa cell gap junctional membrane before meiotic resumption occurs and are consistent with the proposition that reduction in the transfer of a granulosa cell arrester into the oocyte may participate in the regulation of this process.

cAMP and Oocyte Meiotic Status

Following the original observation of Cho et al. (41) that the cyclic nucleotide derivative $N^6,O^{2'}$-dibutyryl cAMP (dbcAMP) prevents spontaneous meiotic maturation in mouse oocytes, numerous studies with various modulators of intracellular cAMP (cAMP phosphodiesterase inhibitors [42, 43], forskolin [44–47], and choleratoxin [31]) in a wide variety of mammalian species have supported the possibility that cAMP is a universal regulator of oocyte meiotic status among mammals. However, conflicting results obtained from laboratories working with different species have led to considerable controversy as to whether species variation exists, whether cAMP is the sole meiotic regulator, and whether the cyclic nucleotide originates from the oocyte itself or from its associated cumulus mass. Attention has focused particularly on attempts to correlate oocyte meiotic status with both intraoocyte cAMP levels and the extent of heterologous coupling in oocyte-cumulus complexes (35).

In order to address possible interspecies variation regarding the role of cAMP in the meiotic process, a comparative investigation was undertaken with oocyte-cumulus complexes of rat, hamster, and pig in which spontaneous maturation for all three species occurred under close to identical conditions. The only difference was that Hanks' salts were substituted for Earle's salts in the hamster cultures since hamster oocytes do not mature spontaneously in Earle's-based medium (46). Oocyte meiotic commitment was taken as the criterion for meiotic status, and separate determinations of cAMP were made in oocytes and cumuli of the same complexes. Oocyte-cumulus coupling was assessed using the [^{3}H]-uridine metabolic coupling assay.

As shown in Figure 2.1, the time for 50% of oocytes to undergo irreversible commitment (t_{50}) was 1 h, 1.25 h, and 5 h for rat, hamster,

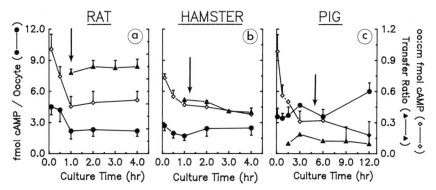

FIGURE 2.1. cAMP content and heterologous coupling during spontaneous meiotic maturation in oocyte-cumulus complexes of rat (*a*), hamster (*b*), and pig (*c*). The data are presented as the mean ± SEM of 6 (rat), 8 (hamster), and 8 (pig) replicates, respectively. The same group of oocyte-cumulus complexes were used for the separate determinations of cAMP within the oocytes (oo) and cumuli (cm). The transfer ratio represents the proportion of radiolabeled uridine marker within a group of oocytes of that present in the corresponding group of cumulus masses. The arrow indicates the time at which 50%-cultured oocytes underwent irreversible commitment to maturation.

and pig oocytes, respectively, indicating interspecies variation with respect to this parameter. In addition, while the t_{50} was preceded by a significant decrease in the intraoocyte cAMP content in the rat (Fig. 2.1a), the level of cAMP in hamster and pig oocytes failed to drop significantly during this or, indeed, any period of culture (Figs. 2.1b and 2.1c, respectively). Nevertheless, there was a significant decrease in the ratio of cAMP in the oocyte:cumulus (oo:cm fmol) during the precommitment period in all three species, which was followed by a much-reduced rate of change during the subsequent culture period. Interestingly, heterologous coupling remained constant through meiotic commitment in the rat oocyte-cumulus complex, while that for the hamster significantly decreased with time ($r = -0.58$, $P < 0.01$) and that for the pig underwent a significant fluctuation during the first 6 h, which was followed by a progressive and significant decrease during the subsequent 9 h ($r = -0.93$, $P < 0.001$).

Collectively, these results indicate that at least within the constraints of the system and the assays employed, there are striking differences among rat, hamster, and pig oocytes regarding the relationships among intraoocyte cAMP levels, heterologous coupling, and meiotic commitment. In the rat, and consistent with observations in the mouse (48), irreversible commitment to meiotic maturation is accompanied by a significant decrease in the oocyte content of cAMP, suggesting that a minimum threshold level of intraoocyte cAMP may be required for maintenance of meiotic arrest. However, the results with hamster and pig show that as in the sheep (49), it is unlikely that a drop in oocyte cAMP triggers meiotic commitment in these oocytes. Taken together, these results suggest that at least in some species, a factor(s) in addition to cAMP is involved in the regulation of meiotic status. As discussed below, possible candidates for an arrester of maturation include one or more purines and/or extradiol, while a candidate for a stimulator of maturation may be progesterone, as shown in the frog (33).

Purines and Oocyte Meiotic Status

Pioneer work regarding a possible role for purines in the regulation of meiotic arrest was undertaken by Eppig, Downs, and their coworkers in the mid-1980s. Their research revealed that various purine compounds are present in mouse (50) and pig (51) follicular fluid that are capable of transiently maintaining meiotic arrest in follicle-free murine oocytes. One of the most active of these follicular purines, hypoxanthine, has been shown to synergize dramatically with the arresting action of cAMP (51) and, at concentrations comparable to those present in murine follicular fluid, to synergize with adenosine in the inhibition of spontaneous meiotic maturation in mouse oocytes (50). It has been suggested that these purines may regulate meiotic status in mouse oocytes by their action on endogenous cAMP levels (50), with hypoxanthine serving as an inhibitor of cAMP phosphodiesterase (52) and phosphophorylated adenosine serving as a substrate for adenylate cyclase. However, since preovulatory LH fails to induce significant decreases in murine intrafollicular levels of either of these purines until after the occurrence of GVBD (50), an LH-induced perturbation of the metabolism, rather than the removal, of hypoxanthine is implicated.

Indeed, several observations support a mediating role for guanyl compounds in hypoxanthine-maintained meiotic arrest. First, guanosine and guanine are more potent meiotic arresters in mouse oocytes than either hypoxanthine or adenosine (51). Second, guanosine is more effective than either hypoxanthine or adenosine in inhibiting cAMP phosphodiesterase activity in mouse oocyte lysates (52). Third, meiotic resumption occurs following exposure of follicle-free cultured oocytes to inhibitors of the enzyme that mediates conversion of hypoxanthine to guanyl compounds (53, 54) or after intraperitoneal injection of such inhibitors to hormonally primed immature mice (55).

Studies with oocytes of other mammalian species have been conducted to investigate the universality of the meiotic-arresting capabilities of purine bases and nucleosides. Indeed, guanosine and hypoxanthine inhibit meiotic maturation in rat oocytes (56), and hypoxanthine maintains meiotic arrest in oocytes of rhesus monkey (57) and cow (58). However, collectively, these studies show that oocytes from different species exhibit different sensitivities to these compounds, although the variety of media employed might account for these differences.

We have recently examined this possibility by exposing hamster cumulus-enclosed and cumulus-free oocytes to hypoxanthine and various other purine bases and nucleosides in a medium identical to that used in all the purine mouse studies conducted by Eppig and Downs. As reported for mouse (50), cumulus-enclosed hamster oocytes were also more sensitive to hypoxanthine than cumulus-free oocytes (ID_{50} = 5.1 mM and 11.4 mM, respectively, where ID_{50} is the dose that inhibited 50% of oocytes from undergoing spontaneous maturation). Interestingly, however, both types of hamster oocytes were much less sensitive to the purine base than their murine counterparts (ID_{50} = 3.37 mM and 4.37 mM, for cumulus-enclosed and cumulus-free mouse oocytes, respectively [50]). In addition, in comparative studies of the relative efficacies of various purine bases and nucleosides to maintain meiotic arrest of cumulus-enclosed hamster oocytes, adenine, adenosine, xanthine, and xanthosine all significantly increased the percentage of oocytes at the GV stage. In contrast, none of these compounds at the same concentrations as used for the hamster study exhibited any significant arresting capability following a 3-h incubation with mouse oocytes (51). These comparative data with hamster and mouse oocytes provide additional support for the concept that the mechanisms that regulate meiotic maturation vary among different mammalian species. In addition, while purines may play some regulatory role in the meiotic process in all mammals, it seems probable that the relative significance of any one of these compounds varies among species.

Steroids and Oocyte Meiotic Status

Since the follicular steroidogenic environment undergoes a dramatic change following the preovulatory surge of LH, many studies have investigated the possible role of steroids in the regulation of meiotic maturation in mammals (reviewed in 58). Since LH induces a decline in estradiol synthesis with a concomitant stimulation in the synthesis of progesterone (59, 60), attention has been particularly paid to possible arresting and stimulating functions for estradiol and progesterone, respectively.

A direct action of estradiol on the spontaneous maturation of cumulus-free oocytes of a number of species has been demonstrated (cow [61], mouse [62, 63], pig [64, 65], and hamster [66]). Furthermore, testosterone has been shown to potentiate both dbcAMP-dependent arrest in mouse oocytes (67, 68) and FSH-dependent arrest in pig oocytes (69). Since granulosa cells aromatize testosterone when stimulated by cAMP (70), it has been suggested that estradiol might mediate these androgenic effects (71). Such a possibility is supported by the finding that cumulus-enclosed pig oocytes cultured in the presence of FSH and testosterone

TABLE 2.1. Relationship between follicular steroid levels and oocyte meiotic commitment in PMSG-hCG-primed golden hamsters.

Follicular compartment	Steroid	n	r-value	P-value
Tissue	E_2	12	−0.712	<0.010
	T	12	0.057	NS
	P_4	12	0.685	<0.020
	$E_2:P_4$	12	−0.782	<0.010
	$E_2:T + P_4$	12	−0.839	<0.001
Fluid	E_2	11	0.396	NS
	T	11	0.135	NS
	P_4	11	0.806	<0.010
	$E_2:P_4$	10	−0.835	<0.001
	$E_2:T + P_4$	10	−0.799	<0.010

undergo GVBD concomitantly with a reduction in the ratio of estradiol: progesterone secreted by the cumulus mass (69).

Collectively, the above studies suggest that estradiol may play a role in modulating cAMP-dependent maintenance of meiotic arrest. Furthermore, the studies with pig oocyte-cumulus complexes exposed to testosterone and FSH suggest that progesterone may play a role in the stimulation of meiotic resumption. Therefore, we have recently undertaken an in vivo study with golden hamsters to identify the relationship between the prevailing follicular steroidogenic environment and the commitment status of the oocyte. For this study, *pregnant mare serum gonadotropin* (PMSG) -primed hamsters were sacrificed at 0.5-h intervals following hCG injection, and dissected follicles were used either for determination of oocyte meiotic commitment status or for assay of steroid content in follicular tissue and in follicular fluid. In this model, meiotic commitment occurs between 2.0 and 2.5 h post-hCG.

Consistent with previous observations (60), preovulatory gonadotropin was found to induce an early and dramatic shift in follicular steroidogenic activity (Table 2.1 and unpublished data). Prior to commitment (between 0 and 1.5 h post-hCG), estradiol levels gradually, but insignificantly, decreased in follicular tissue, but significantly increased in follicular fluid ($P < 0.01$, Newman-Keuls). In contrast, both tissue and fluid testosterone levels significantly increased within 0.5 h post-hCG ($P < 0.01$ in both cases), with those in fluid continuing to increase during the subsequent hour. The kinetics for tissue and fluid progesterone were similar, with progressive rises in both compartments occurring during the period prior to meiotic commitment. While progesterone levels in tissue and fluid and estradiol levels in fluid were maintained during the period of commitment (2.0–2.5 h post-hCG), testosterone levels in tissue and fluid and estradiol levels in tissue declined. During commitment (2.0–2.5 h post-hCG), however, significant decreases in tissue estradiol and testosterone occurred

($P < 0.01$ for both steroids, Newman-Keuls, comparing 2.0 and 2.5 h values) concomitant with a rise, albeit insignificant, in tissue progesterone. In contrast, the fluid levels of both progesterone and estradiol were maintained during this period, while that for testosterone underwent a significant decline ($P < 0.01$).

Analyses of these follicular steroid changes with respect to meiotic commitment revealed that the proportion of oocytes committed to mature was positively correlated with both tissue and fluid progesterone levels, but negatively correlated with tissue estradiol levels (Table 2.1). While no significant correlations between commitment and either tissue or fluid testosterone levels were revealed when all the data were analyzed together, interestingly, a significant positive correlation existed between testosterone levels and commitment during the very early stages of commitment, when 1%–30% of oocytes became meiotically committed. In addition, significant negative correlations were revealed between the proportion of oocytes committed to mature and the ratios of estradiol:progesterone and estradiol:progesterone + testosterone for both follicular tissue and fluid ($P < 0.01$ and $P < 0.001$ for estradiol:progesterone in tissue and fluid, respectively; and $P < 0.001$ and 0.01 for estradiol:progesterone + testosterone in tissue and fluid, respectively).

From a temporal perspective, therefore, the results of the above in vivo steroid/meiotic commitment study are consistent with the possibility that follicular steroidal dynamics may mediate the action of LH to induce meiotic resumption. Failure of the progesterone synthesis inhibitor cyanoketone to interfere with LH-induction of GVBD in explanted rat follicles (72) is not necessarily inconsistent with this proposition. Since cyanoketone binds specifically and irreversibly to 3β-hydroxysteroid dehydrogenase, it inhibits the synthesis of both progesterone and estradiol (73). In addition, the compound has been found to inhibit binding of estradiol to its receptor protein in a competitive manner (74). Thus, the occurrence of GVBD in the presence of cyanoketone in this system may be attributable to an inhibitory effect of cyanoketone on the synthesis and/or the action of estradiol, rather than to an interference with progesterone action. Furthermore, the possible role of estradiol in maintenance of meiotic arrest is not necessarily negated by the lack of an effect of microinjected estradiol upon LH stimulation of maturation in explanted hamster follicles (58). Any elevation in testosterone and/or progesterone may be sufficient to induce meiotic commitment. That progesterone blocked spontaneous GVBD in cumulus-free (63) and cumulus-enclosed (75) mouse oocytes, however, shows that progesterone has meiotic-arresting activity in the extrafollicular environment and that induction of meiotic commitment by progesterone in vivo must require the presence of preovulatory gonadotropin and/or must be mediated by some follicular component that is absent from the oocyte-cumulus system.

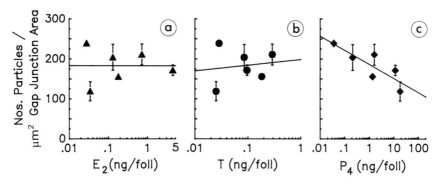

FIGURE 2.2. Relationship between particle density in reticulate regions of gap junctions in hamster membrana granulosa cells and follicular tissue levels of E_2 (a), T (b), and P_4 (c).

The mechanism of action of steroids in regulating oocyte meiotic status remains to be determined. However, since numerous studies indicate that steroids are capable of regulating both the distribution and the integrity of gap junctions (reviewed in 27), the possibility exists that regulation of follicular gap junctional integrity is involved. There is considerable evidence that estradiol up-regulates (76, 77) and progesterone down-regulates (78) gap junctions and that testosterone causes a decrease in gap junction function in androgen-sensitive motorneurons (79). Thus, progesterone and/or testosterone may mediate the action of preovulatory gonadotropin to stimulate meiotic resumption by antagonizing estradiol maintenance of gap junction integrity. Such a possibility is supported by a recent analysis of gap junction particle density in hamster membrana granulosa cells as this relates to follicular steroidal content. The results show that while no correlations existed between any of the tissue steroid levels and the density of particles in nonreticulate gap junctional areas (data not shown) or between estradiol or testosterone and reticulate areas (Figs. 2.2a and 2.2b), progesterone was negatively correlated with particle density in reticulate areas ($P < 0.05$) (Fig. 2.2c). Furthermore, a positive correlation was revealed between this parameter and the ratio of estradiol:progesterone ($P < 0.05$) that was further enhanced upon regressing the data against the ratio of estradiol:testosterone + progesterone ($P < 0.01$) (Fig. 2.3).

One consequence of such testosterone and progesterone antagonism of estradiol up-regulation of gap junction integrity would be an overall disruption in the transfer of cAMP through the follicular syncitium to the oocyte. However, in the majority of mammalian species, including the golden hamster (Fig. 2.1), cAMP levels do not fall significantly prior to GVBD (reviewed in 21). These seemingly contradictory facts may be reconciled by the established LH-induced, dramatic, and very rapid

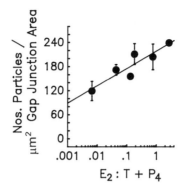

FIGURE 2.3. Relationship between particle density in reticulate regions of gap junctions in hamster membrana granulosa cells and follicular $E_2:T + P_4$ levels.

elevation in endogenous follicular cAMP levels (80, 81) that may precede follicular gap junctional down-regulation to result in increased heterologous transfer of cAMP to the oocyte prior to gap junction down-regulation. Gap junction down-regulation in the follicular syncitium would serve to isolate the oocyte-cumulus complex from the membrana granulosa cells that constitute the larger population of follicular somatic cells. The possibility exists, therefore, that such down-regulation results in an over-all decrease in estradiol in the vicinity of the oocyte below the threshold level required for maintenance of meiotic arrest.

Conclusions

The mammalian oocyte is contained within a dynamic follicular environment that is comprised of a large variety of molecules, any of which may play either independent or interactive roles in the regulation of meiotic maturation. While this chapter has considered research regarding the roles of cAMP, purines, and steroids, it is possible that other follicular components mediate in the meiotic-stimulating activity of LH. Nevertheless, the data reviewed are consistent with a universal role for cAMP in the maintenance of meiotic arrest. However, the mouse and rat are the only species so far examined in which a decrease in the level of intra-oocyte cAMP accompanies meiotic commitment. Therefore, in the majority of species, an additional molecule(s) must be implicated in the regulation of meiotic status.

Steroids may be included in the group of potential meiotic modulators. Results obtained from in vivo experiments using the golden hamster are consistent with a synergism between estradiol and cAMP in the maintenance of meiotic arrest and a role for both testosterone and progesterone in the stimulation of meiotic resumption. The mechanisms underlying such steroidogenic control remain to be determined, although the data show that progesterone effects a rapid decrease in granulosa cell gap junction particle density in response to preovulatory gonadotropin. Such down-regulation may isolate the oocyte-cumulus complex from the follicular syncitium, thereby reducing the intraoocyte level of estradiol below the level required for maintenance of meiotic arrest.

Acknowledgments. The author thanks Larry Nienaber for breeding the hamsters and Milissa Kaufman for assistance in preparing the manuscript.

References

1. Borum K. Oogenesis in the mouse: a study of the meiotic prophase. Exp Cell Res 1961;24:495–507.
2. Baker TG. The control of oogenesis in mammals. In: Midgley AR Jr, Sadler WA, eds. Ovarian follicular development and function. New York: Raven Press, 1979:353–64.
3. Racowsky C. Gamete resources: origin and production of oocytes. In: Pedersen RA, McLaren A, First NL, eds. Animal applications of research in mammalian development. New York: Cold Spring Harbor Laboratory Press, 1991:23–82.
4. Albertini DF, Anderson E. The appearance and structure of intercellular connections during the ontogeny of the rabbit ovarian follicle with particular reference to gap junctions. J Cell Biol 1974;63:234–50.
5. Gilula NB, Epstein ML, Beers WH. Cell-to-cell communication and ovulation: a study of the oocyte-cumulus complex. J Cell Biol 1978;78:58–75.
6. Larsen WJ, Wert SE. Roles of cell gap junctions in gametogenesis and in early embryonic development. Tissue Cell 1988;20:809–48.
7. Amsterdam A, Josephs R, Lieberman ME, Lindner HR. Organization of intramembrane particles in freeze-cleaved gap junctions of rat Graafian follicles: optical diffraction analysis. J Cell Sci 1976;21:93–105.
8. Moor RM, Smith MW, Dawson RMC. Measurement of intercellular coupling between oocyte and cumulus cells using intracellular markers. Exp Cell Res 1980;16:15–29.
9. Herlands RL, Schultz RM. Regulation of mouse oocyte growth: probable nutritional role for intercellular communication between follicle cells and oocytes in oocyte growth. J Exp Zool 1984;229:317–25.
10. Donahue RP. Maturation of the mouse oocyte in vitro, I. Sequence and timing of nuclear progression. J Exp Zool 1968:237–50.
11. Donahue RP. The relation of oocyte maturation to ovulation in mammals. In: Biggers JD, Schuetz AW, eds. Oogenesis. Baltimore: University Park Press, 1972:413–38.
12. Pincus G, Enzmann EV. The comparative behaviour of mammalian eggs in vivo and in vitro. J Exp Med 1935;62:665–75.
13. Thibault C, Szollosi D, Gerard M. Mammalian oocyte maturation. Reprod Nutr Dev 1987;27:865–96.

14. Tsafriri A, Channing CP. Influence of follicular maturation and culture conditions on the meiosis of pig oocytes in vitro. J Reprod Fertil 1975;43:149–52.
15. Sato E, Ishibashi T. Meiotic arresting action of the substance obtained from the surface of porcine ovarian granulosa cells. Jpn J Zootech 1975;48:22–6.
16. Sirard MA, Bilodeau S. Granulosa cells inhibit the resumption of meiosis in bovine oocytes in vitro. Biol Reprod 1990;43:777–83.
17. Tsafriri A. Oocyte maturation in mammals. In: Jones JR, ed. The vertebrate ovary: comparative biology and evolution. New York: Plenum Press, 1978: 409–42.
18. Tsafriri A, Pomerantz SH, Channing CP. Porcine follicular fluid inhibitor of oocyte meiosis: partial characteristics of the inhibitor. Biol Reprod 1978;14:511–6.
19. Eppig JJ, Downs SM. Chemical signals that regulate mammalian oocyte maturation. Biol Reprod 1984;30:1–12.
20. Chang MC. The maturation of rabbit oocytes in culture and their maturation, activation, fertilization, and subsequent development in the fallopian tube. J Exp Zool 1955;128:378–99.
21. Tsafriri A, Channing CP, Pomerantz SH, Lindner HR. Inhibition of maturation of isolated rat oocytes by porcine follicular fluid. J Endocrinol 1977;75:285–91.
22. Jagiello G, Graffeo J, Ducayen M, Prosser R. Further studies of inhibitors of in vitro mammalian oocyte maturation. Fertil Steril 1977;28:476–81.
23. Gwatkin RBL, Anderson OF. Hamster oocyte maturation in vitro: inhibition by follicular components. Life Sci 1976;19:527–36.
24. Leibfried L, First NL. Effect of bovine and porcine follicular fluid and granulosa cells on maturation of oocytes in vitro. Biol Reprod 1980;23:699–704.
25. Racowsky C, McGaughey RW. Further studies of the effects of follicular fluid and membrana granulosa cells on the spontaneous maturation of pig oocytes. J Reprod Fertil 1982;66:505–12.
26. Fleming AD, Kahil W, Armstrong DT. Porcine follicular fluid does not inhibit maturation of rat oocytes in vitro. J Reprod Fertil 1983;69:665–70.
27. Racowsky C, Baldwin KV. In vitro and in vivo studies reveal that hamster oocyte meiotic arrest is maintained only transiently by follicular fluid, but persistently by membrana/cumulus granulosa cell contact. Dev Biol 1989;134:297–306.
28. Larsen WJ, Wert SE, Brunner GD. Differential modulation of rat follicle cell gap junction populations at ovulation. Dev Biol 1987;123:61–71.
29. Larsen WJ, Wert SE, Brunner GD. A dramatic loss of cumulus cell gap junctions is correlated with germinal vesicle breakdown in rat oocytes. Dev Biol 1986;113:517–21.
30. Liebfried L, First NL. Follicular control of meiosis in the porcine oocyte. Biol Reprod 1980;23:705–9.
31. Dekel N, Beers WH. Rat oocyte maturation in vitro: relief of cyclic AMP inhibition by gonadotropins. Proc Natl Acad Sci USA 1978;75:4369–73.
32. Dekel N, Beers WH. Development of the rat oocyte in vitro: inhibition of maturation in the presence or absence of the cumulus oophorus. Dev Biol 1980;75:247–54.

33. Schuetz AW. Action of hormones in germinal vesicle breakdown in frog (*Rana pipiens*) oocytes. J Exp Zool 1967;166:347–54.
34. Kanatani H, Shirai H, Nakanishi K, Kurokawa T. Isolation and identification of meiosis-inducing substance in starfish *Asterias anurensis*. Nature (London) 1969;211:273–7.
35. Racowsky C, Satterlie RA. Metabolic, fluorescent dye and electrical coupling between hamster oocytes and cumulus cells during meiotic maturation in vivo and in vitro. Dev Biol 1985;108:191–202.
36. Racowsky C, Baldwin KV, Larabell CA, DeMarais AA, Kazilek CJ. Down-regulation of membrana granulosa cell gap junctions is correlated with irreversible commitment to resume meiosis in golden Syrian hamster oocytes. Eur J Cell Biol 1989;49:244–51.
37. Moor RM, Osborn JC, Cran DG, Walters DE. Selective effect of gonadotropins on cell coupling, nuclear maturation and protein synthesis in mammalian oocytes. J Embryol Exp Morphol 1981;61:347–65.
38. Eppig JJ. The relationship between cumulus-oocyte coupling, oocyte meiotic maturation, and cumulus expansion. Dev Biol 1982;89:268–72.
39. Salustri A, Siracusa G. Metabolic coupling, cumulus expansion and meiotic resumption in mouse cumuli oophori cultured in vitro in the presence of FSH or dbcAMP, or stimulated in vivo by hCG. J Reprod Fertil 1983;68:335–41.
40. Eppig JJ, Downs SM. Gonadotropin-induced murine oocyte maturation in vivo is not associated with decreased cyclic adenosine monophosphate in the oocyte-cumulus complex. Gamete Res 1988;20:125–31.
41. Cho WK, Stern S, Biggers JD. Inhibitory effect of dibutyryl cAMP on mouse oocytes maturation in vivo. J Exp Zool 1974;187:383–6.
42. Wassarman PM, Josefowicz WJ, Letourneau GE. Meiotic maturation of mouse oocytes in vitro: inhibition of maturation at specific stages of nuclear progression. J Cell Sci 1976;22:531–45.
43. Magnusson C, Hillensjo T. Inhibition of maturation and metabolism of rat oocytes by cyclic AMP. J Exp Zool 1977;201:138–47.
44. Olsiewski PJ, Beers WH. cAMP synthesis in the rat oocyte. Dev Biol 1983;100:287–93.
45. Racowsky C. Effect of forskolin on the spontaneous maturation and cyclic AMP content of rat oocyte-cumulus complexes. J Reprod Fertil 1984;72:107–16.
46. Racowsky C. Effect of forskolin on the spontaneous maturation and cyclic AMP content of hamster oocyte-cumulus complexes. J Exp Zool 1985;234:87–96.
47. Racowsky C. Effect of forskolin on maintenance of meiotic arrest and stimulation of cumulus expansion, progesterone and cyclic AMP production by pig oocyte-cumulus complexes. J Reprod Fertil 1985;74:9–21.
48. Schultz RM, Montgomery RR, Belanoff JR. Regulation of mouse oocyte meiotic maturation: implication of a decrease in oocyte cAMP and protein dephosphorylation in commitment to resume meiosis. Dev Biol 1983;97:264–73.
49. Moor RM, Heslop JP. Cyclic AMP in mammalian follicle cells and oocytes during maturation. J Exp Zool 1981;216:205–9.

50. Eppig JJ, Ward-Bailey PF, Coleman DL. Hypoxanthine and adenosine in murine ovarian follicular fluid: concentrations and activity in maintaining oocyte meiotic arrest. Biol Reprod 1985;33:1041-9.
51. Downs SM, Coleman DL, Ward-Bailey PF, Eppig JJ. Hypoxanthine is the principal inhibitor of murine oocyte maturation in a low molecular weight fraction of porcine follicular fluid. Proc Natl Acad Sci USA 1985;82:454-8.
52. Downs SM, Daniel SAJ, Bornslaeger EA, Hoppe PC, Eppig JJ. Maintenance of meiotic arrest in mouse oocytes by purines: modulation of cAMP levels and cAMP phosphodiesterase activity. Gamete Res 1989;23:323-34.
53. Downs SM, Coleman DL, Eppig JJ. Maintenance of murine oocyte meiotic arrest: uptake and metabolism of hypoxanthine and adenosine by cumulus cell-enclosed and denuded oocytes. Dev Biol 1986;117:174-83.
54. Downs SM, Eppig JJ. Induction of mouse oocyte maturation in vivo by perturbants of purine metabolism. Biol Reprod 1987;36:431-7.
55. Miller JGO, Behrman HR. Oocyte maturation is inhibited by adenosine in the presence of FSH. Biol Reprod 1986;35:833-7.
56. Warikoo PK, Bavister BD. Hypoxanthine and cyclic adenosine 5'-monophosphate maintain meiotic arrest of rhesus monkey oocytes in vitro. Fertil Steril 1989;51:886-9.
57. Sirard MA, First NL. In vitro inhibition of oocyte nuclear maturation in the bovine. Biol Reprod 1988;39:229-34.
58. Racowsky C, Baldwin KV. Modulation of intrafollicular estradiol in explanted hamster follicles does not affect oocyte meiotic status. J Reprod Fertil 1989; 87:409-20.
59. Eiler H, Nalbandov AV. Sex steroids in follicular fluid and blood plasma during the estrous cycle of pigs. Endocrinology 1977;100:331-8.
60. Hubbard CJ, Greenwald GS. Cyclic nucleotides, DNA, and steroid levels in ovarian follicles and corpora lutea of the cyclic hamster. Biol Reprod 1982; 26:230-40.
61. Robertson JE, Baker RD. Role of female sex hormones as possible regulators of oocyte maturation [Abstract S7]. Proc 2nd annu meet Soc for the Study of Reproduction, Davis, CA 1969.
62. Nekola MV, Smith DM. Oocyte maturation and follicle cell viability in vitro. Eur J Obstet Gynecol Reprod Biol 1974;4 (suppl):S125-31.
63. Eppig JJ, Koide SL. Effects of progesterone and oestradiol-17β on the spontaneous meiotic maturation of mouse oocytes. J Reprod Fertil 1978; 52:99-101.
64. Richter JD, McGaughey RW. Specificity of inhibition by steroids of porcine oocyte maturation in vitro. J Exp Zool 1979;209:81-90.
65. Racowsky C, McGaughey RW. In the absence of protein, estradiol suppresses meiosis of porcine oocytes in vitro. J Exp Zool 1982;224:103-10.
66. Racowsky C. Antagonistic actions of estradiol and tamoxifen upon forskolin-dependent meiotic arrest, intercellular coupling and the cyclic AMP content of hamster oocyte-cumulus complexes. J Exp Zool 1985;234:251-60.
67. Schultz RM, Montgomery RR, Ward-Bailey P, Eppig JJ. Regulation of oocyte maturation in the mouse: possible roles of intercellular communication, cAMP and testosterone. Dev Biol 1983;95:294-304.
68. Kaji E, Bornslaeger EA, Schultz RM. Inhibition of mouse oocyte cylic AMP phosphodiesterase by steroid hormones: a possible mechanism for steroid hormone inhibition of oocyte maturation. J Exp Zool 1987;243:489-93.

69. Racowsky C. Androgenic modulation of cyclic adenosine monophosphate (cAMP)-dependent meiotic arrest. Biol Reprod 1983;28:774–87.
70. Erickson GF, Ryan KJ. The effect of LH/FSH, dibutyryl cAMP, and prostaglandins on the production of estrogens by rabbit granulosa cells in vitro. Endocrinology 1975;97:108–13.
71. Rice C, McGaughey RW. The effect of testosterone and dibutyryl cAMP on the spontaneous maturation of pig oocytes. J Reprod Fertil 1981;62:245–56.
72. Lindner HR, Tsafriri A, Lieberman ME, et al. Gonadotropin action on cultured Graafian follicles: induction of maturation division of the mammalian oocyte and differentiation of the luteal cell. Recent Prog Horm Res 1974; 30:79–138.
73. Young G, Kagawa H, Nagahama Y. Inhibitory effect of cyanoketone on salmon gonadotropin-induced estradiol-17β production by ovarian follicles of the Amago salmon (*Oncorhyncus rhodurus*) in vitro. Gen Comp Endocrinol 1982;47:357–60.
74. Wolfson AJ, Richards J, Rotenstein D. Cyanoketone competition with estradiol for binding to the cytosolic estrogen receptor. J Steroid Biochem 1983;19:1817–8.
75. Batten BE, Roh SI, Kim MH. Effects of progesterone and a progesterone antagonist (RU486) on germinal vesicle breakdown in the mouse. Anat Rec 1989;223:387–92.
76. MacKenzie LW, Garfield RE. Hormonal control of gap junctions in the myometrium. Am J Physiol 1985;248:C296–308.
77. Burghardt RC, Gaddy-Kurten D, Burghardt RL, Kurten RC, Mitchell PA. Gap junction modulation in rat uterus, III. Structure-activity relationships of estrogen receptor-binding ligands on myometrial and serosal cells. Biol Reprod 1987;36:741–51.
78. Garfield RE, Gasc JM, Baulieu EE. Effects of the antiprogesterone RU 486 on preterm birth in the rat. Am J Obstet Gynecol 1987;157:1281–5.
79. Matsumoto A, Arnold AP, Zampighi GA, Micevych PE. Androgenic modulation of gap junctions between motoneurons in the rat spinal cord. J Neurosci 1988;8:4177–83.
80. Hubbard CJ. The effect in vitro of alterations in gonadotropins and cyclic nucleotides on oocyte maturation in the hamster. Life Sci 1983;33:1695–702.
81. Hubbard CJ. Cyclic AMP changes in the component cells of Graafian follicles: possible influences on maturation in the follicle-enclosed oocytes of hamsters. Dev Biol 1986;118:343–51.

3

Translation Control in Oocytes: A Critical Role for the Poly(A) Tail of Maternal mRNAs

MARCIA L. O'CONNELL, JOAQUIN HUARTE, DOMINIQUE BELIN,
JEAN-DOMINIQUE VASSALLI, AND SIDNEY STRICKLAND

The majority of mRNAs in a eukaryotic cell are modified at both their 5′ and 3′ ends. A modified guanylate residue or cap structure is present at the 5′ end that, among possible roles, might be required as a barrier to exonucleases. A *polyadenylic acid* (poly(A)) tail is present at the 3′ end of most mRNAs, for which a single essential function has not as yet been identified. It has been postulated that this structure may serve regulatory functions. In particular, the poly(A) tail has been implicated in regulating RNA stability, transport from the nucleus, and translation (1, 2).

While the majority of the data accumulated since the discovery of the poly(A) tail focused on its involvement in RNA stability, attention has shifted recently to its role in translation. As alluded to above, not all transcripts are polyadenylated, so the poly(A) tail cannot be required for the translation of all messages. However, it is possible that in cases in which the translational status of a particular message changes, polyadenylation may be utilized as a mechanism to regulate this transition. In particular, developmental systems from various species have provided examples of transcripts whose 3′ poly(A) tails are modulated in length. These examples complement numerous systems in which polyadenylated transcripts tend to be translated more efficiently (3–5) and emphasize the regulatory role of the poly(A) tail in directing either the initiation or cessation of mRNA translation. This chapter focuses on this regulatory role, as the range of possible functions of the poly(A) tail has been extensively reviewed previously (1, 2).

Polyadenylation During Oogenesis

The maturation of the germ cells of many species provides a unique situation in which the translational status of various transcripts must change dramatically. Many mRNAs are synthesized and stored during the growth of the germ cells and then recruited for translation at specific times during maturation. It is not surprising, therefore, that a correlation between the poly(A) tail length of specific messages and their translational status was first demonstrated in clam oocytes (6). During oogenesis, these messages are transcribed and stored and have a short poly(A) tail. Upon maturation, the poly(A) tail is extended, and one can begin to detect protein. This correlation between extension of a poly(A) tail and translation of the message has since been observed with a number of transcripts in *Xenopus* oocytes (7), the HPRT and *tissue plasminogen activator* (tPA) messages in mouse oocytes (8, 9), the fem-3 mRNA in *C. elegans* germ cells (10), and the Mst87F mRNA in *Drosophila* spermatocytes (11).

The question generated by these observations is whether polyadenylation actually regulates translation or merely occurs concomitantly. For both tPA in mouse oocytes and a series of messages in *Xenopus*, the regulatory role of the poly(A) tail has been investigated by RNA injection into oocytes (7, 12). In primary mouse oocytes, the tPA message is dormant and has a short poly(A) tail of 20–40 As. During oocyte maturation, the message is further adenylated and translated. Injection of synthetic RNAs has demonstrated that a long poly(A) tail is necessary and sufficient for translational activation, and sequences within the tPA 3'*-untranslated region* (UTR) mediate the extension of the poly(A) tail during oocyte maturation (12). Mutational analysis has defined the two sequence elements required, the first being the canonical nuclear cleavage/polyadenylation signal AAUAAA and the second, an A/U-rich *cytoplasmic polyadenylation element* (CPE). The degree of dependence upon the sequence AAUAAA distinguishes this process from nuclear polyadenylation: The nuclear reaction is a two-step process that requires the AAUAAA for the addition of the first 10–30 As, but only this oligo A stretch for the subsequent adenylation (13), while the cytoplasmic process seems to depend upon the presence of the AAUAAA sequence throughout the process.

The regulatory role of the poly(A) tail has also been demonstrated in *Xenopus* oocytes, and the mechanism shares many features with that of the mouse oocyte (7). The B4 mRNA requires a long poly(A) tail for its translation, and the G10 message requires ongoing polyadenylation for its polysomal recruitment. In both cases, polyadenylation is dependent on both the AAUAAA sequence and an A/U-rich CPE. Though Schafer et al. (11) did not determine whether polyadenylation regulated the translation of Mst87F in *Drosophila* germ cells, they did define a 12-nt A/U-rich sequence in the 5'-UTR that is required for both translation and polyadenylation.

Though, in the majority of cases, a long poly(A) tail correlates with increased translation, it is important to note that the opposite has been observed for certain messages. In mouse spermatocytes, the 2 protamine mRNAs and the transition protein 1 transcript are all deadenylated when loaded on polysomes (reviewed in 14). In the case of protamine 1, once again correct temporal expression is dependent on specific sequence elements in the mRNAs, located in the 3'-UTR. A similar correlation exists with the *Xenopus* histone mRNAs in oocytes, which are deadenylated at the time of translational activation (15, 16).

A provocative deduction that has emerged from investigation of transcripts whose polyadenylation promotes their translation is that regulated deadenylation may impart dormancy prior to translational recruitment. It has long been suggested that specific mRNAs are "masked" in immature oocytes until the protein product is required, the presumption being that proteins bound to the messages prevent their being loaded on polysomes (reviewed in 17). It is conceivable that these factors mediate the deadenylation of the transcript, preventing it from being recognized by the translational machinery. This process would be distinct from the default deadenylation process seen in mature *Xenopus* oocytes that is concomitant with the removal of translated RNAs from polysomes (18, 19).

Regulation via Factor Binding

With the identification of specific sequence elements that mediate polyadenylation and perhaps also deadenylation, it will become possible to search for factors that regulate the processes. Richter et al. have identified binding activities in mature oocytes that interact specifically with the CPEs from a number of *Xenopus* transcripts (20). The factors have different affinities for different CPEs, presumably as a result of the variability in CPE sequences, allowing for the differential regulation of the transcripts. Furthermore, they have demonstrated that a 58-Kd factor that binds the B4 CPE is phosphorylated in the presence of a maturation promoting factor (MPF) that itself induces B4 polyadenylation (21).

Finally, the mechanism by which the poly(A) tail regulates translation is largely unknown. It is thought that, in general, the poly(A) tail does not exist as naked RNA, but is bound by either the nuclear or cytoplasmic *poly(A) binding protein* (PABP). In fact, the occupation by, or displacement of, PABP may itself serve a regulatory function, as genetic experiments in yeast have demonstrated that in the absence of PABP, translation is inhibited at the level of initiation (22). This is consistent with the observation that the poly(A) tail can be an enhancer of translation initiation in reticulocyte lysates (23). By analogy with transcriptional regulation, perhaps the poly(A) tail with or without associated PABP interacts with proteins bound to the 5'- and 3'-UTRs, thus stimulating or inhibiting their activity. The identification and purification of these factors has been initiated and will allow the investigation of the interactions that mediate translational regulation by the extent of 3' polyadenylation.

References

1. Jackson RJ, Standart N. Do the poly(A) tail and 3' untranslated region control mRNA translation? Cell 1990;62:15–24.
2. Brawerman G. The role of the poly(A) sequence in mammalian messenger RNA. Crit Rev Biochem 1981;10:1–38.
3. Huez G, Marbaix G, Hubert E, et al. Role of the polyadenylate segment in the translation of globin messenger RNA in *Xenopus* oocytes. Proc Natl Acad Sci 1974;71:3143–6.
4. Drummond DR, Armstrong J, Colman A. The effect of capping and polyadenylation on the stability, movement, and translation of synthetic RNAs in *Xenopus* oocytes. Nucl Acids Res 1985;13:7375–92.
5. Galili G, Kawata EE, Smith LD, Larkins BA. Role of the 3' poly(A) sequence in translational regulation of mRNAs in *Xenopus* oocytes. J Biol Chem 1988;623:5764–70.
6. Rosenthal ET, Tansey TR, Ruderman JV. Sequence-specific adenylations and deadenylations accompany changes in the translation of maternal mRNA after fertilization of *Spisula* oocytes. J Mol Biol 1983;166:309–27.
7. McGrew LL, Dworkin-Rastl E, Dworkin M, Richter J. Poly(A) elongation during *Xenopus* oocyte maturation is required for translational recruitment and is mediated by a short sequence element. Genes Dev 1989;3:803–15.
8. Paynton BV, Rempel R, Bachvarova R. Changes in state of adenylation and time course of degradation of maternal mRNAs during oocyte maturation and early embryonic development in the mouse. Dev Biol 1988;129:304–14.
9. Huarte J, Belin D, Vassalli A, Strickland S, Vassalli J-D. Meiotic maturation of mouse oocytes triggers the translation and polyadenylation of dormant tissue-type plasminogen activator mRNA. Genes Dev 1987;1:1201–11.
10. Ahringer J, Kimble J. Control of the sperm-oocyte switch in *C. elegans* hermaphrodites by the fem-3 3' untranslated region. Nature (London) 1991;349:346–8.
11. Schafer M, Kuhn R, Bosse F, Schafer U. A conserved element in the leader mediates post-meiotic translation as well as cytoplasmic polyadenylation of a *Drosophila* spermatocyte mRNA. EMBO J 1990;9:4519–25.
12. Vassalli J-D, Huarte J, Belin D, et al. Regulated polyadenylation controls mRNA translation during meiotic maturation of mouse oocytes. Genes Dev 1989;3:2163–71.
13. Sheets MD, Wickens M. Two phases in the addition of a poly(A) tail. Genes Dev 1989;3:1401–12.
14. Hecht N. Regulation of "haploid expressed genes" in male germ cells. J Reprod Fertil 1990;88:679–93.
15. Ruderman JV, Woodland HR, Sturgess EA. Modulation of histone messenger RNA during early development of *Xenopus laevis*. Dev Biol 1979;71:71–82.
16. Woodland HR. Histone synthesis during the development of *Xenopus*. FEBS Lett 1980;121:1–7.
17. Richter J. Translational control during early development. Bioessays 1991;13:179–83.

18. Fox CA, Wickens M. Poly(A) removal during oocyte maturation: a default reaction selectively prevented by specific sequences in the 3' UTR of certain maternal mRNAs. Genes Dev 1990;4:2287-98.
19. Varnum SM, Wormington WM. Deadenylation of maternal mRNAs during *Xenopus* oocyte maturation does not require specific cis- sequences: a default mechanism for translational control. Genes Dev 1990;4:2278-86.
20. McGrew LL, Richter J. Translational control by cytoplasmic polyadenylation during *Xenopus* oocyte maturation: characterization of cis and trans elements and regulation by cyclin/MPF. EMBO J 1990;9:3743-51.
21. Paris J, Swenson K, Piwnica-Worms H, Richter JD. Maturation-specific polyadenylation: in vitro activation by p34 (cdc2) and phosphorylation of a 58 kD CPE-binding protein. Genes Dev 1991;5:1697-1708.
22. Sachs AB, Davis RW. The poly(A) binding protein is required for poly(A) shortening and 60S ribosomal subunit-dependent translation initiation. Cell 1989;58:857-67.
23. Munroe D, Jacobson A. mRNA poly(A) tail, a 3' enhancer of translational initiation. Mol Cell Biol 1990;10:3441-5.

4

Establishment of Competence for Preimplantation Embryogenesis During Oogenesis in Mice

JOHN J. EPPIG

Shortly after the initiation of meiosis in oocytes during fetal life, this process becomes arrested at the diplotene stage. This arrest is sustained by the presence of meiosis-arresting substances within the oocyte (1) and by the absence of factors essential for the progression of meiosis (2). As the oocytes near completion of their growth phase, about the time of follicular antrum formation, they become competent to resume meiosis but remain arrested within the follicle because of the meiosis-arresting action of the somatic cells comprising the follicle wall (3–5). Meiosis normally resumes in response to the preovulatory surge of *luteinizing hormone* (LH) (6) but also resumes spontaneously, in the absence of gonadotropins, when the oocyte is removed from the follicle and cultured in a supportive medium (7, 8). Some oocytes in early antral follicles are competent to resume meiosis but are unable to progress to metaphase II (9). Therefore, *germinal vesicle breakdown* (GVBD) and subsequent completion of meiosis I are distinctly separable in oocytes and are sequentially acquired. Furthermore, oocytes must become capable of producing additional factors to progress beyond the initial stages of oocyte maturation.

The final goal of oocyte development is to produce a gamete competent to become fertilized and initiate embryogenesis. Maternal regulatory factors persist in preimplantation embryos and are critical for successful development. Messenger RNA coding for Oct-3, a transcription factor, is detectable in growing and mature oocytes, but not in primordial (resting) oocytes (10, 11). Zygotes failed to cleave to the 2-cell stage after fertilized eggs were injected with oligonucleotides antisense to Oct-3 mRNA, suggesting that Oct-3 is a maternal factor essential for first cleavage (12). These results show that regulatory molecules essential for early embryogenesis are synthesized in oocytes and support the hypothesis that translation products of maternal messages are required for first cleavage (13).

Oocyte culture systems are extremely valuable experimental tools for studying the mechanisms regulating oogenesis and subsequent embryogenesis. For example, in the classic experiments of Pincus and his colleagues (7, 14), oocytes were isolated from Graafian follicles and placed in culture, where they subsequently underwent spontaneous GVBD. This led to the hypothesis that somatic follicular components function to maintain the oocytes in meiotic arrest, a hypothesis supported by research in several laboratories. Observations on spontaneous maturation in vitro also led to the conclusion that oocytes sequentially acquire competence to initiate and, subsequently, to complete the first meiotic division (9). In early studies, oocytes matured in culture could not be fertilized, and this led to the opinion that oocyte maturation in vitro did not lead to the production of a normal gamete. It seemed that oocytes matured outside the follicle were deficient in some way; although nuclear maturation was apparently normal, cytoplasmic maturation was not. The cytoplasmic deficiency appeared to involve a failure to produce a male pronuclear growth factor (15). Since those early studies, however, it has been shown that oocytes of mice (16), rats (17, 18), sheep (19), cattle (20, 21), and cats (22) can be successfully fertilized after maturation in vitro. These results demonstrated that spontaneous maturation may produce a gamete competent to undergo fertilization and embryogenesis if maturation is carried out using appropriate culture conditions. This does not imply, however, that the mechanisms initiating spontaneous maturation in vitro and in vivo are the same.

Embryonic Developmental Competence of Mouse Oocytes Matured In Vitro: Correlation with Follicular Development

Since a large group of follicles develops almost synchronously in neonatal and juvenile mice, oocytes and follicles at increasing stages of development can be isolated from mice of increasing ages until the animals are about 1 month old (23, 24) (Fig. 4.1). In vitro studies with such oocytes recovered from immature mice have shown that the percentage of in vitro-matured ova that cleave to the 2-cell stage after insemination increases with advancing follicular development (Fig. 4.2) (24). Likewise, the competence of 2-cell stage embryos to develop to the blastocyst stage also increases with follicular development (Fig. 4.2) (24). The highest percentage of ova that cleave to the 2-cell stage is achieved by oocytes isolated from 20-day-old mice, but the highest percentage of development of 2-cells to the blastocyst stage is not achieved until the oocyte donors are 24 days old. Accordingly, the ratio of ova that cleave to the 2-cell

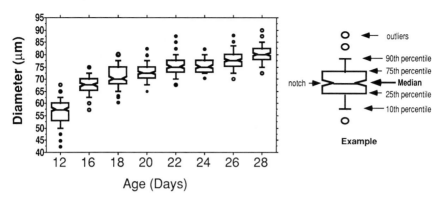

FIGURE 4.1. Diameter of oocytes 12–28 days old. Diameters are presented using notched box-and-whisker plots as the percentile distribution of oocytes (see example). Nonoverlapping notches between groups indicate a significant difference (P < 0.05) in the populations, as described in reference 29. Data adapted from Eppig and Schroeder (24).

stage to embryos that develop from the 2-cell stage to blastocyst is high when the oocyte donors are 18 days old; more than 3 times as many ova reach the 2-cell stage as complete the 2-cell stage to blastocyst transition (Fig. 4.2). In contrast, the 2-cell:blastocyst ratio reaches a minimum when the donors are 24 days old; only 1.3 times as many ova reach the 2-cell stage as complete the 2-cell stage to blastocyst transition (Fig. 4.2). These results suggest that developing oocytes acquire competence to complete the 2-cell stage to blastocyst transition after becoming competent to cleave to the 2-cell stage.

The differing developmental capacities in oocytes isolated from donors of increasing age probably reflect differences in content of regulatory molecules essential for successful embryogenesis. What are the factors that determine whether oocytes contain or can produce these regulatory molecules? This question was investigated with regard to the effect of stimulation with gonadotropin in vivo and in vitro.

To assess the effect of gonadotropins on the developmental competence of oocytes, *pregnant mare serum gonadotropin* (PMSG), which stimulates follicular development in vivo, was injected into 20- and 24-day-old mice, and the *germinal vesicle* (GV)-stage oocytes were isolated 48 h later when the mice were either 22 or 26 days old. The rationale for comparing the developmental capacity of oocytes from 22- and 26-day-old mice is that overall follicular development is slightly more advanced in the older group, thereby enabling comparison of the influence of follicles at slightly different stages of development. Oocytes were matured and fertilized in vitro. The frequency of cleavage to the 2-cell stage was the same in all groups: 80%–90% (Fig. 4.3A). The percentage that com-

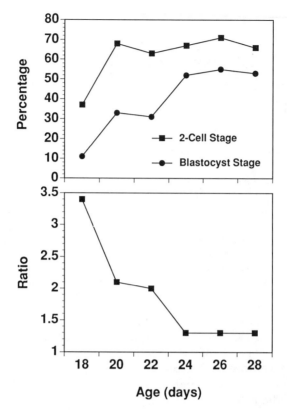

FIGURE 4.2. Development of embryos derived from in vitro-matured oocytes isolated from mice 18–28 days old. The top panel shows the percentage of oocytes that cleaved to the 2-cell stage after insemination and the percentage of 2-cell stage oocytes that developed to blastocysts. These percentages were used to calculate the ratio of embryos that cleaved to the 2-cell stage to the embryos that completed the 2C to B transition shown in the bottom panel. Data adapted from Eppig and Schroeder (24).

pleted the *2-cell stage to blastocyst* (2C to B) transition was also determined. There was no difference between the untreated and gonadotropin-treated 22-day-old groups, but gonadotropin injection resulted in an increase in the percentage of oocytes from the 26-day-old group that completed the 2C to B transition.

In the next experiment, the effect of FSH treatment of oocyte-cumulus cell complexes maturing in vitro on competence to complete the 2C to B transition was determined. In experiments using mice not treated with PMSG, FSH had no effect on this competency in oocytes isolated from 22-day-old mice but increased this capacity when the complexes were isolated from 26-day-old mice (Fig. 4.3B). In contrast, FSH did not

4. Establishment of Competence for Preimplantation Embryogenesis   47

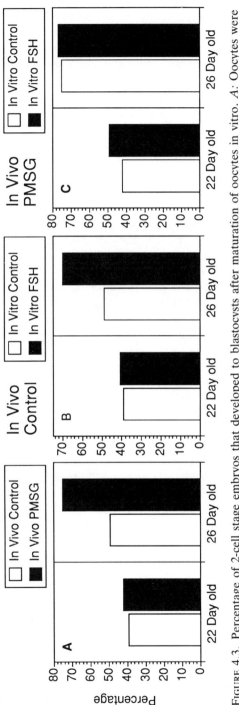

FIGURE 4.3. Percentage of 2-cell stage embryos that developed to blastocysts after maturation of oocytes in vitro. A: Oocytes were isolated at the GV stage from mice 48h after injection with PMSG or from uninjected control mice. B: Oocytes were isolated at the GV stage from mice not treated with PMSG and matured in the presence or absence of FSH (1-μg/mL oFSH^{-17}, NIAMDD). C: Percentage of 2-cell stage embryos that developed to blastocysts after maturation in vitro. Oocytes were isolated at the GV stage from mice injected 48h previously with PMSG and matured in the presence or absence of FSH. Data adapted from Eppig, Schroeder, and O'Brien (25).

increase the developmental competence of oocytes isolated from PMSG-primed 26-day-old mice above that already increased by PMSG in vivo (Fig. 4.3C). However, FSH treatment of complexes from 26-day-old mice not treated with PMSG increased the developmental capacity to approximately the same level as that resulting from PMSG treatment (Figs. 4.3B and 4.3C).

Taken together, therefore, the above results indicate that gonadotropins can have profound effects on the preimplantation developmental competence of oocytes and that these effects are dependent upon the age of the donor mice. Although the age-dependent effect is probably related to differing stages of follicular development, PMSG treatment dramatically accelerates follicular development in mice of both ages. In fact, the distribution of follicles of various sizes is about the same in PMSG-treated mice of both ages (25). Therefore, although the development of the somatic components of the follicles is accelerated by PMSG, the developmental capacity of the oocytes contained therein does not appear to be coordinately accelerated. Nevertheless, 2 to 3 times more oocytes are recovered from the ovaries of both ages after PMSG treatment, so gonadotropins accelerate oocyte development coordinately with follicular development at some stage(s) of follicular development, probably the later stages.

FSH treatment of in vitro-maturing oocytes promoted the acquisition of increased developmental competence only when the oocytes were isolated from 26-day-old mice not previously treated with PMSG in vivo. In this way, FSH treatment in vitro mimicked PMSG treatment in vivo, but although the result of both gonadotropin treatments had the same end result (increased frequency of completion of the 2C to B transition), it is not known whether the processes leading to that end are the same. Oocytes affected by PMSG treatment in vivo are in the GV stage and are situated in the complex system of cell-to-cell communications existing in the intact follicle. In contrast, oocytes affected by FSH treatments in vitro are undergoing maturation and are isolated from these complex cellular interactions of the intact follicle. It is possible that PMSG treatment in vivo initiates some events of maturation before GVBD in some oocytes; they may in fact be "more mature" in some ways than many, but not all, of the oocytes present in the follicles of the untreated mice. This more advanced level of maturation must be related in some way to the oocytes' capacity to carry out processes associated with the 2C to B transition.

This leads to the question of whether the oocytes contain the critical molecules for this transition before undergoing GVBD or whether the molecules are produced during GVBD. The observation that FSH treatment of maturing oocytes can also increase the frequency of 2C to B transition seems to argue that the critical molecules for this transition are synthesized during maturation. Oocytes in less-developed follicles, as are many of those in 22-day-old mice, are probably incompetent to produce the critical molecules that promote the acquisition of the capacity to complete the 2C to B transition. Thus, it appears that there are at least two critical steps in oocyte development that relate to competence to complete the 2C to B transition: (i) production of molecules that regulate the synthesis of transition factor(s) and (ii) the synthesis of the transition factor(s).

Because the cumulus cells of complexes isolated from the 22-day-old mice did not promote the synthesis of transition factor(s) in response to FSH during maturation in vitro, it was important to establish whether the cumulus cells of these complexes are capable of responding to FSH. That they can respond to FSH is evident from the observation that virtually all of the cumuli oophori from both 22- and 26-day-old mice underwent expansion in response to FSH whether they were isolated from PMSG-treated or untreated mice. In addition, FSH stimulated elevated *cyclic adenosine monophosphate* (cAMP) levels in complexes from all groups (Fig. 4.4). Levels of cAMP attained after treatment of complexes from PMSG-treated mice with FSH were about 10-fold greater than those from untreated mice, but there was no difference in the cAMP levels attained in complexes treated with FSH, whether the complexes were obtained from 22- or 26-day-old mice. The level of cAMP attained by complexes in response to FSH, therefore, seems to be unrelated to the ability of the complexes to produce transition factor(s); the factor was produced in response to FSH in complexes isolated from 26-day-old mice not treated with PMSG, but not in those isolated from 22-day-old mice, yet there was no difference in the levels of cAMP produced in response to FSH.

Based on these observations, the author hypothesizes that many of the follicles in 22-day-old mice are restricted in their ability to complete the 2C to B transition because they are incompetent to produce the transition factor(s) and/or because the somatic components of the follicles are incompetent to signal the oocyte appropriately to become competent. The author also hypothesizes that somatic cells of more advanced follicles, such as those more commonly found in the older mice, signal the oocyte in response to gonadotropin to produce the critical transition factor(s). Obviously, the molecules that regulate the acquisition of the capacity to complete the 2C to B transition, as well as those that actually promote the transition, must be identified, and the mechanisms by which the follicular somatic cells participate in this aspect of oocyte development are critical questions for developmental and reproductive biology.

The author proposes that oocytes restricted in their ability to complete the 2C to B transition exist in a distinct stage of oocyte development. A classification of GV-stage mouse oocytes based on characteristics of competence to complete meiosis and preimplantation development is presented in Table 4.1. According to this system, embryogenesis-restricted oocytes are classified as stage IV oocytes. Although stage IV oocytes are able to complete the nuclear events of meiosis and progress to metaphase

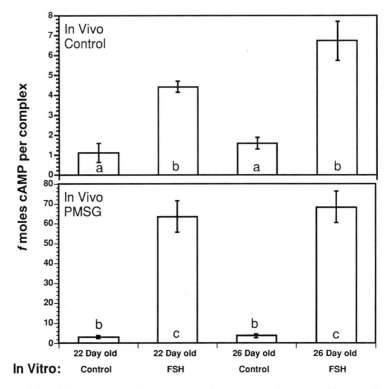

FIGURE 4.4. cAMP content of oocyte-cumulus cell complexes. cAMP was determined using radioimmunoassay of complexes cultured for 2h in medium with or without FSH (1µg/mL), both in the presence of the cAMP-phosphodiesterase inhibitor isobutyl methylxanthine (50µM). Error bars indicate the SEM. The significance of differences between means were evaluated using the Student-Newman-Keuls Multiple Range Test of log-transformed data; different letters on the bars indicate $P < 0.05$ (N = 4).

TABLE 4.1. Classification of GV-stage oocytes based on competence to complete meiosis and undergo preimplantation development.

		Stage				
		I	II	III	IV	V
Competence	GVBD	No	No	Yes	Yes	Yes
	Metaphase II	No	No	No	Yes	Yes
	2-cell stage	No	No	No	Yes	Yes
	Blastocyst	No	No	No	No	Yes
Other characteristics	Growth	"Resting"	Growing	Growing	Nearly grown	Fully grown
	Follicular origin	Primordial	Preantral	Early antral	Antral	Antral

II, producing the first polar body, and although they are able to cleave to the 2-cell stage after fertilization, they are unable to support development from the 2-cell stage to the blastocyst stage. Many of the embryos from stage IV oocytes that fail to complete the 2C to B transition block at or near the 2-cell stage, although some reach the morula stage. Failure to activate zygotic transcription results in rapid termination of embryo development. In mice, zygotic transcription is initiated in the 2-cell stage (26, 27), and inhibition of this transcription results in developmental arrest at this stage (28). The author proposes that stage IV oocytes are deficient in a factor(s) necessary for the normal sequence of activation of the zygotic genome and, therefore, embryos derived from stage IV oocytes are unable to develop significantly beyond the 2-cell stage. Accordingly, it is hypothesized that 2C to B transition factors participate in activation of the zygotic genome.

Acknowledgments. This paper is dedicated to Al Schroeder, who died in September, 1990. Al was a good friend and a valuable colleague who was an important participant in the studies reviewed here. This research was supported by the National Cooperative Program on Non-Human In Vitro Fertilization and Preimplantation Development, NIH, through Cooperative Agreement HD-21970. I thank Marilyn O'Brien for her highly skilled and dedicated technical assistance in these studies.

References

1. Fulka J. Maturation-inhibiting activity of growing mouse oocytes. Cell Differ Dev 1985;17:45–8.
2. Balakier H. Induction of maturation in small oocytes from sexually immature mice by fusion with meiotic or mitotic cells. Exp Cell Res 1978;112:137–41.
3. Tsafriri A, Channing CP. An inhibitory influence of granulosa cells and follicular fluid upon oocyte meiosis in vitro. Endocrinology 1975;96:922–7.
4. Leibfried L, First N. Effect of bovine and porcine follicular fluid and granulosa cells on maturation of oocytes in vitro. Biol Reprod 1980;23:699–704.
5. Racowsky C, Baldwin KV. In vitro and in vivo studies reveal that hamster oocyte meiotic arrest is maintained only transiently by follicular fluid, but persistently by membrana/cumulus granulosa cell contact. Dev Biol 1989; 134:297–306.
6. Lindner HR, Tsafriri A, Lieberman ME, et al. Gonadotropin action on cultured Graafian follicles: induction of maturation division of the mammalian oocyte and differentiation of the luteal cell. Recent Prog Horm Res 1974; 30:79–138.
7. Pincus G, Enzmann EV. The comparative behavior of mammalian eggs in vivo and in vitro, I. The activation of ovarian eggs. J Exp Med 1935;62: 655–75.
8. Edwards RG. Maturation in vitro of mouse, sheep, cow, pig, rhesus monkey and human ovarian oocytes. Nature (London) 1965;208:349–51.

9. Sorensen RA, Wassarman PM. Relationship between growth and meiotic maturation of the mouse oocyte. Dev Biol 1976;50:531–6.
10. Rosner MH, Vigano MA, Ozato K, et al. A POU-domain transcription factor in early stem cells and germ cells of the mammalian embryo. Nature (London) 1990;345:686–92.
11. Schöler HR, Hatzopoulos AK, Balling R, Suzuki N, Gruss P. A family of octamer-specific proteins present during mouse embryogenesis: evidence for germline-specific expression of an Oct factor. EMBO J 1989;8:2543–50.
12. Rosner MH, De Santo RJ, Arnheiter H, Staudt LM. Oct-3 is a maternal factor required for the first mouse embryonic division. Cell 1991;64:1103–10.
13. Johnson MH. The molecular and cellular basis of preimplantation development. Biol Rev 1981;56:463–98.
14. Pincus G, Saunders B. The comparative behavior of mammalian eggs in vivo and in vitro, IV. The maturation of human ovarian ova. Anat Rec 1939;75:537–45.
15. Thibault C. Are follicular maturation and oocyte maturation independent processes? J Reprod Fertil 1977;51:1–15.
16. Schroeder AC, Eppig JJ. The developmental capacity of mouse oocytes that matured spontaneously in vitro is normal. Dev Biol 1984;102:493–7.
17. Shalgi R, Dekel N, Kraicer PF. The effect of LH on the fertilizability and developmental capacity of rat oocytes matured in vitro. J Reprod Fertil 1979;55:429–35.
18. Vanderhyden BC, Armstrong DT. Role of cumulus cells and serum on the in vitro maturation, fertilization, and subsequent development of rat oocytes. Biol Reprod 1989;40:720–8.
19. Staigmiller RB, Moor RM. Effect of follicle cells on the maturation and developmental competence of ovine oocytes matured outside the follicle. Gamete Res 1984;9:221–9.
20. Trounson AO, Willadsen SM, Rowson LEA. Fertilization and developmental capacity of bovine follicular oocytes matured in vitro and in vivo and transferred to the oviducts of rabbits and cows. J Reprod Fertil 1977;51:321–7.
21. Sirard MA, Parrish JJ, Ware CB, Leibfried-Rutledge ML, First NL. The culture of bovine oocytes to obtain developmentally competent embryos. Biol Reprod 1988;39:546–52.
22. Johnston LA, O'Brien SJ, Wildt DE. In vitro maturation and fertilization of domestic cat follicular oocytes. Gamete Res 1989;24:343–56.
23. Mangia F, Epstein CJ. Biochemical studies of growing mouse oocytes: preparation of oocytes and analysis of glucose-6-phosphate dehydrogenase and lactate dehydrogenase activities. Dev Biol 1975;45:211–20.
24. Eppig JJ, Schroeder AC. Capacity of mouse oocytes from preantral follicles to undergo embryogenesis and development to live young after growth, maturation and fertilization in vitro. Biol Reprod 1989;41:268–76.
25. Eppig JJ, Schroeder AC, O'Brien MJ. Developmental capacity of mouse oocytes matured in vitro: effects of gonadotropic stimulation, follicular origin, and oocyte size. J Reprod Fertil 1992;95:119–127.
26. Flach G, Johnson MH, Braude PR, Taylor RAS, Bolton VN. The transition from maternal to embryonic control in the 2-cell mouse embryo. EMBO J 1982;1:681–6.

27. Taylor KD, Piko L. Patterns of mRNA prevalence and expression of B1 and B2 transcripts in early mouse embryos. Development 1987;101:877–92.
28. Golbus MS, Calarco PG, Epstein CJ. The effects of inhibitors of RNA synthesis (a-amanitin and actinomycin D) on preimplantation mouse embryogenesis. J Exp Zool 1973;186:207–16.
29. Kafadar K. Notched box-and-whisker plot. Encyclopedia of statistical sciences; 6. 1985:367–70.

Part II

Epigenetic Control of Embryo Development

5

Regulation of Hamster Embryo Development In Vitro by Amino Acids

BARRY D. BAVISTER AND SUSAN H. MCKIERNAN

Considerable attention has been directed towards determining carbohydrate energy substrates for supporting in vitro development of preimplantation embryos, mostly in studies with mice (1, 2). In contrast, there has been little interest in examining amino acid requirements, doubtless because no regulatory role for amino acids was found in studies with mouse embryos. Although glycine as the sole fixed-nitrogen source was able to support 8-cell mouse embryo development (3), later studies showed that a fixed-nitrogen source was not essential for development of 2-cell mouse embryos (4).

However, there does appear to be a covert role for amino acids in cultured mouse embryos: Inclusion of amino acids in the culture medium enhanced postimplantation embryo development (5). That amino acids might play a role in embryo development in vitro was indicated in an early study that examined nutrient requirements of cultured rabbit embryos (6). Deletion of different groups of components from the culture medium revealed that only omission of the amino acids severely restricted embryo development. The precise role of amino acids in supporting rabbit embryo development was not established. Development of rat 8-cell embryos in vitro was also improved by the presence of amino acids (7), and, recently, the in vitro "block to development" of 1-cell rat embryos was overcome using a simple culture medium (HECM-1) containing 20 amino acids (8).

The impetus for the studies described here came from initial attempts to support development of golden hamster embryos in culture. Hamster embryos have been particularly difficult to sustain in vitro, exhibiting blocks to development at the 2-, 4-, and 8-cell stages (9–14). Although the golden hamster has been used for many years for research on fertilization (15–18), the refractoriness of the preimplantation embryos to culture has severely restricted use of this species for developmental studies. However, because hamster embryos are very stringent in their requirements for in vitro development, they could be a more sensitive model for probing epigenetic factors influencing early development.

As a first step towards overcoming the blocks to development and perhaps explaining their cause(s), we cultured hamster 8-cell embryos using a modified Tyrode's solution containing lactate, pyruvate, and bovine serum albumin (9). Almost none (2%) of the "early" 8-cell embryos—that is, those collected approximately 54 h post-egg activation by sperm—could develop into blastocysts in vitro, although embryos collected slightly later (~61 h) fared better (22% blastocysts). After many unsuccessful attempts to improve embryo development, such as by altering the pH of the culture medium and adding serum, we found that amino acids stimulated development: 18-fold (36% blastocysts) with early 8-cell embryos and 3-fold (66%) with slightly older embryos. The amino acids used were those found to be important for in vitro maturation of hamster oocytes: glutamine, phenylalanine, methionine, and isoleucine (19). These amino acids were not tested separately in our early experiments (9). A few of the blastocysts grown from 8-cell embryos in the presence of amino acids were able to develop into late fetuses following embryo transfer.

These early experiments have led to more detailed studies, some of which are described below, that confirm the importance of specific amino acids in modulating the development of hamster embryos in vitro. Moreover, the huge concentrations of amino acids both within eggs and embryos and in the female reproductive tract secretions (mice, rabbits: 20–23) indicate a physiological function for these compounds in vivo beyond any metabolic substrate role. In this chapter, we address the questions: What are the amino acid requirements for hamster embryo development in vitro, and what are possible functional roles of these amino acids?

Requirements for Amino Acids by Cultured Hamster Embryos

Initial Studies on the Importance of Amino Acids

Before early cleavage stage hamster embryos could be cultured to morulae and blastocysts, studies were restricted to embryos collected at the 8-cell stage. We found that glutamine alone stimulated blastocyst development more effectively than a large group of amino acids (all those found in Ham's F-10 medium), although vitamins were also needed to support blastocyst hatching (24). Subsequently, glutamine was found to be a key component of the culture medium for supporting the transition from

8-cells to morula, as well as for blastocyst formation. Glutamine at 1.0 mM (together with lactate, pyruvate, and glucose) could support ~60% blastocyst development (25).

Other amino acids were examined in groups of 4, together with suboptimal (0.1 mM) glutamine. All the groups stimulated development to the blastocyst stage, except for the proline, cysteine, arginine, and phenylalanine group, which was inhibitory; however, the group comprising alanine, hydroxyproline, isoleucine, and lysine further stimulated development to late (expanded) blastocysts. In one study with in vitro fertilized hamster eggs, we found that the presence of cumulus cells enhanced development to the 2-cell stage (further development was blocked), but the addition of "Gwatkin's" amino acids (glutamine, phenylalanine, methionine, and isoleucine) replaced the need for cumulus cells (26). This result was our first hint that amino acids in the oviductal environment might play an important role in embryo development during the first cell cycle (see section entitled *Effect of Hypotaurine on Development of 1-Cell Embryos*).

These early studies may have been somewhat confounded by the inclusion of BSA in the culture medium, which could obscure a requirement for amino acids by developing embryos. In addition, this protein preparation is notoriously variable and contaminated in a variety of ways with other compounds (27). We therefore replaced BSA with a chemically defined polymer, *polyvinylalcohol* (PVA), which had proved useful for supporting hamster gametes in studies on sperm motility and fertilization (28, 29). This change eliminated any possibility that BSA was indirectly supplying amino acids to the cultured embryos.

Studies with Chemically Defined Culture Media

Using a modified Tyrode's solution containing PVA as the only macromolecule, we confirmed that amino acids could support development of 8-cell hamster embryos to the blastocyst stage (30). An important advance in these studies was the discovery that development of 2-cell hamster embryos to the morula and blastocyst stages could be supported in chemically defined (protein-free) culture media when glucose and phosphate were deleted (11–13). The generic medium lacking these compounds was designated *hamster embryo culture medium* (HECM), although there are now several modifications of the original medium, HECM-1. When development of embryos collected at the 2-cell and 8-cell stages was compared, significantly more 2-cell embryos developed to the 8-cell stage or further when 20 amino acids were used versus only 4 ("Gwatkin's") amino acids; conversely, when freshly collected 8-cell embryos were cultured, more developed into late blastocysts with 4 amino acids than with 20 (Fig. 5.1) (12). This indicated that the amino acid preferences of hamster embryos change during development: Either a group of amino

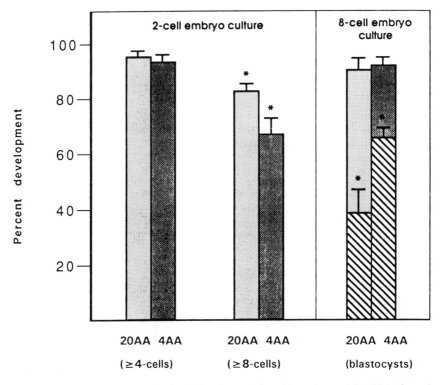

FIGURE 5.1. Comparison of HECM-1 (containing 20 amino acids, 20AA) and HECM-2 (4 amino acids, 4AA) for supporting in vitro development of hamster 2-and 8-cell embryos. Development of cultured 2-cell embryos (left panel) to ≥4-cells was scored at 24 h and to ≥8-cells at 48 h; embryos cultured from the 8-cell stage to blastocysts (right panel) were scored at 18 h. Hatched portions of histogram (right panel) equal late blastocysts (blastocele cavity ≥2/3 of embryo diameter). Values are % development ± SEM; *differ significantly (P ≤ 0.001). Results are from 12 replicate experiments, each with ~20 embryos. Reprinted with permission from Seshagiri and Bavister (12), © by Wiley-Liss, a division of John Wiley & Sons, Inc., 1991.

acids, comprising more than 4 but less than 20, might be optimal for hamster embryo development or media containing different amino acids might be required for culture of early and late cleavage stage embryos. This question is still being examined in our laboratory.

Modulation of 1-Cell Embryo Development by Single Amino Acids

Development of 1-cell hamster embryos was also modulated by amino acids. When 1-cell embryos were cultured in HECM-1, which contains 20 amino acids, 55% reached the 4-cell stage, although very few developed further (31). This was the first demonstration of hamster 1-cell development in vitro through 2 cleavage divisions. With glutamine and hypotaurine (see *Effect of Hypotaurine on Development of 1-Cell Embryos*) included in all media, effects of adding single amino acids were studied on 1-cell development. Results were similar to our earlier study with 8-cell embryos (25) in that some amino acids were inhibitory (phenylalanine, valine, isoleucine, tyrosine, arginine, and tryptophan), while others were stimulatory (glycine, cystine, and lysine). This study was limited by the difficulty of obtaining more than 2 cleavage divisions in vitro, which was later attributed in part to the presence of pyruvate (14).

Encouraged by the finding that single amino acids could modulate development of 1-cell embryos, a more complex experiment was designed in which 20 amino acids were examined, each at 2 concentrations (0.5 and 0.05 mM), for a total of 40 experimental treatments. The medium used was HECM-3, which contains no pyruvate, and reduced lactate and glutamine (14). Standard conditions were: culture for 72 h in 50-µL drops of medium in a 60-mm petri dish (Falcon Plastics) overlaid with 10-mL washed silicone oil and equilibration and embryo culture in 10% CO_2/5% O_2/85% N_2 at 37°C. Each replicate was split into 3 sections done on separate days, with 14, 14, and 12 experimental treatments per section; each section had its own control treatment (basic medium with glutamine only). Four replicates were performed; within each, treatments were coded and randomly assigned to sections.

One-cell embryos were collected from superovulated hamsters (usually 40–60 per donor) ~10 h post-egg activation (9) and distributed to treatments within a section using *controlled pooling*. That is, embryos, segregated by donor, were held in equilibrated control medium prior to distribution until all were collected; then, the embryos from one donor were distributed equally to all treatments, the same number of embryos from the second donor were added to the same treatment drops, and so on, using 4–5 donors per day. On each day, the number of embryos available from the donor that gave the smallest collection was used as the standard for all other donors. In this way, embryos from each donor were equally represented in all treatments. After culturing the embryos undisturbed for 72 h, the numbers of morulae and blastocysts were recorded, and all embryos were examined using Hoechst nuclear stain (32, 33) to estimate *mean cell number* (= total nuclei counted within a treatment divided by the number of embryos cultured). Data were analyzed by 2-way ANOVA on: % (morula + blastocyst) × block (replicate) and mean nuclei number × block (replicate). Significant differences were determined by a protected L.S.D.

The results of this study are shown in Figure 5.2. Values for percent development (morula + blastocyst) are ranked. Although the 95% confidence interval is large, due both to inherent biological variability and to

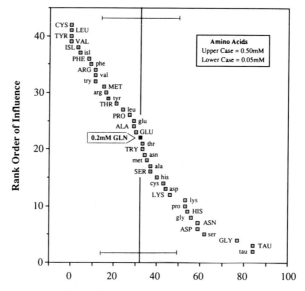

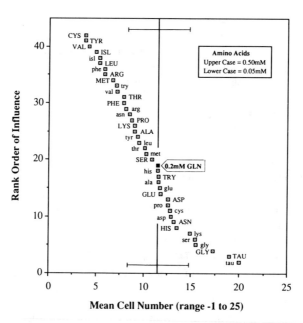

FIGURE 5.2. Effect of single amino acids on 1-cell hamster embryo development. Embryos collected 10 h postactivation of eggs were cultured for 72 h in medium HECM-3 containing glutamine (0.2 mM) and one of the amino acids shown at 0.5 or 0.05 mM. Mean cell number was determined by nuclear (Hoechst) staining. In each panel, the vertical line marks the position of the control mean, glutamine only (arrow); horizontal bars denote the 95% confidence interval of the mean. Results are from 4 replicate experiments, with a total of 31–50 embryos per experimental data point and 127 for the control. Values falling outside the bars differ significantly from the control ($P < 0.05$). Data from McKiernan and Bavister (unpublished).

the complexity of the experimental design, there is no doubt that some amino acids exerted a profound effect on embryo development. In particular, taurine (the product of hypotaurine oxidation) at both high and low concentrations increased development almost 3-fold over the control (containing 0.2-mM glutamine), while glycine (0.5 mM), serine (0.05 mM), and aspartic acid and asparagine (both 0.5 mM) all appeared to be stimulatory. In contrast, cysteine, leucine, tyrosine, valine, isoleucine, and phenylalanine were all inhibitory at 0.5 mM, consistent with the results of our preliminary study with 1-cell embryos (31).

A similar picture emerged when mean cell number was used as the end point: Taurine and glycine were again stimulatory, while cysteine, leucine, tyrosine, valine, and isoleucine were all inhibitory at 0.5 mM (Fig. 5.2). This study, showing a modulatory effect of single amino acids on embryo development from 1-cell to the blastocyst stage, strongly encourages us to continue investigating the role of amino acids. Current experiments are aimed at refining the protocol so that interactions between amino acids can be examined.

Effect of Hypotaurine on Development of 1-Cell Embryos

Two circumstantial lines of argument persuaded us to examine the role of hypotaurine in early embryo development. One was the demonstration that hypotaurine is essential in vitro for maintenance of functional viability (motility) in hamster sperm (29). The other was the huge concentrations of (hypo)taurine found in the female reproductive tract and in eggs and embryos (20–23). Since the male and female gametes, as well as very early embryos (zygotes), are exposed to the same environment in vivo—that is, the oviductal ampullary fluid—we hypothesized that (hypo)taurine might also be necessary for supporting development of the zygote in vitro. This tentative hypothesis was proved to be correct.

First, in vivo fertilized hamster eggs were collected from superovulated donors 3, 6, or 9 h after the (estimated) time of egg activation (sperm penetration) in vivo, which is 4 A.M. (9). These eggs were cultured in medium HECM-3 with or without hypotaurine (1 mM). Two effects on development were seen. First, in the absence of hypotaurine, the percentage of embryos that reached ⩾8-cells (Fig. 5.3) or morula/blastocyst stages (not shown) after 72-h culture increased as the time of collection was delayed. Only 7% of eggs collected 3 h post-egg activation reached morula/blastocyst stages, while 64% of 9-h eggs developed to those stages (not shown). (All embryos were examined for development at the same time of day so that the total incubation time [in vivo + in vitro] was equal for all treatments.) Second, addition of hypotaurine at 3 h post-egg activation more than doubled the percentage of embryos reaching ⩾8-cells (Fig. 5.3) and quadrupled the percentage reaching

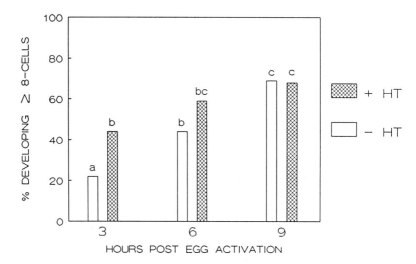

FIGURE 5.3. Stimulation of early 1-cell hamster embryo development by hypotaurine (HT). Embryos collected 3, 6, or 9h postactivation by sperm were cultured in HECM-3 ± hypotaurine 1 mM for 72h, then scored for development to ≥8-cells. Arcsin-transformed percentages are shown. Means with different superscripts differ significantly (P ≤ 0.05, Fisher's L.S.D.). Data from Barnett and Bavister (unpublished).

morulae/blastocysts (not shown). The effect of hypotaurine diminished with increasing age of eggs postactivation until there was no difference plus or minus hypotaurine for eggs collected 9h postactivation (Fig. 5.3).

Since hypotaurine exerted an effect that was more notable with eggs collected earlier postactivation (6-h versus 3-h data, Fig. 5.3), we hypothesized that if eggs were collected from the oviduct at 0h—that is, immediately preceding activation—then fertilized in vitro, hypotaurine would have an even more marked effect on their development. *In vitro fertilization* (IVF) was carried out as described (34), except that eggs were coincubated with capacitated sperm for 2h in HECM-3, then cultured for 96h ± hypotaurine under 10% CO_2 and 10% O_2. In the absence of hypotaurine, no IVF embryos reached the 8-cell stage, whereas with hypotaurine, 14% of IVF embryos developed into morulae or blastocysts (35, 36). Thus, hypotaurine was essential for advanced development of IVF embryos—that is, when egg activation takes place in vitro.

We conclude from these studies that (i) very early 1-cell hamster embryos have reduced capability for subsequent development in vitro, indicating an important role for the oviductal environment during the first few hours postactivation; (ii) hypotaurine can partially compensate for the premature loss of oviductal influence, suggesting that provision of this amino acid may be one way in which the oviduct supports early embryo development; and (iii) IVF eggs are exceptionally sensitive to the culture environment, and their subsequent development may be compromised because they have been exposed only briefly to the oviductal milieu. A corollary of this latter point is that embryos derived from IVF *follicular* eggs (e.g., as in human or bovine IVF) may be even more compromised because they have received no contribution from the oviduct.

Production of Living Young from Embryos Cultured in Chemically Defined Media Containing Amino Acids

Studies on the requirements of embryos for amino acids, or for any other environmental constituent, are incomplete until the viability of the cultured embryos is demonstrated. In the present context, evidence for modulation of development by amino acids might not be helpful towards understanding regulatory mechanisms without proof that embryos produced under these conditions are viable. At different stages during the progress of our studies, we have transferred embryos to recipient hamsters to evaluate viability.

Embryos collected at the 2-cell stage and cultured up to the 8-cell/morula stages in HECM were transferred to uteri of pseudopregnant hamsters. About 25% of the transferred embryos produced live offspring, the same percentage as when noncultured embryos were transferred (37). This result clearly shows that hamster embryos can not only develop, but also maintain viability in chemically defined culture medium. The relatively low success rate appears to be due to factors involved with the transfer procedure itself, such as inadequate synchronization between embryos and recipients. In another study (38), hamster embryos cultured from 8-cells to blastocysts in HECM were also able to develop into normal offspring following embryo transfer. Most recently, IVF hamster embryos were cultured in HECM up to the 8-cell/morula stages (72 h), then transferred to recipients; approximately 17% developed into term offspring (36). Incorporation of hypotaurine in the culture medium was essential for development of these IVF embryos past the 4-cell stage.

Physiological Role of Amino Acids

In numerous studies, amino acid pools and/or uptake kinetics have been examined in preimplantation mouse embryos (20–23, 39–44). A complex picture emerges of changing amino acid transport characteristics as the embryo develops from 1-cell to blastocyst, which is highly indicative of important physiological roles for particular amino acids during this phase of development. However, there seems to be little understanding of the specific functions of these changes in developing embryos.

There are several ways in which amino acids could contribute to early embryo development. Among the most obvious roles are serving as metabolic substrates for energy production, since several amino acids can indirectly enter the TCA cycle, and as anabolic substrates for protein synthesis. However, a variety of other roles, direct or indirect, have been proposed for amino acids, including serving as osmoregulators (45) and chelators of heavy metal ions (5). The importance of amino acids for embryo development in vivo is indicated by the high concentrations of particular amino acids in eggs/embryos and in reproductive tract fluids, ranging up to 20 mM or 30 mM (values for glycine and taurine, respectively, in mouse eggs), 9–10 mM (glycine in mouse and rabbit morulae), and 11 mM or 18 mM (glycine in oviduct and uterus) (20–23). Since these concentrations are far in excess of any metabolic requirements, they point to some other regulatory role for amino acids in early embryo development that remains to be determined. Unfortunately, the concentrations of amino acids in reproductive tract secretions and embryos of the hamster that would help assessment of their physiological functions in light of developmental effects described above are not known.

One possibility that has not been considered is that specific amino acids function as regulators of *intracellular pH* (pH_i). This concept is strengthened by recent reports (46, 47; and see Chapter 8, this volume) that mouse embryos lack the conventional array of mechanisms found in virtually all somatic cells, predominantly the Na^+/H^+ antiporter, for reducing intracellular acid load. If such fundamental pH_i-buffering systems are absent from early embryos, then some other mechanism, as yet unknown, must be present. In the environment of the early preimplantation embryo, all the elements are in place for a proton-transporting mechanism based on amino acid gradients across the cell membrane. This proposed role in pH_i buffering does not exclude other important physiological/metabolic roles for amino acids, and it is suggested as a possible addition to the pH_i-buffering mechanism proposed by Baltz et al. (Chapter 8, this volume).

Amino Acids as Regulators of pH_i: A Hypothetical Model

This model is predicated on the idea that amino acids, present at high concentrations in mammalian eggs and reproductive tract fluids, can shuttle into and out of the egg/embryo, exporting protons. For simplicity, only taurine and glycine are considered since these amino acids are present at the highest concentrations in the egg/embryo and in the reproductive tract secretions, although this does not preclude participation of other amino acids. In addition, (hypo)taurine and glycine are the most important modulators of hamster embryo development in vitro (Figs. 5.2 and 5.3) (48). Other essentials for operation of this mechanism include the abundance of Na^+ in the reproductive tract fluids and the presence of carbonic anhydrase in oviduct cells (49).

Taurine and glycine have both very low (~2–3) and high (~9) pK's; at pH 6–7, they exist predominantly as zwitterions, but in equilibrium with small amounts of the acid form ($-CH_2SO_3^-$, $-CH_2COO^-$), which is capable of accepting H^+ (buffering). The ratio of acid form:zwitterion form, ≤1:100, at first seems too low for an effective buffer system, except for two properties: (i) the very high concentration of the parent molecules (~20 mM) and (ii) the constant removal/disposal of accepted protons, exactly the characteristics that permit (for example) bicarbonate ion to buffer effectively in vivo in spite of its poor buffering properties in vitro. In the proposed scheme (Fig. 5.4), the large concentration of zwitterions actually serves as a reservoir to maintain the acid form at effective levels. In this hypothetical model, sodium glycinate or taurinate is actively transported into the embryo (50–52); sodium salt formation is favored by the high $[Na^+]_o$ in *oviduct fluid* (OF) and the alkaline pH of OF. In the embryo cytoplasm, Na^+ dissociates (due to low $[Na^+]_i$, ~5 mM, and lower pH_i), and protons generated by embryo metabolism are accepted by SO_3^- or $-COO^-$; the acid (associated) form of the amino acid is then transported and/or diffuses down its concentration gradient out of the cell, promptly exchanging H^+ for Na^+ in the more alkaline, Na^+-rich OF; the cycle repeats itself continuously. The H^+ in OF does not accumulate because it is rapidly converted to H_2O (with CO_2 formation) by the action of carbonic anhydrase in the oviductal epithelium (49); CO_2 is eliminated in the usual way via the systemic circulation.

The existence of such a "glycine/taurine shuttle" mechanism could help to explain why embryos exhibit slow development and reduced viability in culture. Removing embryos from the reproductive tract and placing them into culture media lacking adequate concentrations of, for example, glycine and taurine (Figs. 5.2–5.4) (21–23) would cause efflux of these amino acids, resulting in loss of their high intracellular levels (53) and failure of pH_i control, to the detriment of key cellular functions (54–56). According to the model depicted in Figure 5.4, loss of the intracellular amino acids would result in excessive export of protons and abnormal elevation of pH_i. Although there are no direct supportive data, consistent with this suggestion are observations that incorporation of weak acids (CO_2 or DMO) into the culture environment, which would reduce pH_i, strikingly increases hamster embryo development in vitro (13, 57). To assess their possible roles in regulating pH_i, effects of different concentrations of amino acids on embryo development (Fig. 5.2) need to be correlated with pH_i measurements (46, 47).

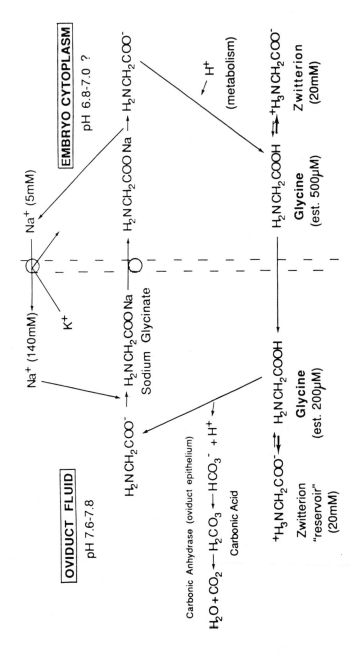

FIGURE 5.4. Hypothetical model for amino acid pH_i-buffering mechanism in mammalian embryos. Glycine (or taurine) "shuttles" across the cell membrane as the sodium salt (influx) and as the protonated free acid (efflux), thereby exporting H^+ and buffering pH_i. See text for details.

Conclusions

Detailed studies have shown that amino acids (particularly glutamine, glycine, and [hypo]taurine) are important modulators of embryo development in vitro in hamsters, and there are indications that this may be true in general for embryos of some other species. Individual amino acids can have powerful stimulatory or inhibitory effects on embryo development. Some of the most potent stimulators of embryo development in vitro (taurine and glycine) are also found in huge concentrations in vivo, arguing for some important physiological role(s) other than as metabolic substrates. The exact roles of amino acids in supporting early embryo development are not well understood, but regulation of pH_i is one possibility. The data obtained from in vitro studies indicate a need to evaluate the contribution of individual amino acids to preimplantation embryonic development and to assess their utility in improving culture media for supporting normal development in vitro. The fact that hamster embryos are very stringent in their requirements for in vitro development could make them a highly sensitive model for probing the epigenetic factors influencing early development.

Acknowledgments. We are grateful to our colleagues for their help in evaluation of amino acid effects on hamster embryo development, particularly Deborah Barnett, Scott Schini, and Polani Seshagiri. We thank Dorothy Boatman for helpful suggestions on preparation of the manuscript. Studies described in this chapter were done as part of the National Cooperative Program on Non-Human In Vitro Fertilization and Preimplantation Development and were funded by the National Institute of Child Health and Human Development, NIH, through Cooperative Agreement HD-22023.

References

1. Brinster RL. Studies on the development of mouse embryos in vitro, IV. Interaction of energy sources. J Reprod Fertil 1965;10:227–40.
2. Biggers JD. Pioneering mammalian embryo culture. In: Bavister BD, ed. The mammalian preimplantation embryo. Regulation of growth and differentiation in vitro. New York: Plenum Press, 1987:1–22.
3. Whitten WK. Culture of tubal ova. Nature (London) 1957;179:1081–2.
4. Cholewa JA, Whitten WK. Development of two-cell mouse embryos in the absence of a fixed-nitrogen source. J Reprod Fertil 1970;22:553–5.
5. Mehta TS, Kiessling AA. Development potential of mouse embryos conceived in vitro and cultured in ethylenediaminetetraacetic acid with or without amino acids or serum. Biol Reprod 1990;43:600–6.
6. Kane MT, Foote RH. Culture of two- and four-cell rabbit embryos to the expanding blastocyst stage in synthetic media. Proc Soc Exp Biol Med 1970;133:921–5.

7. Zhang X, Armstrong DT. Presence of amino acids and insulin in a chemically defined medium improves development of 8-cell rat embryos in vitro and subsequent implantation in vivo. Biol Reprod 1990;42:662–8.
8. Kishi J, Noda Y, Narimoto K, Umaoka Y, Mori T. Block to development in cultured rat one-cell embryos is overcome using medium HECM-1. Hum Reprod 1991;6:1445–8.
9. Bavister BD, Leibfried ML, Lieberman G. Development of preimplantation embryos of the golden hamster in a defined culture medium. Biol Reprod 1983;28:235–47.
10. Schini SA, Bavister BD. Development of hamster embryos through the "two-cell block" in chemically-defined medium. J Exp Zool 1988;245:111–5.
11. Schini SA, Bavister BD. Two-cell block to development of cultured hamster embryos is caused by phosphate and glucose. Biol Reprod 1988;39:1183–92.
12. Seshagiri PB, Bavister BD. Relative developmental abilities of hamster 2- and 8-cell embryos cultured in HECM-1 and HECM-2. J Exp Zool 1991;257:51–7.
13. McKiernan SH, Bavister BD. Environmental variables influencing in vitro development of hamster 2-cell embryos to the blastocyst stage. Biol Reprod 1990;43:404–13.
14. McKiernan SH, Bavister BD, Tasca RJ. Energy substrate requirements for in vitro development of hamster 1- and 2-cell embryos to the blastocyst stage. Hum Reprod 1991;6:64–75.
15. Yanagimachi R, Chang MC. Fertilization of hamster eggs in vitro. Nature (London) 1963;200:281–2.
16. Yanagimachi R, Chang MC. In vitro fertilization of golden hamster ova. J Exp Zool 1964;156:361–76.
17. Yanagimachi R. Mammalian fertilization. In: Knobil E, Neill JD, Ewing LL, Markert CL, Greenwald GS, Pfaff DW, eds. Physiology of reproduction. New York: Raven Press, 1988:135–86.
18. Cherr GN, Drobnis EZ. Fertilization in the golden hamster. In: Dunbar BS, O'Rand MG, eds. A comparative overview of mammalian fertilization. New York: Plenum Press, 1991:217–43.
19. Gwatkin RB, Haidri AA. Requirements for the maturation of hamster oocytes in vitro. Exp Cell Res 1973;76:1–7.
20. Leese HJ, Aldridge S, Jeffries KS. The movement of amino acids into rabbit oviductal fluid. J Reprod Fertil 1979;56:623–6.
21. Schultz GA, Kaye PL, McKay DJ, Johnson MH. Endogenous amino acid pool sizes in mouse eggs and preimplantation embryos. J Reprod Fertil 1981;61:387–93.
22. Miller JGO, Schultz GA. Amino acid content of preimplantation rabbit embryos and fluids of the reproductive tract. Biol Reprod 1987;36:125–9.
23. Kaye PL. Metabolic aspects of the physiology of the preimplantation embryo. In: Rossant J, Pedersen RA, eds. Experimental approaches to mammalian embryonic development. Cambridge, UK: Cambridge University Press, 1986:267–92.
24. Kane MT, Carney EW, Bavister BD. Vitamins and amino acids stimulate hamster blastocysts to hatch in vitro. J Exp Zool 1986;239:429–32.
25. Carney EW, Bavister BD. Stimulatory and inhibitory effects of amino acids on development of hamster eight-cell embryos in vitro. J In Vitro Fertil Embryo Transfer 1987;4:162–7.

26. Juetten J, Bavister BD. The effects of amino acids, cumulus cells and bovine serum albumin on in vitro fertilization and first cleavage division of hamster eggs. J Exp Zool 1983;227:487–90.
27. McKiernan SH, Bavister BD. Different lots of bovine serum albumin inhibit or stimulate in vitro development of hamster embryos. In Vitro Cell Dev Biol 1992;28A:154–6.
28. Bavister BD. Evidence for a role of post-ovulatory cumulus components in supporting fertilizing ability of hamster spermatozoa. J Androl 1982;3:365–72.
29. Boatman DE, Bavister BD, Cruz E. Addition of hypotaurine can reactivate immotile golden hamster spermatozoa. J Androl 1990;11:66–72.
30. Kane MT, Bavister BD. Protein-free culture medium containing polyvinylalcohol, vitamins and amino acids supports development of eight-cell hamster embryos to hatching blastocysts. J Exp Zool 1988;247:183–7.
31. Bavister BD, Arlotto TM. Influence of single amino acids on the development of hamster 1-cell embryos in vitro. Mol Reprod Dev 1990;25:45–51.
32. Purcel VG, Wall RJ, Rexroad CE Jr, Hammer RE, Brinster RL. A rapid whole-mount staining procedure for nuclei of mammalian embryos. Theriogenology 1985;24:687–91.
33. Boatman DE, Andrews JC, Bavister BD. A quantitative assay for capacitation: evaluation of multiple sperm penetration through the zona pellucida of salt-stored hamster eggs. Gamete Res 1988;19:19–29.
34. Bavister, BD. A consistently successful procedure for in vitro fertilization of golden hamster eggs. Gamete Res 1989;23:139–58.
35. Barnett DK, Bavister BD. Development of in vitro fertilized hamster embryos to morulae and blastocysts in a chemically defined culture medium. Biol Reprod 1991;44 (suppl 1):155.
36. Barnett DK, Bavister BD. Hypotaurine requirement for in vitro development of golden hamster one-cell embryos into morulae and blastocysts, and production of term offspring from in vitro fertilized ova. Biol Reprod 1992;47:297–304.
37. Schini SA, Bavister BD. Normal offspring produced after transfer of hamster embryos grown from two- to eight-cells in a chemically-defined culture medium. Theriogenology 1990;33:1255–62.
38. Seshagiri PB, Bavister BD. Assessment of hamster blastocysts derived from eight-cell embryos cultured in medium HECM-2: cell numbers and viability following embryo transfer. J In Vitro Fertil Embryo Dev 1990;7:229–35.
39. Brinster RL. Uptake and incorporation of amino acids by the preimplantation mouse embryo. J Reprod Fertil 1971;27:329–38.
40. Borland RM, Tasca RJ. Activation of a Na^+-dependent amino acid transport system in preimplantation mouse embryos. Dev Biol 1974;36:169–82.
41. Borland RM, Tasca RJ. Na^+-dependent amino acid transport in preimplantation mouse embryos, II. Metabolic inhibitors and nature of the cation requirement. Dev Biol 1975;46:192–201.
42. Kaye PL, Schultz GA, Johnson MH, Pratt HPM, Church RB. Amino acid transport and exchange in preimplantation mouse embryos. J Reprod Fertil 1982;65:367–80.
43. Van Winkle LJ. Amino acid transport in developing animal oocytes and early conceptuses. Biochim Biophys Acta 1988;947:173–208.

44. Van Winkle LJ, Campione AL, Gorman JM, Weimer BD. Changes in the activities of amino acid transport systems bo$^+$ and L during development of preimplantation mouse conceptuses. Biochim Biophys Acta 1990;1021:77–84.
45. Van Winkle LJ, Haghighat N, Campione AL. Glycine protects preimplantation mouse conceptuses from a detrimental effect on development of the inorganic ions in oviductal fluid. J Exp Zool 1990;253:215–9.
46. Baltz JM, Biggers JD, Lechene C. Apparent absence of Na^+/H^+ antiport activity in the two-cell mouse embryo. Dev Biol 1990;138:421–9.
47. Baltz JM, Biggers JD, Lechene C. Two-cell stage mouse embryos appear to lack mechanisms for alleviating intracellular acid loads. J Biol Chem 1991;266:6052–7.
48. Bavister BD, Barnett DK, McKiernan SH. Hypotaurine stimulates in vitro development of early one-cell hamster embryos to the morula/blastocyst stage. Biol Reprod 1991;44 (suppl 1):155.
49. Lutwak-Mann C. Carbonic anhydrase in the female reproductive tract: occurrence, distribution and hormonal dependence. J Endocrinol 1955; 13:26–38.
50. Hobbs JG, Kaye PL. Glycine and Na^+ transport in preimplantation mouse embryos. J Reprod Fertil 1986;77:61–6.
51. Van Winkle LJ, Campione AL, Kester SE. A possible effect of the Na^+ concentration in oviductal fluid on amino acid uptake by cleavage-stage mouse embryos. J Exp Zool 1985;235:141–5.
52. Van Winkle LJ, Haghighat N, Campione AL, Gorman JM. Glycine transport in mouse eggs and preimplantation conceptuses. Biochim Biophys Acta 1988; 941:241–56.
53. Sellens MH, Stein S, Sherman MI. Protein and free amino acid content in preimplantation mouse embryos and in blastocysts under various culture conditions. J Reprod Fertil 1981;61:307–15.
54. Gillies RJ. Intracellular pH and growth control in eukaryotic cells. In: Cameron, Pool TB, eds. The transformed cell. New York: Academic Press, 1981:353–95.
55. Busa WB, Nuccitelli R. Metabolic regulation via intracellular pH. Am J Physiol 1984;246:R409–38.
56. Boron WF. Intracellular pH regulation in epithelial cells. Annu Rev Physiol 1986;48:377–88.
57. Carney EW, Bavister BD. Regulation of hamster embryo development in vitro by carbon dioxide. Biol Reprod 1987;36:1155–63.

6

Energy Metabolism in Preimplantation Development

HENRY J. LEESE

Energy metabolism is ultimately concerned with the generation and utilization of ATP. These two aspects will be considered in turn.

ATP Generation by Preimplantation Embryos

Evidence from a number of sources indicates that early mammalian embryos carry out oxidative phosphorylation. Glycolysis becomes important in some species in the later preimplantation stages. Some evidence for aerobic metabolism follows. First, preimplantation mouse embryos fail to develop in the absence of oxygen (1) or in the presence of cyanide or 2,4-dinitrophenol, which both inhibit oxidative metabolism (2). Second, measurements of the partial pressure of oxygen in oviduct and uterine fluids (3, 4) indicate that preimplantation embryos inhabit an aerobic environment. Third, Mills and Brinster (5) showed that early mouse embryos consume oxygen in vitro; the rate of consumption is low up to the 8-cell stage, but then increases dramatically. Using values for the protein content of preimplantation mouse embryos (6), the data may be converted into conventional units of QO_2 (i.e., μL/mg dry wt/h). When this is done, it is apparent that prior to the 8-cell stage, mouse embryos have a QO_2 (~4) equivalent to a relatively quiescent adult tissue such as the skin, but that by the blastocyst stage (QO_2 = 24), they have become as metabolically active as the adult heart. Fourth, preimplantation embryos contain abundant mitochondria whose structure is modified during development in a manner consistent with the changes in oxygen uptake (7).

This chapter is dedicated to John D. Biggers.

Origin of Fuels Used by Preimplantation Embryos

It is appropriate to inquire as to the origin of the fuel(s) being oxidized. There are two possibilities: Embryos could utilize endogenous sources of energy or take up nutrients from the medium in which they are suspended. A third possibility is that both endogenous and exogenous sources could contribute. At the present time, the evidence (which is all derived from in vitro experiments) is in favor of embryos taking up exogenous nutrients, but a contribution from endogenous sources is by no means ruled out.

Evidence for Consumption of Endogenous Nutrients

Oocytes are large cells by mammalian standards, and a nuclear: cytoplasmic ratio characteristic of the adult is not established until late in preimplantation development. The size of mammalian eggs may perhaps be explained on evolutionary grounds in that true mammals are descended from amphibians and reptiles, which have large, yolk-laden eggs. Although the yolk has largely been lost in mammalian eggs, its influence may have persisted in the form of substantial cytoplasmic reserves of energy. An alternative explanation for the size of mammalian eggs is proposed at the end of this chapter.

Simple calculations show that at least in the mouse embryo, for which there is the most data, the amount of stored glycogen is only sufficient to supply the embryo's oxidative substrate needs for about 7h (8). Similarly, the amount of fat in the mouse embryo is insufficient to sustain oxidative metabolism for more than a few hours. The situation in the embryos of other species could be different. It is well known that cattle embryos, particularly those fertilized and cultured in vitro, contain substantial amounts of fat (9), which gives them a dark, granular appearance.

The situation with regard to protein is intriguing. Mouse embryos lose dry mass and protein as they develop from zygotes to blastocysts (6). If the protein lost, amounting to about 20% of the total, were oxidized, it could sustain the mouse embryo for at least 24h. This possibility is more likely in the preimplantation rabbit embryo, which can develop for 3–4 cleavage divisions in the absence of exogenous substrates (10).

Evidence for Consumption of Exogenous Nutrients

There are four strong lines of evidence in favor of the proposition that preimplantation embryos consume nutrients derived from oviduct and uterine fluids. First, in order to grow mouse embryos in vitro, it is necessary to supplement the culture medium with an energy source(s) (11). In the mouse, pyruvate is obligatory for the survival of unfertilized and fertilized oocytes and for the first cleavage division. Lactate acts synergistically with pyruvate and, as sole substrate, can support the cleavage of 2-cell mouse embryos. Glucose as sole substrate cannot support development until the 4–8 cell stage. Amino acids are generally unnecessary, at least during the early stages; lipids and nucleic acid precursors are also not required. These nutrient requirements are not necessarily true of other rodent species. For example, in the hamster, pyruvate cannot support the development of 1-cell embryos in vitro, and development from the 2-cell stage requires the deletion of glucose and phosphate. Furthermore, hamster embryo development in culture has a general requirement for amino acids (see Chapter 5, this volume). Until more is revealed of embryo metabolism in this and other species, it will be difficult to interpret these findings. A macromolecule is advisable, but not essential. The most common is bovine serum albumin. There is the intriguing possibility that exogenous albumin may be taken up by pinocytosis, hydrolyzed, and utilized by the embryo, either for protein synthesis or as an energy source (12).

It is interesting to speculate why lactate cannot support the development of the earliest stages of the mouse embryo. If supplied at high concentrations, lactate is likely to enter the embryo and be converted to pyruvate, since *lactate dehydrogenase* (LDH) is a highly active enzyme that is thought to bring about an equilibrium between the cytosolic pools of pyruvate and lactate. Why cannot the pyruvate formed in this way substitute for that added exogenously? One possibility is that the embryo cannot adequately dispose of the NADH generated in the conversion of lactate to pyruvate (D.H. Williamson, personal communication). It could be the case, for example, that the shuttle by which cytosolic NADH enters the mitochondria is not fully developed at the 1-cell stage. Since lactate rather than pyruvate can support the cleavage of 1-cell hamster embryos, it is possible that the NADH shuttle is better developed in this species.

Numerous permutations on the nutrient requirement theme have been devised. They have recently been reviewed by Leese (8). The possibility has to be faced that there is no single, optimal culture medium for the preimplantation development of a given species (13). Glucose, for example, is only required at certain stages of mouse preimplantation development (14), and embryos probably have some capacity to adapt to whatever nutrient(s) is available (15). Lawitts and Biggers (16), using a Simplex optimization technique, found that a range of media could successfully support mouse embryo development.

Perhaps the most obvious example of the capacity of embryos to adapt to the environment in which they find themselves in vivo comes from the practice of human in vitro fertilization, in which cultured embryos taken approximately 48 h following insemination are transferred into the uterus, the lumen of which they would not normally encounter until about 2 days later. The second line of evidence is that nutrients have been shown to be taken up by embryos in vitro. The original work, using radioisotopes, was done mainly by Brinster, Biggers, Wales, and Whittingham (17) using mouse and rabbit embryos. The results have now been extended to domestic species, particularly by Rieger and colleagues (18), and the early results on the mouse confirmed using a noninvasive approach that does not involve the use of radioisotopes (19, 20) and that can be extended to the human and the rat (21, 22).

Third, nutrients that oocytes and embryos are known to consume in vitro are present in follicular, oviduct, and uterine fluids in vivo (15, 23–25). Eggs and embryos are exposed to different environments during their preimplantation stages, which raises the question whether this should be acknowledged in the design of culture and experimental media (13, 15).

The fourth line of evidence is that it is possible to calculate the amounts of oxygen required by preimplantation mouse embryos to oxidize completely the pyruvate and glucose whose uptake has been determined experimentally (19, 26). When this is done (8), there is good agreement with the values for QO_2 obtained by Mills and Brinster (5), suggesting that nutrient consumption in vitro does represent a physiological phenomenon.

Conclusion

The evidence that preimplantation embryos within the female tract generate their ATP from substrates in oviduct and uterine fluids is very strong, but not conclusive. Proof would require the demonstration of nutrient consumption in vivo, and methods to test this are not at present available.

ATP Consumption by Preimplantation Embryos

The two major energy-consuming processes in adult mammalian cells are protein synthesis and ion pumping, mainly by Na^+,K^+ ATPase. The contribution of protein synthesis to energy expenditure is thought to lie between about 15% and 25%, depending on the tissue, its physiological status, and the species concerned. The figure for the Na^+,K^+ ATPase is generally higher, between 15% and 50%. Other energy-consuming processes include the synthesis of macromolecules other than protein (e.g., nucleic acids, triglycerides, and glycogen), the maintenance of intracellular pH, and substrate cycling (27). It is theoretically possible to arrive at an estimate of the energy cost of protein synthesis in early mouse embryos. There are two approaches.

The first approach exploits data on protein degradation. Merz et al. (28) found that the half-life of newly synthesized protein for 2-cell mouse embryos was 12.2 h. The figures for 8-cell embryos and blastocysts were

13.8 and 13.3 h, respectively. Wiebold and Anderson (29) found a higher figure at the 2–4 cell stage (20.4 h), but a comparable figure for morulae and blastocysts (12.2 h). Assuming an average protein content of 25 ng per embryo (6) and given the figures of Merz et al., 12.5-ng protein needs to be synthesized every 13 h to replace that degraded, or approximately 1 ng/h. Using the data of Wiebold and Anderson (29) for the 2–4 cell stage, the figure is about 0.5 ng/embryo/h. If the energy cost of protein synthesis is assumed to be 5 mole ATP per mole peptide bond formed, with 1 mole ATP responsible for the transport of an amino acid across the plasma membrane and 4 mole ATP for the actual synthetic process (27), this equates to 4.5 kJ/g protein formed, or 4.5×10^{-9} kJ/h (for 1 ng formed). Irrespective of which substrate is being oxidized, the energy equivalent of 1-L oxygen is the same, at about 20 kJ. Using the data of Mills and Brinster (5) for the oxygen consumption of preimplantation mouse embryos, the following intriguing conclusions are reached.

On the basis of the protein half-lives calculated by Merz et al. (28), the oxygen consumption at the 2- and 8-cell stages is insufficient to account for the energy requirement of protein synthesis, but at the morula, blastocyst, and late blastocyst stages, protein synthesis accounts for 64%, 49%, and 24%, respectively, of the energy available from oxidative metabolism. Using the data of Wiebold and Anderson (29), 75% of the energy available at the 2-cell stage would be consumed in the synthesis of protein, with similar figures to those quoted above at the morula and blastocyst stages.

Obviously, these calculations have to be treated with extreme caution; they are crucially dependent on the value chosen for protein half-life, where there are methodological and species differences (30). The values of Merz et al. (28) are, in all likelihood, underestimates since newly synthesized proteins tend to be unstable and turn over relatively rapidly. A further problem concerns the energetics of protein degradation, which are ill defined. They indicate, however, that the early preimplantation stages appear to use virtually all their ATP to synthesize protein, but that by the time the blastocyst stage has been reached, the proportion has fallen to about 25%. It is surely no coincidence that the second major energy-consuming process—that is, ion pumping associated with blastocoel cavity formation and mediated by the Na^+, K^+ ATPase— becomes significant at this time (31).

The second approach to calculating the energy cost of protein synthesis is to use data from experiments on the incorporation of amino acids into preimplantation mouse embryos. The hourly rates of leucine incorporation for 8–16 cell embryos and blastocysts were 0.035 and 0.125 pM/embryo, respectively (32). Assuming leucine comprises 8.2% of protein amino acid residues, it may be calculated that the cost of protein synthesis is only 6% and 9% of the total energy available at the 8-cell and blastocyst stages, respectively. These figures obviously are considerably below those calculated on the basis of protein degradation. The true figures probably lie between the estimates given by the two approaches. Until the issue is resolved, conclusions on the quantitative fate of ATP in the preimplantation embryo have to be provisional.

Nutrient Utilization by Preimplantation Embryos: The Role of Glutamine

The best-characterized species with respect to preimplantation embryo energy substrate utilization are the mouse, the rabbit and the human, where information is available for all developmental stages. In other species, the focus of attention has usually been on the blastocyst stage since this is of the greatest commercial interest. As indicated earlier, embryos in general utilize such aerobic substrates as pyruvate, lactate, and amino acids during their early preimplantation stages (8), with the question of fat oxidation, particularly in the domestic species, warranting further study.

Of potential amino acid substrates, the one of most interest is glutamine, which is consumed by preimplantation mouse and cattle embryos (18, 33, 34) and is a component of the medium CZB (35) that can overcome the *2-cell block* in mouse embryos (see also Chapter 5, this volume). Glutamine uptake increases with development and, hence, cell division, a characteristic shared with adult tissue, such as enterocytes, thymocytes, colonocytes, fibroblasts, and tumor cells (36). Also coincident with the later stages of preimplantation development, at least in the mouse, rat, and human, is a high rate of aerobic glycolysis (15, 22, 37).

Newsholme et al. (36) have proposed an ingenious explanation for the role of high rates of glycolysis and glutaminolysis in rapidly dividing cells. Glycolysis and glutaminolysis are the sources of metabolic intermediates for biosynthetic pathways: glucose 6-phosphate for the formation of ribose 5-phosphate required for DNA and RNA synthesis and glycerol 3-phosphate for phospholipid synthesis. Glutaminolysis provides carbon and nitrogen atoms for the biosynthesis of purines, pyrimidines, GTP, and NAD^+. In each case, the flux rates through glycolysis and glutaminolysis far exceed the requirement for these biosynthetic precursors. Newsholme et al. (36) have reconciled this by the notion that DNA, RNA, phospholipids, and the like, will only be required at precise times in the cell cycle and that the "maintenance of high rates of glycolysis and glutaminolysis at all times can be seen...as a device to allow intermediates to be 'tapped off' at the precise rate required whenever they are required for biosynthesis."

The hypothesis has a further feature that makes it attractive in the case of the late preimplantation embryo: It provides an answer to the question of why the embryo ceases to consume pyruvate. The switch from pyruvate to glucose as preferred substrate (given a mixture of the two) is quite marked and takes place between 94 and 104 h post-HCG for mouse embryos removed at the 2-cell stage (Martin, Leese, unpublished). According to Newsholme et al. (36), continuing pyruvate oxidation would provide large amounts of energy and an ATP concentration sufficient to inhibit glycolysis and glutaminolysis, by direct inhibition of 6-phosphofructokinase or via elevated citrate (38) and by inhibition of oxoglutarate dehydrogenase, respectively. Such inhibition would render glycolysis and glutaminolysis incapable of acting as "dynamic buffers."

Unifying Principles

Preimplantation development may, in all mammalian species, be divided into two phases on biochemical grounds: that which precedes and that which follows the activation of the embryonic genome. The requirement of preimplantation embryos for ATP may be seen in the light of this transition (8), the period controlled by maternal message being characterized by relatively low metabolic activity, with a switch to greater energy availability following activation of the embryonic genome in readiness for blastocoel cavity formation. This dramatic change in metabolism was well summarized by Epstein (39) (Fig. 6.1). The timing of the switch in different species and the molecular biology involved have recently been admirably reviewed (40).

These considerations and the fact, discussed earlier, that embryos initially lose a small amount of dry mass as they develop may be used to account for the size of the egg and early preimplantation embryo. By expending a considerable amount of energy in the ovary to produce a large oocyte, the energy cost of successive embryonic divisions to form many smaller cells within the same fixed volume will be much less than the alternative of beginning with an adult-sized cell and having to double its volume, weight, and zona pellucida at each division.

A second and much less emphasized division of preimplantation development can be made in terms of the environment of the embryo, contrasting the period spent in the oviduct with that spent in the uterus. The sojourn in the oviduct lumen is remarkably constant across the whole range of mammals, while that in the uterus varies between species, depending on the time of implantation.

It may be significant, at least in the mouse, that the dramatic increase in metabolic activity (Fig. 6.1) culminates in the passage of the embryo from the oviduct to the uterus. Apart from some speculation on the role of aerobic glycolysis in the uterine lumen prior to implantation (41), there has been little attempt to account for this shift in metabolism in terms of the embryo's changing environment. Such attempts at linking biochemistry with physiology could be a fruitful research area and have implications for our ability to grow early embryos satisfactorily in culture.

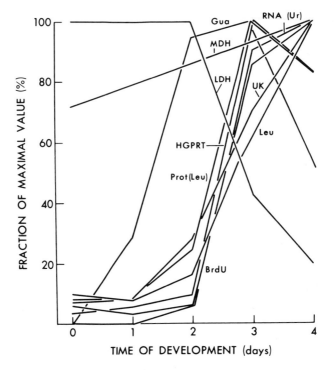

FIGURE 6.1. Summary of biochemical changes taking place in preimplantation mouse embryos. Values are expressed as percent of maximal value attained during the preimplantation period. While the most pronounced alterations appear to be occurring between days 2 (8–16 cells) and 3 (early blastocyst), important changes also occur both prior to and following this time. (Enzyme activities: HGPRT = hypoxanthine guanine phosphoribosyltransferase; LDH = lactate dehydrogenase; MDH = malate dehydrogenase; UK = uridine kinase. Transport capacities: Gua = guanine; BrdU = 5-bromodeoxyuridine; Leu = leucine. Macromolecular synthetic rates: Prot(Leu) = incorporation of leucine into protein; RNA(Ur) = incorporation of uridine into RNA). Reprinted with permission from Epstein (39).

Acknowledgments. The author's studies were supported by the UK Science and Engineering Council, The Medical Research Council, and The Wellcome Trust.

References

1. Auerbach S, Brinster RL. Effect of oxygen concentration on the development of two-cell mouse embryos. Nature (London) 1968;217:465–6.
2. Thomson JL. Effect of inhibitors of carbohydrate metabolism on the development of preimplantation mouse embryos. Exp Cell Res 1967;46:252–62.

3. Leese HJ. The formation and function of oviduct fluid. J Reprod Fertil 1987;82:843–56.
4. Yochim JM. Intrauterine oxygen tension and metabolism of the endometrium during the preimplantation period. In: Blandau RJ, ed. The biology of the blastocyst. University of Chicago Press, 1971:363–82.
5. Mills RM, Brinster RL. Oxygen consumption of preimplantation mouse embryos. Exp Cell Res 1967;47:337–44.
6. Sellens MH, Stein S, Sherman MI. Protein and free amino acid content in preimplantation mouse embryos and in blastocysts under various culture conditions. J Reprod Fertil 1981;61:307–15.
7. Stern S, Biggers JD, Anderson E. Mitochondria and early development of the mouse. J Exp Zool 1971;176:179–92.
8. Leese HJ. Metabolism of the preimplantation mammalian embryo. In: Milligan SR, ed. Oxford reviews of reproductive biology; 13. Oxford University Press, 1991:35–72.
9. Betteridge KJ, Fléchon J-E. The anatomy and physiology of preattachment bovine embryos. Theriogenology 1988;29:155–87.
10. Kane MT. Minimal nutrient requirements for culture of one-cell rabbit embryos. Biol Reprod 1987;37:775–8.
11. Biggers JD. Pioneering mammalian embryo culture. In: Bavister BD, ed. The mammalian preimplantation embryo: regulation of growth and differentiation in vitro. New York: Plenum Press, 1987:1–22.
12. Pemble LB, Kaye PL. Whole protein uptake and metabolism by mouse blastocysts. J Reprod Fertil 1986;78:149–57.
13. Leese HJ. The environment of the preimplantation embryo. In: Edwards RG, ed. Establishing a successful human pregnancy. Serono Symposia Publications; 66. New York: Raven Press, 1990:143–54.
14. Brown JJG, Whittingham DG. The roles of pyruvate, lactate and glucose during preimplantation development of embryos from F hybrid mice in vitro. Development 1991;112:99–105.
15. Gardner DK, Leese HJ. Concentrations of nutrients in mouse oviduct fluid and their effects on embryo development and metabolism in vitro. J Reprod Fertil 1989;88:361–8.
16. Lawitts JA, Biggers JD. Optimization of mouse embryo culture media using simplex methods. J Reprod Fertil 1991;91:543–56.
17. Brinster RL. Nutrition and metabolism of the ovum, zygote and blastocyst. In: Greep RO, Astwood EB, eds. Handbook of physiology; vol II. Washington, D.C.: American Physiological Society, 1973:165–85.
18. Rieger D, Guay P. Measurement of the metabolism of energy substrates in individual bovine blastocysts. J Reprod Fertil 1988;83:585–91.
19. Leese HJ, Barton AM. Pyruvate and glucose uptake by mouse ova and preimplantation embryos. J Reprod Fertil 1984;32:9–13.
20. Butler JE, Lechene C, Biggers JD. Noninvasive measurement of glucose uptake by two populations of murine embryos. Biol Reprod 1988;39:779–86.
21. Hardy K, Hooper MAK, Handyside AH, Rutherford AJ, Winston RML, Leese HJ. Non-invasive measurement of glucose and pyruvate uptake by individual human oocytes and preimplantation embryos. Hum Reprod 1989; 4:188–91.
22. Brison DR, Leese HJ. Energy metabolism in late preimplantation rat embryos. J Reprod Fertil 1991;93:245–51.

23. Leese HJ, Lenton EA. Glucose and lactate in human follicular fluid: concentrations and inter-relationships. Hum Reprod 1990;5:915–9.
24. Gott AL, Hardy K, Winston RML, Leese HJ. The nutrition and environment of the early human embryo. Proc Nutr Soc 1990;17:546–7.
25. Edirisinghe WR, Wales RG. Influence of environmental factors on the metabolism of glucose by preimplantation mouse embryos in vitro. Aust J Biol Sci 1985;38:411–20.
26. Gardner DK, Leese HJ. The role of glucose and pyruvate transport in regulating nutrient utilization by preimplantation mouse embryos. Development 1988;194:423–9.
27. Kelly JM, McBride BW. The sodium pump and other mechanisms of thermogenesis in selected tissues. Proc Nutr Soc 1990;49:185–202.
28. Merz EA, Brinster RL, Brunner S, Chen HY. Protein degradation during preimplantation development of the mouse. J Reprod Fertil 1981;61:415–8.
29. Wiebold JL, Anderson GB. Lethality of a tritiated amino acid in early mouse embryos. J Embryol Exp Morphol 1985;88:209–17.
30. Jung T. Protein synthesis and degradation in non-cultured and in-vitro cultured rabbit blastocysts. J Reprod Fertil 1989;86:507–12.
31. Biggers JD, Bell JE, Benos DJ. The mammalian blastocyst: transport functions in a developing epithelium. Am J Physiol 1988;255:419–32.
32. Epstein CJ, Smith SA. Amino acid uptake and protein synthesis in preimplantation mammalian embryos. Dev Biol 1973;33:171–84.
33. Gardner DK, Clarke RN, Lechene CP, Biggers JD. Development of a noninvasive ultramicrofluorometric method for measuring the uptake of glutamine by single preimplantation mouse embryos. Gamete Res 1989;24:427–38.
34. Chatot CL, Tasca RJ, Ziomek CA. Glutamine uptake and utilization by preimplantation mouse embryos in CZB medium. J Reprod Fertil 1990;89:335–46.
35. Chatot CL, Ziomek CA, Bavister BD, Lewis JL, Torres I. An improved culture medium supports development of random-bred 1-cell mouse embryos in vitro. J Reprod Fertil 1989;86:679–88.
36. Newsholme EA, Newsholme P, Curi R, Crabtree B, Ardawi MSM. Glutamine metabolism in different tissues: its physiological and pathological importance. In: Kinney JM, Borum PR, eds. Perspectives in clinical nutrition. Baltimore-Munich: Urban & Schwarzenberg, 1989:71–98.
37. Gott AL, Hardy K, Winston RML, Leese HJ. Non-invasive measurement of pyruvate and glucose uptake and lactate formation by single human preimplantation embryos. Hum Reprod 1990;5:104–8.
38. Barbehenn EK, Wales RG, Lowry OH. The explanation for the blockade of glycolysis in early mouse embryos. Proc Natl Acad Sci USA 1974;71:1056–60.
39. Epstein CJ. Gene expression and macromolecular synthesis during preimplantation embryonic development. Biol Reprod 1975;12:82–105.
40. Telford NA, Watson AJ, Schultz GA. Transition from maternal to embryonic control in early mammalian development: a comparison of several species. Mol Reprod Dev 1990;26:90–100.
41. Leese HJ. Energy metabolism of the blastocyst and uterus at implantation. In: Yoshinaga K, ed. Blastocyst implantation. Serono Symposia, USA. Boston: Adams Publishing, 1989:39–44.

7

Growth Factors as Regulators of Mammalian Preimplantation Embryo Development

BERND FISCHER

Low-molecular weight peptides with a wide range of different actions are collectively called *growth factors* (GF). Their biological action is multifunctional (1, 2). Based on structural homology and shared receptor binding activity, GF are grouped into families, such as the *epidermal growth factor* (EGF), *insulin* and *insulin-like growth factor* (IGF), *transforming growth factor β* (TGFβ), *fibroblast growth factor* (FGF), and *platelet-derived growth factor* (PDGF) families (3). Binding of a GF to its cell surface receptor initiates, via activation of second messengers, cellular responses related either to cell proliferation (growth) or to cell differentiation. GF mainly act locally by paracrine or autocrine mechanisms. Oncogenes can be part of the intracellular signal transduction (reviewed in 4, 5). The exact mechanisms and substances involved in the signaling process and function, however, are still unknown.

Most of our information regarding GF action and involvement of oncogenes in GF-stimulated cellular responses was gained in studies employing cell lines. A more stringent biological approach, however, is to study GF action in intact and interacting tissues. Developing embryos offer such a biological system.

GF and Mammalian Embryogenesis

Most work on the involvement of GF in embryo development has been performed in nonmammalian species or in postimplantation stages in mammals (3, 6–8). The interest in GF action on mammalian preimplantation development has significantly increased during the last few years. The reason for this increase may be twofold. First, the general effects of GF—that is, stimulation of cell proliferation and cell differentiation—are classical features of preimplantation embryo development. Thus, regulatory effects by GF seem as likely to occur on preimplantation as on postimplantation development. Second, sensitive techniques have been established in recent years to investigate GF and GF actions in preimplantation embryos—that is, in specimens characterized by limitations in respect to the amount of material available for investigation. A landmark was the work of Rappolee and colleagues on the differential expression of GF genes in preimplantation mouse embryos (9) (see also, 10). Their results, showing expression (i) of some but not all GF studied and (ii) in a stage-specific manner, pointed to a subtle role of GF in controlling early embryogenesis. However, verification of specific GF functions is required to prove their involvement in the regulation of mammalian preimplantation embryo development (11). Recent functional studies have provided the first evidence for such an involvement.

GF and Mammalian Preimplantation Embryo Development

Reviewing recent publications on GF and mammalian preimplantation embryo development (inevitably incomplete in such a fast-growing field), two observations have to be emphasized. First, the majority of studies have been performed in the mouse, and second, these have focused on a few GF, mainly on the IGF and EGF families (Table 7.1). Our knowledge is still fragmentary and just beginning to disclose the first regulatory actions. Present experimental findings are summarized below (see Table 7.1).

Developmentally Regulated GF Gene Expression

Expression of GF and GF binding sites/receptors is developmentally regulated. Some GF and GF receptors are not expressed during preimplantation development in mammals. Others are present as maternal transcripts and/or zygotic transcripts, the latter being first expressed during cleavage or after blastocyst formation (9, 10). GF and GF receptors specifically expressed during cleavage have not yet been detected. GF found in cleavage stage embryos are also transcribed in blastocysts. Stage-specific effects are known from several GF. Insulin (12–15) and EGF (16), for example, exert their actions on compacted morulae and blastocyst stages, but not on early cleavage stages.

Differential GF Expression and Binding to Trophoblast and Embryoblast Cells

After differentiation of blastomeres into *embryoblast* (ICM) and *trophoblast* (TR) cells, expression and function of some GF are cell

TABLE 7.1. Studies on GF/GF receptors in mammalian preimplantation embryos.

Species	GF	Embryonic stage	Main results	Reference
Mouse	TGFα TGFβ$_1$ EGF PDGF-A bFGF NGF CSF	Unfertilized oocytes, 2-c, 4-c, 16-c, blast	Stage-specific expression of GF genes during preimplantation development (mRNA detection by RT-PCR) Translation of transcripts into GF protein (immunofluorescence detection)	9
Mouse	IGF-I IGF-II Insulin kFGF	Unfertilized oocytes, 2-c, 4-c, 16-c, blast	Stage-specific expression of GF genes during preimplantation development Stage-specific expression of IGF-I, IGF-II, and insulin receptor genes Increase in protein synthesis by IGF-II and insulin in cultured blast	10
Mouse	EGF TGFα TGFβ$_1$ IGF-I	Cultured 2-c, 8-c	Increase in cell number, blastocyst formation, and zona hatching in cultured embryos by EGF, TGFα, and TGFβ$_1$	18
Mouse	EGF	Blast, cultured 2-c, 8-c	EGF receptor with M$_r$ of 170,000 and tyrosine kinase activity Tyrosine phosphorylation-mediated signal transduction is important for normal preimplantation development	48
Mouse	TGFα EGF	Mor, early blast	Dose-dependent stimulation of blastocyst expansion Can be inhibited by inhibitors of EGF receptor kinase activity Stimulation of methionine uptake 30% nonresponders to GF stimulation	23
Mouse	TGFβ$_2$	2-c, 4-c, mor, blast	GF protein detectable (by immunofluorescence) from 4-c stage onwards Specific staining in TR, not in ICM	17
Mouse	IGF-II	Blast	No gene expression demonstrable by in situ hybridization (in contrast to slightly older postimplantation stages)	54
Mouse	Insulin	2-c cultured to mor and blast, noncompacted and compacted 8-c	Dose-dependent increase in protein synthesis in all stages ≥ compacted 8-c, but not in noncompacted 8-c	12

TABLE 7.1. Continued.

Species	GF	Embryonic stage	Main results	Reference
Mouse	Insulin	2-c cultured to noncompacted and compacted mor, blast; in vivo blast	Increase in cell number Increased mor and blast formation	22
Mouse	Insulin	Oocytes, 2-c, 4-c, 8-c, blast	Detection of insulin receptors in compacted 8-c, mor, blast, and isolated ICM	55
Mouse	Insulin	2-c cultured to noncompacted mor, mor or blast	Specific stimulation of ICM proliferation, no effect on TR Increased blast formation	19
Mouse	Insulin	Oocyte, 2-c, 4-c, 8-c, mor, blast	Stage-specific binding by mor and blast	56
Mouse	Insulin	2-c, 4-c, 8-c, mor, blast	Stage-specific binding and receptor-mediated endocytosis of insulin from mor stage onwards Binding by TR and ICM Insulin receptor (localization by immunocytochemistry) on TR (especially polar TR) and ICM Increase in RNA and DNA synthesis in mor and blast Probably no de novo synthesis of insulin by preimplantation mouse embryos	13–15
Mouse	Insulin IGF-I	Blast	Stimulation of protein synthesis in TR and ICM by insulin and IGF-I Crossreaction of IGF-I with the insulin receptor	57
Mouse	Insulin IGF-I IGF-II	Oocyte, 2-c, 4-c, 8-c, mor, blast	Dose-dependent specific binding of insulin by mor and blast, not by younger stages Binding by TR and ICM Displacement of insulin binding by IGFs Binding of IGF-I and IGF-II by TR outgrowths	58
Mouse	Insulin IGF-I IGF-II	2-c, 8-c, mor, blast	Increase in protein synthesis in mor and blast by insulin, not by IGF-I, IGF-II Increase in RNA and DNA synthesis by insulin in 8-c, clearly in mor and blast; only slight or no effects by IGF-I, IGF-II	20

7. Mammalian Preimplantation Embryo Development

Species	GF	Stage	Observation	Ref
Mouse	IGF-II Insulin EGF TGFα	Cultured 1-/2-c	Increased protein synthesis in blast Increased cell number	21
Mouse	TGFα	Blast	GF protein detectable in virtually every cell	21
Mouse	EGF	2-c, mor, blast	Cell- (TR) and stage-specific (mor/blast transition) increase in protein synthesis	16
Mouse	EGF	Blast	Stimulated appearance of *fos* protein	27
Rat	IGF-II	Blast (periimplantation)	No gene expression	59
Rabbit	EGF	Blast (periimplantation)	Immunohistochemical detection of EGF receptors in TR	41
Rabbit, golden hamster	EGF	Blast	No specific binding	32
Cow	TGFβ bFGF	Cultured fertilized oocytes	GF interaction to promote development beyond 8-c in vitro block	60
Cow	TGFα	16-c	Increased blast formation Shortened 4th-cell cycle	28
Cow	PDGF	Cultured 2-c and 8-c	May induce expression of *c-myc* and *c-fos*	28
Pig	Insulin	Isolated blastomeres	Increased blast formation	61
Pig	IGF-I	Spherical to elongated blast	Stage-specific amount of GF protein No clear stage specificity in detected GF mRNA, probably low number of transcripts	62
Pig	IGF-I EGF	TR from D15–19 blast	Specific binding by freshly isolated and cultured TR EGF receptor with M_r of 160,000 EGF stimulated cell proliferation IGF-I/insulin receptor	49
Pig	EGF	Elongated blast	Specific GF binding in a time-dependent and saturable manner	33
Pig	EGF	Preimplantation embryos	EGF receptor with M_r of 170,000	50
Cow, pig, sheep	EGF	Spherical to elongated blast	Specific binding TGFα competes for EGF binding sites	32
Horse	IGF-II	Blast (periimplantation)	GF gene expression	63

Note: GF: See text for full name. (2-c, 4-c, etc. = number of blastomeres in cleavage stage embryos; mor = morulae; blast = blastocysts; TR = trophoblast cells; ICM = embryoblast cells.)

specific. Examples are the specific expression of TGFβ$_2$ in TR (17), the binding of EGF to TR (18), and the specific effects reported for EGF on TR (16) and insulin on ICM (19). However, not all specific effects could be confirmed in other studies. Insulin, for example, was found to bind to ICM and TR (13–15). Therefore, current information is not conclusive. A differential action of GF on the various embryonic cell lines and cell line derivates may apply to some, but not to all, GF.

Main Function: Cell Proliferation and Cell Differentiation

As expected from their known functional properties in other tissues, GF can stimulate cell proliferation and differentiation in preimplantation embryos. Stimulation of growth-related processes, such as increased RNA and DNA (13–15), protein synthesis (10, 12, 16, 20, 21), or rise in cell number (18, 22), are typical embryonic responses to GF. It has already been emphasized that some GF initiate these reactions only in blastocysts. However, the described GF effects may not simply reflect growth. They may be attributed to specialized functions of GF during the second part of preimplantation embryo development, such as supporting special nutritive functions; for example, transport of nutrients (insulin [13–15]), blastocyst expansion (EGF, TGFα [23]), or cellular changes directed towards implantation (24 and Chapter 19, this volume).

Some GF are produced by the pregnant uterus (see below). GF of embryonic origin (Table 7.1) may be directed towards both the maternal and embryonic systems. By controlling subtle developmental processes within the embryo and endometrium, GF would be potent candidates for generating synchronized embryo-maternal interactions during the pre-implantation period (25).

GF Action in Preimplantation Embryos May Involve Oncogene Expression

Cellular binding of GF—for example, of EGF—is known to induce expression of oncogenes (26). There are preliminary reports (27, 28) that GF may also induce oncogene expression in preimplantation embryos. It is most likely that certain oncogenes will turn out to be integral components of the intracellular growth control network/cascade (5) following GF binding. The interest for investigating oncogene expression in pre-implantation embryos is twofold. First, as in postimplantation stages (29–31), the expression and biological significance of oncogenes during mammalian preimplantation development are still largely unknown. Second, as in other tissues, interactions of oncogenes with GF are likely and may help to understand GF function.

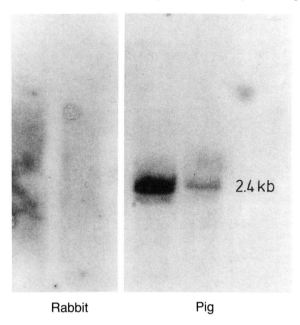

Rabbit   Pig

FIGURE 7.1. Expression of c-*myc* in pig D13, but not in rabbit D6 preimplantation embryos. The blot was hybridized with a mouse c-*myc* cDNA probe (second exon) in phosphate buffer with 7% SDS at 65°C for 20 h. It was washed under stringent conditions (2 × 20 min incubations each, in 5% and 1% SDS, all at 65°C) and exposed to a preflashed X-ray film for 7 to 9 days. Based on the migration of the standards, pig c-*myc* mRNA size was calculated to be ~2.4 kb, corresponding well with data for human, described in reference 51, and mouse, described in reference 52, c-*myc* mRNA. The hybridized Northern blot was kindly provided by S.J. Nass, University of Wisconsin–Madison (32).

In somatic cells, EGF induces the expression of c-*myc* and c-*fos* (26)—that is, the expression of oncogenes closely associated with cell growth and cell differentiation. It may not be coincidental that elongated blastocysts of the pig, which bind EGF (32, 33), express c-*fos* (31) and c-*myc* (32) (Fig. 7.1). In preliminary assays (positive control data from adult rabbit tissue are still missing), we could detect neither a specific EGF binding (Fig. 7.2; see below) nor c-*myc* mRNA (32) (Fig. 7.1) in day 6 rabbit blastocysts.

Clarification of Relevant GF and Possible Species Differences Needed

Experimental proof for an effect on preimplantation embryo development has been provided largely for members of only two families, namely,

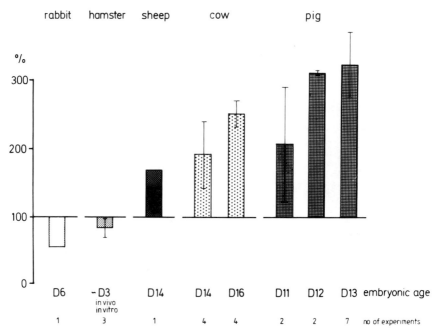

FIGURE 7.2. EGF binding in mammalian preimplantation embryos. EGF binding is expressed as total binding in percent of nonspecific binding; nonspecific binding = 100% ($\bar{x}$ ± SEM). Total binding was determined by incubating homogenized embryonic samples with ^{125}I-EGF (mouse, 0.5 nM) for 60 min after 15 min pretreatment with diluent. Pretreatment with unlabeled EGF (mouse, receptor-grade, 167 nM) instead of diluent was employed to measure nonspecific binding. Incubation was performed in Hepes-buffered modified Tyrode solution at 37°C. Binding affinities and receptor numbers given in the text were evaluated by Woolf plot analysis, as described in reference 53.

insulin/IGF and EGF/TGFα. Other GF may also affect mammalian preimplantation embryos but have not yet or have only in a few cases been investigated (TGFβ, PDGF, and bFGF). In addition, attention has to be paid to species differences. Preimplantation development of mammalian embryos is much more divergent than generally assumed, indicated by differences ranging from energy and amino acid metabolism (Chapter 5, this volume) to cell proliferation and numbers (34 and Chapter 13, this volume) and morphology. In such species as the cow (35, 36), sheep (37, 38), and pig (39, 40), blastocysts undergo a drastic change in morphology from a spherical blastocyst to an elongated blastocyst of, in the pig, up to several meters in length within a few days. This remodeling is associated with a considerable increase in protein concentration (40). In other species, embryos hardly expand at all (mouse and hamster) and implant as spherical blastocysts (primates, equids, lagomorpha, and rodents). Such diverse development obviously must be differently regulated.

TABLE 7.2. K_d of EGF binding in various tissues/cells (pM).

Tissues/cell lines	K_d	Reference
Human epidermoid cancer cells (A 431) with overexpression of the EGF receptor	5000, 700	64
Human epidermoid cancer cells (A 431) with overexpression of the EGF receptor	400	65
Human breast cancer	200	66
Mouse mammary tissue	10,000, 4000	67
Human placenta syncytiotrophoblast	200	68
Mouse uterine tissue	1000	43
Pig D12 and D13 embryos	200–12	32
Cow D16 embryos	230–80	32

EGF Binding in Mammalian Preimplantation Embryos

EGF and its receptor are expressed in the pregnant uterus of several species (41–45 and Chapter 19, this volume), most likely influenced by estrogens (42, 46). EGF is a potent mitogen (47). It is therefore one of the possible factors involved in the rapid and substantial increase in size during blastocyst elongation in ruminants and the pig.

We started to test this hypothesis by examining EGF binding in elongating (cow, sheep, and pig) and nonelongating blastocysts (rabbit and golden hamster) (32). We could demonstrate a specific binding of EGF by cow, sheep, and pig embryos, but not by rabbit and hamster embryos (Fig. 7.2). In the pig and cow, younger, not yet elongated stages (cow D14, pig D11) also bound EGF (Fig. 7.2). The binding affinities ranged from 200–12 pM and 230–80 pM in pig (D12 and D13) and cow embryos (D16), respectively, and lie in the range of other tissues known for high EGF binding (Table 7.2). The number of binding sites was estimated to be 8–20 pM/mg protein in pig D12, 1–5 in pig D13, and ~0.4 in cow D16 embryos. Specificity of binding was proved in competitive binding assays with insulin. Insulin did not compete with EGF for the binding sites. TGFα, however, was a strong competitor, especially if human recombinant TGFα was employed (32).

It has to be emphasized, however, that EGF binding has been shown in species with nonelongating embryos, autoradiographically (18) and in crosslinking studies (48) in the mouse, and immunohistochemically in the rabbit (41). The differences between these findings need further study; for example, investigation of binding affinities, binding sites/receptor numbers, and functional properties of the binding sites in embryos of these species. The biochemical characteristics of the EGF binding sites in preimplantation embryos of the mouse (23, 48) and pig (49, 50) closely resemble the properties of the classical EGF receptor known from adult tissues. The molecular weight is approximately 160,000 (49) to 170,000 (48, 50), and the receptor-associated tyrosine kinase activity was proven biochemically (48) as well as biologically (23, 48). It has now to be investigated whether preimplantation embryos in other mammalian species also express the EGF receptor and when the receptor becomes functional during ontogeny.

Conclusions

In the last few years, experimental evidence has accumulated indicating that GF are involved in the regulation of preimplantation embryo development in mammals. A new, unique function of GF related specifically to preimplantation embryo development has not been found. Clear evidence for a role of GF in the control of cell differentiation and the initiation of specific developmental processes is still lacking. However, developmentally regulated expression of GF and GF receptor genes; a possible differential expression and binding of GF by trophoblast and embryoblast cells; and the stimulation of embryonic cell proliferation, protein synthesis, and blastocyst expansion by GF are strong indicators for a fundamental involvement of GF in mammalian preimplantation development. GF function seems more related to the second part of preimplantation development (blastocyst formation and development) rather than to cleavage. Expression of specific oncogenes may be part of the GF-initiated growth-signaling pathway in embryos. Effects of some GF on preimplantation development have already been documented; effects of others will surely be discovered in the future. Embryos of different species will probably differ in their need for specific GF. GF are produced by the embryo itself and/or delivered by the maternal system. Investigation of GF-generated maternal-embryonic interactions promises insights of basic importance for our understanding of early pregnancy in mammals.

Acknowledgments. The author thanks Armin Schumacher for helpful comments on the manuscript and Jacques Beckman and Gisela Mathieu for typing the manuscript. The work on EGF binding and expression of c-*myc* was performed in close cooperation with Barry D. Bavister, Terri A. Rose, Lewis G. Sheffield, Sharryl J. Nass, and Paul J. Bertics, University of Wisconsin, Madison. This research was made possible by financial support of the Max Kade Foundation, New York, NY.

References

1. Sporn MB, Roberts AB. Peptide growth factors are multifunctional. Nature (London) 1988;332:217-9.
2. Sporn MB, Roberts AB. The multifunctional nature of peptide growth factors. In: Sporn MB, Roberts AB, eds. Peptide growth factors and their receptors, I. Handbook of experimental pharmacology; vol 95/I. Berlin: Springer-Verlag, 1990:3-15.
3. Mercola M, Stiles CD. Growth factor superfamilies and mammalian embryogenesis. Development 1988;102:451-60.
4. Studzinski GP. Oncogenes, growth, and the cell cycle: an overview. Cell Tissue Kinet 1989;22:405-24.
5. Leutz A, Graf T. Relationships between oncogenes and growth control. In: Sporn MB, Roberts AB, eds. Peptide growth factors and their receptors, II. Handbook of experimental pharmacology; vol 95/II. Berlin: Springer-Verlag, 1990:655-703.
6. Whitman M, Melton DA. Growth factors in early embryogenesis. Annu Rev Cell Biol 1989;5:93-117.
7. Heath JK, Smith AG. Growth factors in embryogenesis. Br Med Bull 1989; 2:319-36.
8. Nilsen-Hamilton M. Growth factor signaling in early mammalian development. In: Rosenblum IY, Heyner S, eds. Growth factors in mammalian development. Boca Raton, FL: CRC Press, 1989:135-65.
9. Rappolee DA, Brenner CA, Schultz R, Mark D, Werb Z. Developmental expression of PDGF, TGF-α, and TGF-β genes in preimplantation mouse embryos. Science 1988;241:1823-5.
10. Rappolee DA, Sturm KS, Schultz GA, Pederson RA, Werb Z. The expression of growth factor ligands and receptors in preimplantation mouse embryos. In: Heyner S, Wiley LM, eds. Early embryo development and paracrine relationships. New York: Alan R. Liss, 1990:11-25.
11. Simmen FA, Simmen RCM. Peptide growth factors and proto-oncogenes in mammalian conceptus development. Biol Reprod 1991;44:1-5.
12. Harvey MB, Kaye PL. Insulin stimulates protein synthesis in compacted mouse embryos. Endocrinology 1988;122:1182-4.
13. Heyner S, Rao LV, Jarett L, Smith RM. Preimplantation mouse embryos internalize maternal insulin via receptor-mediated endocytosis: pattern of uptake and functional correlations. Dev Biol 1989;134:48-58.
14. Heyner S, Mattson BA, Smith RM, Rosenblum IY. Insulin and insulin-like growth factors in mammalian development. In: Rosenblum IY, Heyner S, eds. Growth factors in mammalian development. Boca Raton, FL: CRC Press, 1989:91-112.
15. Rao LV, Farber M, Smith RM, Heyner S. The role of insulin in preimplantation mouse development. In: Heyner S, Wiley LM, eds. Early embryo development and paracrine relationships. New York: Alan R. Liss, 1990:109-24.
16. Wood SA, Kaye PL. Effects of epidermal growth factor on preimplantation mouse embryos. J Reprod Fertil 1989;85:575-82.
17. Slager HG, Lawson KA, van den Eijnden-van Raaij AJM, de Laat SW, Mummery CL. Differential localization of TGF-β2 in mouse preimplantation and early postimplantation development. Dev Biol 1991;145:205-18.

18. Paria BC, Dey SK. Preimplantation embryo development in vitro: cooperative interactions among embryos and role of growth factors. Proc Natl Acad Sci USA 1990;87:4756–60.
19. Harvey MB, Kaye PL. Insulin increases the cell number of the inner cell mass and stimulates morphological development of mouse blastocysts in vitro. Development 1990;110:963–7.
20. Rao LV, Wikarczuk ML, Heyner S. Functional roles of insulin and insulinlike growth factors in preimplantation mouse embryo development. In Vitro Cell Dev Biol 1990;26:1043–8.
21. Werb Z. Expression of EGF and TGF-α genes in early mammalian development. Mol Reprod Dev 1990;27:10–15.
22. Gardner HG, Kaye PL. Insulin increases cell numbers and morphological development in mouse pre-implantation embryos in vitro. Reprod Fertil Dev 1991;3:79–91.
23. Dardik A, Schultz RM. Blastocoel expansion in the preimplantation mouse embryo: stimulatory effect of TGF-α and EGF. Development 1991;113:919–30.
24. Tamada H, Das SK, Andrews GK, Dey SK. Cell-type-specific expression of transforming growth factor-α in the mouse uterus during the peri-implantation period. Biol Reprod 1991;45:365–72.
25. Fischer B. Effects of asynchrony on rabbit blastocyst development. J Reprod Fertil 1989;86:479–91.
26. McCaffrey P, Ran W, Campisi J, Rosner MR. Two independent growth factor-generated signals regulate c-*fos* and c-*myc* mRNA levels in Swiss 3T3 cells. J Biol Chem 1987;262:1442–5.
27. Adamson ED. EGF receptor activities in mammalian development. Mol Reprod Dev 1990;27:16–22.
28. Larson RC, Ignotz GG, Currie WB. Platelet derived growth factor (PDGF) initiates completion of the fourth cell cycle of bovine embryo development. J Reprod Fertil Abstr Ser 1991;7:6.
29. Zimmerman KA, Yancopoulos GD, Collum RG, et al. Differential expression of *myc* family genes during murine development. Nature (London) 1986;319:780–3.
30. Downs KM, Martin GR, Bishop JM. Contrasting patterns of *myc* and N-*myc* expression during gastrulation of the mouse embryo. Genes Dev 1989;3:860–9.
31. Whyte A, Stewart HJ. Expression of the proto-oncogene *fos* (c-*fos*) by preimplantation blastocysts of the pig. Development 1989;105:651–6.
32. Fischer B, Rose TA, Nass SJ, Sheffield LG, Bavister BD. Specific binding of epidermal growth factor (EGF) and expression of c-*myc* oncogene in mammalian preimplantation embryos. J Reprod Fertil Abstr Ser 1991;7:7.
33. Letcher LR, Simmen RCM, Pope WF. Characterization of epidermal growth factor receptors in preimplantation swine embryos. J Anim Sci 1989;suppl 67:106.
34. Kane MT. In vitro growth of preimplantation rabbit embryos. In: Bavister BD, ed. The mammalian preimplantation embryo. Regulation of growth and differentiation in vitro. New York: Plenum Press, 1987:193–217.

35. Chang MC. Development of bovine blastocyst with a note on implantation. Anat Rec 1952;113:143–61.
36. Gustafsson H, Plöen L. The morphology of 16 and 17 day old bovine blastocysts from virgin and repeat breeder heifers. Anat Histol Embryol 1986;15:277–87.
37. Chang MC, Rowson LEA. Fertilization and early development of Dorset Horn sheep in the spring and summer. Anat Rec 1965;152:303–16.
38. Rowson LEA, Moor RM. Development of the sheep conceptus during the first fourteen days. J Anat 1966;100:777–85.
39. Perry JS, Rowlands IW. Early pregnancy in the pig. J Reprod Fertil 1962;4:175–88.
40. Anderson LL. Growth, protein content and distribution of early pig embryos. Anat Rec 1978;190:143–54.
41. Hofmann GE, Anderson TL. Immunohistochemical localization of epidermal growth factor receptor during implantation in the rabbit. Am J Obstet Gynecol 1990;162:837–41.
42. DiAugustine RP, Petrusz P, Bell GI, et al. Influence of estrogens on mouse uterine epidermal growth factor precursor protein and messenger ribonucleic acid. Endocrinology 1988;122:2355–63.
43. Brown MJ, Zogg JL, Schultz GS, Hilton FK. Increased binding of epidermal growth factor at preimplantation sites in mouse uteri. Endocrinology 1989;124:2882–8.
44. Huet-Hudson YM, Andrews GK, Dey SK. Epidermal growth factor and pregnancy in the mouse. In: Heyner S, Wiley LM, eds. Early embryo development and paracrine relationships. New York: Alan R Liss, 1990:125–36.
45. Simmen RCM, Simmen FA. Regulation of uterine and conceptus secretory activity in the pig. J Reprod Fertil 1990;suppl 40:279–92.
46. Lingham RB, Stancel GM, Loose-Mitchell DS. Estrogen regulation of epidermal growth factor receptor messenger ribonucleic acid. Mol Endocrinol 1988;2:230–5.
47. Fisher DA, Lakshmanan J. Metabolism and effects of epidermal growth factor and related growth factors in mammals. Endocr Rev 1990;11:418–42.
48. Paria BC, Tsukamura H, Dey SK. Epidermal growth factor-specific protein tyrosine phosphorylation in preimplantation embryo development. Biol Reprod 1991;45:711–8.
49. Corps AN, Brigstock DR, Littlewood CJ, Brown KD. Receptors for epidermal growth factor and insulin-like growth factor-I on preimplantation trophoderm of the pig. Development 1990;110:221–7.
50. Zhang Y, Paria DC, Dey SK, Davis DL. Characterization of the epidermal growth factor receptor in preimplantation conceptuses. Dev Biol 1992;151:617–21.
51. Watt R, Stanton LW, Marcu KB, et al. Nucleotide sequence of cloned cDNA of human c-*myc* oncogene. Nature (London) 1983;303:725–8.
52. Bernard O, Cory SC, Gerondakis S, Webb E, Adams JM. Sequence of the murine and human cellular *myc* oncogenes and two modes of *myc* transcription resulting from chromosome translocation in B lymphoid tumors. EMBO J 1983;2:2375–83.

53. Keightley DD, Fisher RJ, Creessie NAC. Properties and interpretation of the Woolf and Scatchard plots in analyzing data from steroid receptor assays. J Steroid Biochem 1983;19:1407–12.
54. Lee JE, Pintar J, Efstratiadis A. Pattern of the insulin-like growth factor II gene expression during early mouse embryogenesis. Development 1990; 110:151–9.
55. Harvey MB, Kaye PL. Visualization of insulin receptors on mouse pre-embryos. Reprod Fertil Dev 1991;3:9–15.
56. Rosenblum IY, Mattson BA, Heyner S. Stage-specific insulin binding in mouse preimplantation embryos. Dev Biol 1986;116:261–3.
57. Harvey MB, Kaye PL. Mouse blastocysts respond metabolically to short-term stimulation by insulin and IGF-1 through the insulin receptor. Mol Reprod Dev 1991;29:253–8.
58. Mattson BA, Rosenblum IY, Smith RM, Heyner S. Autoradiographic evidence for insulin and insulin-like growth factor binding to early mouse embryos. Diabetes 1988;37:585–9.
59. Florance RSK, Senior PV, Byrne S, Beck F. The expression of IGF-II in the early post-implantation rat conceptus. J Anat 1991;175:169–79.
60. Larson RC, Ignotz GG, Currie WB. Defined medium containing TGFβ and bFGF permits development of bovine embryos beyond the "8-cell block." J Reprod Fertil Abstr Ser 1990;5:16.
61. Saito S, Niemann H. Effects of extracellular matrices and growth factors on the development of isolated porcine blastomeres. Biol Reprod 1991; 44:927–36.
62. Letcher R, Simmen RCM, Bazer FW, Simmen FA. Insulin-like growth factor-I expression during early conceptus development in the pig. Biol Reprod 1989;41:1143–51.
63. Stewart F, Allen WR. Gene expression in equine conceptus around the time of implantation. J Reprod Fertil Abstr Ser 1991;7:39.
64. Kawamoto T, Sato JD, Le A, et al. Growth stimulation of A431 cells by epidermal growth factor: identification of high-affinity receptors for epidermal growth factor by an anti-receptor monoclonal antibody. Proc Natl Acad Sci USA 1983;80:1337–41.
65. Defize LHK, Arndt-Jovin DJ, Jovin TM, et al. A431 cell variants lacking the blood group A antigen display increased high affinity epidermal growth factor-receptor number, protein-tyrosine kinase activity, and receptor turnover. J Cell Biol 1988;107:939–49.
66. Fitzpatrick SL, La Chance MP, Schultz GS. Characterization of epidermal growth factor receptor and action on human breast cancer cells in culture. Cancer Res 1984;44:3442–7.
67. Taketani Y, Oka T. Biological action of epidermal growth factor and its functional receptors in normal mammary epithelial cells. Proc Natl Acad Sci USA 1983;80:2647–50.
68. Richards RC, Beardmore JM, Brown PJ, Molloy CM, Johnson PM. Epidermal growth factor receptors on isolated human placental syncytiotrophoblast plasma membrane. Placenta 1983;4:133–8.

8

Intracellular pH Regulation by the Preimplantation Embryo

JAY M. BALTZ, JOHN D. BIGGERS, AND CLAUDE LECHENE

It has become clear that the preimplantation embryo possesses a number of rapidly developing and sometimes unique systems for transmembrane transport. For example, mouse embryos possess amino acid transport systems that change from one set to a different set, found only in embryos, over the course of only a few days' development (1); excitable calcium channels are found in abundance in the early mouse and hamster embryo, but disappear completely by the 16-cell stage (2–4); and Na^+ transport into the rabbit blastocoel is completely altered between the 5th and 7th day of development, with the development of amiloride and furosemide sensitivity (5, 6).

In order to analyze changes in transport occurring in the preimplantation embryo, we have been examining in the mouse embryo a set of mechanisms well known to be involved in growth and differentiation: those that regulate and specifically alter *intracellular pH* (pH_i). Intracellular pH is regulated mainly by plasma membrane proteins that transport ions across the membrane. All ion transport occurs via three general mechanisms: (i) pumps that directly use ATP hydrolysis to energize transport, (ii) channels and carriers that mediate the diffusion of ions down their concentration gradients, and (iii) exchangers that translocate one (or more) species of ion down its concentration gradient in order to energize the transport of another species. All transport ultimately derives its energy from ATP—the ATPases do so directly, while the other mechanisms use ion gradients maintained by the Na^+,K^+ ATPase; indeed, ion gradients energize most transport in cells (7). The most common mechanisms that mediate pH_i regulation in almost all mammalian cells are ion exchangers, including the Na^+/H^+ antiport and the HCO_3^-/Cl^- exchanger.

The Na^+/H^+ antiport relieves intracellular acidosis. When pH_i is too low, the antiport exchanges Na^+ running into the cell down its gradient for H^+ removed from the cell (Fig. 8.1). The exchange is 1:1 and, thus, electroneutral (8). The antiport has a pH *set point* below which it is

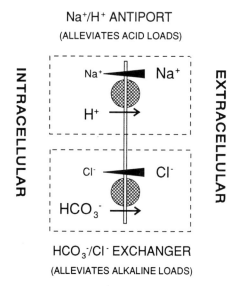

FIGURE 8.1. Two pH_i regulatory mechanisms common in mammalian cells. The Na^+/H^+ antiport (top) normally functions to relieve intracellular acidosis by exporting protons (H^+). Protons are exchanged for Na^+, which flows down its gradient into the cell. The HCO_3^-/Cl^- exchanger (bottom) can effect relief of intracellular alkalosis. HCO_3^- is exported from the cell in exchange for Cl^-. In the presence of both mechanisms, pH_i is kept within a narrow range, above which the HCO_3^-/Cl^- exchanger would drive pH_i down and below which the Na^+/H^+ antiport is activated to drive pH_i up. These mechanisms are nearly ubiquitous in mammalian cells.

active, with activity increasing with decreasing pH_i (9); the antiporter's set point is modulated by phosphorylation in response to stimulation by growth factors (10). The HCO_3^-/Cl^- exchanger can relieve intracellular alkalosis. When pH_i is too high, the exchanger removes intracellular HCO_3^- in exchange for extracellular Cl^- running down its gradient into the cell (Fig. 8.1). Several homologues are known, designated AE1, AE2, and AE3 (11) (AE for *anion exchanger*). AE1 is the red blood cell band 3 protein; AE2 is found in many cells, including epithelia; and AE3 is found in neuronal cells. Like the Na^+/H^+ antiport, the HCO_3^-/Cl^- exchanger is electroneutral. The HCO_3^-/Cl^- exchanger also probably possesses a set point, with activity increasing with increasing pH_i (12, 13).

These two mechanisms are found in virtually every mammalian cell type, maintaining pH_i within a narrow range (typically within a tenth of a pH unit). They are also responsible for physiologically important alterations in pH_i: For example, in a number of cells, their level of activity and active ranges are altered by growth factors, resulting in a shift in the resting pH_i (14–16); pH_i shifts have also been found to accompany differentiation and the switch between proliferation and quiescence (17–19); and the sea urchin egg Na^+/H^+ antiport is responsible for the marked fertilization-induced alkalinization that initiates the late fertilization events (20). Nothing, however, was known about the regulation of pH_i in the mammalian preimplantation embryo. This area was a good candidate for exploration, given the unusual aspects of other embryo transport systems, the central role of ions in the formation and maintenance of the blastocoel, and the implication of these systems and pH_i alterations in cell division and differentiation.

Measuring pH_i and Ion Transport

The general scheme for assessing pH_i regulation is to perturb pH_i, making the cytoplasm either more acidic or more basic, and follow its recovery or the concomitant flux of any other ions after the perturbation.

Perturbing pH_i

Mouse 2-cell embryos were acid loaded by a method that is used routinely in other cell types—the *ammonium pulse*—wherein a pulse exposure to an ammonium salt in the medium leaves the cells with a net acid load. The mechanism is as follows: Ammonia in solution is an equilibrium mixture of NH_3 and NH_4^+, with NH_3 being much more membrane permeant. Thus, immediately upon exposure, the cells are alkalinized by the influx of NH_3, which combines with H^+ in the cytoplasm until the equilibrium concentration of NH_4^+ is reached. However, since NH_4^+ is not completely impermeant, a much slower, but significant, influx of this ion subsequently drives the equilibrium back towards NH_3, releasing protons and thus producing a slow fall of pH_i from the initial high value. When the ammonia-containing external medium is replaced, all of the internal NH_3 and NH_4^+ exits as the much more permeant NH_3, including that which entered as NH_4^+. Thus, each NH_4^+ that entered leaves behind an H^+. This net gain of H^+ by the cell results in a significant acid load (21, 22). We used 25 mM ammonium chloride or sulfate to accomplish acid loading. This method has the great advantage that there are no foreign substances in the cell while the pH_i recovery is being measured since all of the ammonia has been removed.

There is, unfortunately, no similar method for alkalinization that we have found to work in embryos. Thus, we alkaline-loaded embryos by exposing them to ammonium salts that then remained in the bath for the duration of the experiment. The acidification due to NH_4^+ influx discussed above was slow enough compared to recovery from alkalosis by the cells that the specific component of recovery could be measured.

Measuring pH_i

We measured pH_i in preimplantation embryos using a pH-sensitive fluorophore: 2′,7′-bis(2-carboxyethyl)-5(and 6)-carboxy fluorescein (BCECF, Molecular Probes, Eugene, OR). The dye is loaded as the membrane-permeant acetoxymethyl ester form, which is cleaved intracellularly and becomes trapped in the cell. The dye exhibits a linear response to pH in the range of approximately 6.3 to 7.8 (23). The fluorescence is measured from an image collected with an extremely sensitive camera (ISIT, Dage-MTI, Michigan City, IN) by an image analysis and quantification system (Interactive Video Systems, Concord, MA). The fluorescence intensities are recorded for each of a number of embryos at 2 wavelengths—one pH sensitive and one pH insensitive. The ratio of the 2 intensities is dependent only on pH (24). Thus, the pH_i of a number of embryos can be determined simultaneously as a function of time; we generally had 10 embryos in each experiment (25). The concentration of dye and level of illumination used have no effect on the viability of 2-cell mouse embryos as assayed by their ability to develop to the blastocyst stage in culture (25).

Measuring Intracellular Ions

Intracellular ion concentrations were determined using electron probe X-ray microanalysis (26). Briefly, embryos are flash frozen and freeze dried. The electron probe is then used to assay for a number of elements (e.g., Na, K, and Cl) simultaneously. When a randomly selected subset of embryos from a given experimental condition are frozen at different times, the change in concentration of the ion(s) of interest over time can be determined.

pH_i Regulatory Systems of the 2-Cell Mouse Embryo

We chose to concentrate initially on the 2-cell stage embryo because its cells are still easily (optically) accessible, but cells within the same embryo can be compared. The idea was to characterize fully the regulatory mechanisms at this stage for a basis of comparison with the other stages.

Recovery from Acid Loads

Two-cell embryos were acid loaded as described above. The result of a typical experiment in which the pH_i of a group of 2-cell embryos is followed through acid loading and recovery is shown in Figure 8.2. As with other types of cells, pH_i is relatively stable in the absence of any

8. Intracellular pH Regulation by the Preimplantation Embryo   101

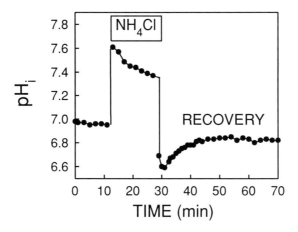

FIGURE 8.2. Acid loading and recovery in 2-cell mouse embryos. The embryos were acid loaded by an NH$_4$Cl pulse (in this case, 15 min) as detailed in the text. Following removal of the NH$_4$Cl, the cytoplasm is significantly acidified relative to baseline pH$_i$. The recovery can then be observed to occur over approximately the next 15 min. The points represent the mean pH$_i$ of the 20 blastomeres of ten 2-cell embryos that was measured as described in the text.

perturbations. It is, however, lower than that normally seen in other cells for reasons discussed below. The effect of the ammonium pulse can be seen as a marked alkalinization lasting for the duration of the pulse, as can the slower acidification during this period. After the ammonium-containing medium was replaced, the cells are left with a net acid load. Recovery then proceeds and is completed in about 15 min. It is this recovery phase that is of particular interest here: What are the mechanisms driving it? This can be answered by comparing the recoveries observed under various conditions designed to inhibit the mechanisms thought to be responsible: If the recovery is partially or totally abolished under conditions designed specifically to inhibit one mechanism, then it is fairly certain that the mechanism participates in the recovery.

We first investigated the contribution of the ubiquitous Na$^+$/H$^+$ antiport to the recovery from acid loading in the 2-cell mouse embryo. The scheme of experiments is as follows: Comparison was made of the recovery seen in a normal culture medium with those seen either in the presence of the Na$^+$/H$^+$ antiport inhibitor amiloride (or its derivatives) or in the absence of external Na$^+$. Both of these conditions should inhibit any component of the recovery due to the antiport; if all of the recovery were mediated by the antiport, then the curve would remain entirely flat and at a much lower level after acid loading, with no recovery seen. Recoveries were compared by fitting a single exponential to the recovery phase. The rate of the recovery can be expressed quantitatively as a *rate constant*—that is, the coefficient of time in the exponent of the

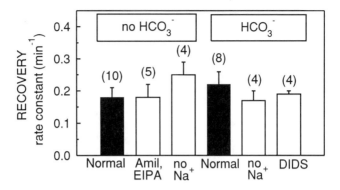

FIGURE 8.3. Rate of recovery from acid loads under various conditions. The rate of recovery following acid loading is given by the first-order rate constant of the exponential recovery; a larger rate constant corresponds to a faster recovery. Under "normal" conditions, pH_i regulatory mechanisms should be functional (filled bars). In the absence of HCO_3^- (left), only the Na^+/H^+ antiport would function, while in its presence (right), both the antiport and any HCO_3^--dependent mechanisms would function (e.g., the Na^+-dependent HCO_3^-/Cl^- exchanger and/or the Na^+,HCO_3^- cotransporter). The open bars show recoveries under conditions that would inhibit recovery via these mechanisms (for details, see text). The observation that the recovery is not slowed by any of these inhibitory conditions indicates that recovery from acid loads in 2-cell mouse embryos is not mediated by the mechanisms found in other cells. Numbers above bars indicate number of experiments; error bars are standard errors.

exponential. The method and the analysis are explained in detail in reference 25; the most important feature is that a larger rate constant corresponds to a faster recovery, with a rate constant of zero indicating no recovery at all. All these experiments were done in bicarbonate-free media buffered with HEPES. This eliminates any possible contribution by bicarbonate-dependent mechanisms and makes interpretation simpler.

In control acid-loading experiments, recoveries were indeed observed, similar to that seen in Figure 8.2. They proceeded with a rate constant that averaged $0.18\,\text{min}^{-1}$ in HCO_3^--free medium (Fig. 8.3). Therefore, the question is, How much is this rate affected by conditions that inhibit the Na^+/H^+ antiport? This was investigated, as described above, by addition of the inhibitor amiloride (1 mM), or its more potent derivative ethylisopropylamiloride (EIPA, 100 μM), or by the absence of Na^+ during recovery. Figure 8.3 shows the results of such experiments: The recovery in the presence of amiloride or EIPA (data pooled) was $0.18\,\text{min}^{-1}$, while in the absence of Na^+, it was $0.25\,\text{min}^{-1}$. These values are not significantly different from the control ("normal" medium), with the rate of recovery in the absence of Na^+ being, if anything, faster. Our surprising results showed that the recovery from acid loading in the 2-cell mouse embryo did not involve the Na^+/H^+ antiport—a mechanism that is ubiquitous in other cells (25). Furthermore, other experiments confirm that there is no role for the antiport in pH_i regulation in these cells: Removing Na^+ from the medium (which reverses the Na^+ gradient and should cause the antiport to run backwards and acidify the cell) had no effect on the baseline pH_i (25). Thus, the embryo, at least at the 2-cell stage, is unlike other cells in its pH_i regulation, with the recovery proceeding without apparent activity of the Na^+/H^+ antiport.

There are several other pH_i regulatory mechanisms that have been found in some types of cells, usually in conjunction with the antiport, that are able to effect relief of acid loads. The other known acid-load relieving systems in mammalian cells are an Na^+-dependent HCO_3^-/Cl^- exchanger that imports HCO_3^- to neutralize acid, in conjunction with Na^+ entering the cell down its gradient and Cl^- exiting up its gradient (27, 28); an Na^+,HCO_3^- cotransporter that imports HCO_3^- along with Na^+ (29, 30); and an H^+ ATPase that uses the energy of ATP hydrolysis to pump protons directly out of the cell (31, 32). Of these, the former two depend on bicarbonate, which was nominally absent from the solutions in these experiments. The latter—the H^+ ATPase—could mediate recovery under HCO_3^--free conditions.

In experiments similar to those in which we assayed for the antiport, we examined whether an ATPase was active. This ATPase is inhibited by *N-ethyl maleimide* (NEM) or by depletion of intracellular ATP. However, in the presence of NEM or with ATP depleted by incubation of the cells in cyanide, the recoveries were again not slowed and were, indeed, faster than the control recoveries (the mean rate constants were 0.36 and 0.31, respectively). Therefore, an ATPase was not effecting the recovery (33).

We also tested for the presence of either of the two HCO_3^--dependent systems that relieve acid loads by following recoveries from acid loads in HCO_3^-/CO_2-buffered media (18 mM HCO_3^-, 5% CO_2). The rates can be directly compared since the cytoplasmic buffering capacity of the embryo does not change upon addition of HCO_3^-/CO_2 (33). These mechanisms—the Na^+-dependent HCO_3^-/Cl^- exchanger and the Na^+,HCO_3^- cotransporter—both depend on Na^+, so we again examined recoveries in the absence of Na^+, but this time in HCO_3^-/CO_2-buffered media. Again, there was no effect. The rate constant for recovery in Na^+ containing HCO_3^-/CO_2-buffered medium was $0.22\,min^{-1}$, while in Na^+-free medium, it was $0.17\,min^{-1}$ (Fig. 8.3) (33). These means were not significantly different, nor were they different from the recoveries in the HCO_3^--free media. Also, the Na^+-dependent HCO_3^-/Cl^- exchanger, and sometimes the Na^+,HCO_3^- cotransporter, is inhibited by the anion transport inhibiting stilbenes, such as *4,4'-diisothiocyanostilbene 2,2'-disulfonic acid* (DIDS). The presence of DIDS (100 μM), however, had no effect, with the recovery rate constant with DIDS being $0.19\,min^{-1}$ (Fig. 8.3) (33).

Thus, we are left with a lack of detectable mechanisms for the relief of acid loads in the 2-cell mouse embryo since the recovery proceeds at essentially the same rate in the presence or absence of HCO_3^-, in the absence of external Na^+, or in the presence of such inhibitors as amiloride and DIDS. While the cells of these embryos do, indeed, recover from acid loads, it is clear that, at least under these conditions, they do not behave like somatic cells when they are challenged by intracellular acidosis. The process by which we believe they recover is discussed below.

Recovery from Alkaline Loads

As explained above, alkaline loading was accomplished by introducing ammonium and leaving it there, so that the entire experiment takes place in the alkaline phase of the curve shown in Figure 8.2; the ammonia-containing medium is never removed. Thus, the embryos can be alkaline loaded, and the recovery, now in the acid direction, followed. The only known mechanism for the relief of alkaline loads is the HCO_3^-/Cl^- exchanger (this is not dependent on Na^+ and should not be confused with the Na^+-dependent exchanger that relieves acidosis, as discussed above). We therefore investigated whether an HCO_3^-/Cl^- exchanger was active in the 2-cell embryo.

These experiments were done in HCO_3^-/CO_2-buffered media (18 mM HCO_3^-, 5% CO_2). First, it was necessary to show that there was, in fact, recovery from alkaline loading. This is shown on the left in Figure 8.4, where a quite rapid fall in pH_i is seen in HCO_3^-/CO_2-buffered medium after the cells were alkaline loaded. In contrast to the situation with acid loading, this recovery is mediated by a specific mechanism. The recovery, quantified as the initial rate of change of intracellular H^+ concentration following alkaline loading, is inhibited to the same extent by DIDS, lack of HCO_3^- (in HCO_3^--free HEPES-buffered medium), or by lack of Cl^- (Fig. 8.5) (26). The right panel in Figure 8.4 shows an example of a recovery in the absence of HCO_3^-. Note that the initial recovery portion is much slower than the corresponding portion in the presence of HCO_3^-. These observations imply the existence of an HCO_3^-/Cl^- exchanger mediating the recovery.

There was a residual recovery that was about 50% as fast as the control (Figs. 8.4 and 8.5) under the conditions that inhibited HCO_3^-/Cl^- exchange. This is mostly due to the slow entry of NH_4^+ into the cells (as discussed above) and, possibly, direct entry of H^+ into the cells (see below) (26). Since ammonium ion is not normally present in appreciable quantities in vivo or in culture, this residual recovery may not be physiologically relevant so that the HCO_3^-/Cl^- mediates all of the recovery from alkaline loading under in vivo or culture conditions.

We also confirmed that as expected for an HCO_3^-/Cl^- exchange mechanism, Cl^- was being transported into the embryos during recovery.

8. Intracellular pH Regulation by the Preimplantation Embryo 105

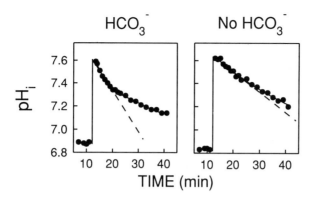

FIGURE 8.4. Alkaline loading and recovery in 2-cell mouse embryos. The embryos were alkaline loaded by the introduction of NH₄Cl. After this alkalinization, recovery was observed. On the left, this is done in the presence of HCO_3^-; on the right, in its absence. The initial recovery proceeds much faster with HCO_3^-, as can be seen by comparing the initial rates (dashed lines). In the pH_i range above about 7.3, the recovery rate is much enhanced by the presence of HCO_3^-, while below pH 7.2, the recoveries are similar. This is due to the HCO_3^-/Cl^- exchanger, which is active above about pH_i 7.2 in the 2-cell mouse embryo.

Electron probe analysis was used to measure Cl^- uptake into Cl^--depleted cells following alkaline loading. There was, indeed, a marked Cl^- uptake that was completely inhibited by DIDS. External HCO_3^-, which is a competitive inhibitor with Cl^- for inward transport on the exchanger, inhibited the Cl^- uptake into the 2-cell embryo following alkaline loading (26). Thus, we concluded that by using an HCO_3^-/Cl^- exchanger, the 2-cell embryo is just like somatic cells in relieving alkaline loads.

In addition, we were able to determine the kinetic parameters of the exchanger in the 2-cell mouse embryo by combining the results of the experiments determining the concentration dependence of inhibition by external HCO_3^- with a series of experiments in which pH_i was followed to determine the rate of recovery with various concentrations of Cl^- in the medium. The exchanger was found to exhibit Michaelis-Menten-type kinetics with respect to Cl^- transport, with the K_m for external Cl^- being about 3 mM; the inhibition constant K_I for external HCO_3^- was about 2 mM (26). In addition, the activity of the exchanger decreases with decreasing pH_i, and it is effectively inactive below a pH_i of about 7.2 (26). This can be seen in Figure 8.4, where the curves with and without HCO_3^- have very different slopes above about pH 7.3, but approximately the same slopes below about pH 7.2. These characteristics are similar to those seen for this system in other cells (12, 34, 35, 36). Therefore, we have concluded that the 2-cell mouse embryo possesses a mechanism found in differentiated cells for the specific relief of intra-

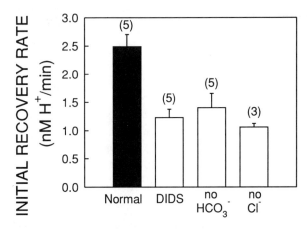

FIGURE 8.5. Rate of recovery from alkaline loads under various conditions. The rate of recovery is given by the initial rate of change of pH_i (dashed lines in Fig. 8.4). The filled bar shows the rate of recovery under "normal" conditions; that is, with the HCO_3^-/Cl^- exchanger functional. The open bars show the recovery rates with the exchanger inhibited by various means: the anion exchange inhibitor, DIDS, absence of HCO_3^-, or depletion of Cl^-. In each case, the rate of recovery is inhibited by about the same extent, ~50%. The remaining 50% is probably due to the passive entry of NH_4^+ and/or H^+ (see text). Numbers above bars indicate number of experiments; error bars are standard errors.

cellular alkalosis—the HCO_3^-/Cl^- exchanger. This result is in marked contrast to the unique situation for acidosis, where the usual mechanisms are undetectable.

How Protons Are Exported from Two-Cell Embryos

Despite there being no detectable activity of pH_i regulatory systems in the pH range below about 7.2, embryos do recover from acid loads. Therefore, some mechanism exists whereby excess protons (or the equivalent) are removed from the cytoplasm of these cells. There were several clues as to the possible mechanism: (i) No other extracellular ions seemed to be involved, and (ii) the baseline (unperturbed) pH_i of these cells was fairly low compared to other cells (overall mean pH_i around 6.9, with means of individual experiments as low as 6.6 [25, 33]). The former points towards a mechanism that is not based on countertransport or cotransport, but one in which protons (or the equivalent) are moved directly. The latter implies that it is possible that nothing maintains pH_i above the equilibrium value for protons across the membrane since the equilibrium pH_i would be well below the baseline pH_i that is normally found in other cells (22). The simplest explanation compatible with these observations was that the cells were merely permeable to protons so that protons were in equilibrium across the plasma membrane and the recoveries from an acid load that we observed were the manifestation of the reestablishment of this equilibrium after a perturbation. There are several predictions that can be made from this hypothesis. One is that the baseline value of pH_i would be dependent on the membrane potential V_m, which in turn is governed largely by the gradient of K^+ across the membrane. Another is that pH_i would be determined by the extracellular pH (since equilibrium is determined by the proton *gradient*). Indeed, when tested, these predictions were validated. Replacing the external medium with one containing 100 mM K^+ (instead of the usual 6 mM), and thus virtually collapsing the membrane potential, resulted in a rapid, marked rise in pH_i to approximately the extracellular pH value (25). Also, an alteration in extracellular pH produced a similar change in pH_i (25). These results led us to conclude that the 2-cell preimplantation embryo probably has a rather large permeability to protons that both determines baseline pH_i and governs the recovery from perturbations to pH_i in the acid and neutral pH range. This is a situation unlike that found in other cells.

In light of the proton permeability, none of the work described thus far would preclude the presence of other pH_i regulatory systems at the 2-cell stage. (See Chapter 5, this volume, for another possible mechanism affecting pH_i). An Na^+/H^+ antiport, for example, could be running at maximal rate and still not be a detectable component of recovery from an acid load if the proton permeability pathway supports a very much larger flux of protons than the antiport. In this case, inhibiting the antiport would decrease the recovery rate by only an undetectably small amount; the antiport flux would be overwhelmed by the permeability flux. However, such a "silent" antiport would not be without significance. Since it would probably be chronically active, given the low resting pH_i of these cells (at least under culture conditions), the antiport could be the main pathway for Na^+ entry into the cell. We are currently investigating Na^+ flux pathways in the embryo, using the electron probe.

We are also currently examining the properties of the proton permeability in the hope of confirming its existence and eventually determining its molecular basis. We are exploiting a phenomenon that we have observed in these embryos: When they are flushed from the oviduct at room temperature and maintained at room temperature in bicarbonate-free medium, they have a remarkably high pH_i (7.7–8.0). The reason for this is unknown, but it is convenient for looking at the proton permeability since the cells appear to become proton permeable upon warming so that we can "turn on" (or at least greatly increase) the permeability when we wish; when the temperature is raised, pH_i immediately starts falling until it reaches the presumed equilibrium value. Preliminary data show that the proton permeability pathway monitored under these conditions behaves as predicted; the rate of decrease of pH_i after warming is highly dependent on the external pH_i and is markedly slowed if the V_m is decreased by high external K^+. It can be reversed by a combination of high external pH and high K^+ so that the cells actually become more alkaline when warmed. These experiments seem to confirm that there is indeed a pathway through the plasma membrane for significant proton flux.

pH_i Regulation and Development

Development of Na^+/H^+ Antiporter Activity

Presumably, Na^+/H^+ antiporter activity in pH_i regulation must become detectable at some point, as it is certainly present in the differentiated tissues of the mouse. Indeed, there is some indication that it is present by the blastocyst stage, as blastocoel formation and maintenance of the early blastocoel is dependent on Na^+ and can be disrupted by ethylisopropylamiloride (37). We are currently addressing this problem using electron probe analysis to determine Na^+ uptake into embryos of different preimplantation stages under conditions that support or inhibit Na^+/H^+ antiporter activity (as was mentioned above for the 2-cell stage). Since a proton permeability pathway would not affect the Na^+ flux, this methodology avoids the complications arising from the putative proton permeability that could mask Na^+/H^+ antiport-mediated pH_i changes. Thus, we should be able to detect any Na^+ fluxes due to an Na^+/H^+ antiport even when the concomitant pH_i changes are undetectable.

In Vivo pH_i Regulation

The in vivo environment of the 2-cell embryo—the oviduct—differs in some important respects from the in vitro conditions. First, the concentration of K^+ in mouse oviductal fluid is high: 25 mM (38). Second, while the pH and bicarbonate concentrations have not been measured for mouse oviductal fluid, they have been found to be high in a number of other species where they have been determined (39). Both these conditions—high K^+ and high pH—would tend to elevate pH_i by lowering the equilibrium value for intracellular proton concentration. Therefore, it is possible that in vivo, the equilibrium value for pH_i is near to or above 7.2, rather than approximately 6.9 as we find in vitro. Since this higher pH_i approaches the active range of the HCO_3^-/Cl^- exchanger, an increase of pH_i above 7.2 caused by the high K^+/high pH oviductal fluid would be opposed by this mechanism, and pH_i would be effectively maintained within a narrow range below 7.2.

At any rate, it is clear that embryo plasma membrane transport systems seem, in many cases, to function differently from those performing the same functions in differentiated cells. It remains to try to answer fully the two main questions: What are these differences and what aspects of embryo physiology during development require these novel mechanisms to perform functions that are accomplished by common systems in almost every differentiated mammalian cell type?

Acknowledgments. This work was supported by NIH Grant HD-21581. We would like to thank Stephen Smith and Khiem Oei for technical assistance.

References

1. Van Winkle LJ. Amino acid transport in developing animal oocytes and early conceptuses. Biochim Biophys Acta 1988;947:173–208.
2. Eusebi F, Colonna R, Mangia F. Development of membrane excitability in mammalian oocytes and early embryos. Gamete Res 1983;7:39–47.
3. Mitani S. The reduction of calcium current associated with early differentiation of the murine embryo. J Physiol 1985;363:71-86.
4. Yoshida S. Action potentials dependent on monovalent cations in developing mouse embryos. Dev Biol 1985;110:200–6.
5. Powers RD, Borland RM, Biggers JD. Amiloride-sensitive rheogenic Na^+ transport in rabbit blastocyst. Nature (London) 1977;270:603–4.
6. Benos DJ, Biggers JD. Sodium and chloride co-transport by preimplantation rabbit blastocysts. J Physiol 1983;342:23–33.
7. Cohen BJ, Lechene C. Na,K pump: cellular role and regulation in non-excitable cells. Biol Cell 1989;66:191–5.
8. Montrose MH, Murer H. Kinetics of Na^+/H^+ exchange. In: Grinstein S, ed. Na^+/H^+ exchange. Boca Raton, FL: CRC Press, 1988:57–75.
9. Aronson PS, Nee J, Suhm MA. Modifier role of internal H^+ in activating the Na^+/H^+ exchanger in renal microvillus membrane vesicles. Nature (London) 1982;299:161–3.
10. Sardet C, Counillon L, Franchi A, Pouyssegur J. Growth factors induce phosphorylation of the Na^+/H^+ antiporter, a glycoprotein of 110 kD. Science 1990;247:723–6.
11. Kopito RR, Lee BS, Simmons DM, Lindsey AE, Morgans CW, Schneider K. Regulation of intracellular pH by a neuronal homolog of the erythrocyte anion exchanger. Cell 1989;59:927–37.
12. Olsnes S, Tonnessen TI, Sandvig K. pH-regulated anion antiport in nucleated mammalian cells. J Cell Biol 1986;102:967–71.
13. Vaughan-Jones RD. Regulation of intracellular pH in cardiac muscle. In: Bock G, Marsh J, eds. Proton passage across cell membranes. CIBA Foundation Symposium 139, Feb. 9–11, 1988. London: Wiley and Sons, 1988:23–46.
14. Moolenaar WH, Tsien RY, van der Saag PT, de Laat SW. Na^+/H^+ exchange and cytoplasmic pH in the action of growth factors in human fibroblasts. Nature (London) 1983;304:645–8.

15. Moolenaar WH, Tertoolen LGJ, de Laat SW. Phorbol esters and diacylglycerol mimic growth factors in raising cytoplasmic pH. Nature (London) 1984;312: 371–3.
16. Ganz MB, Boyarsky G, Sterzel RB, Boron WF. Arginine vasopressin enhances pH_i regulation in the presence of HCO_3^- by stimulating three acid-base transport systems. Nature (London) 1989;337:648–51.
17. Biermann AJ, Tertoolen LGJ, de Laat SW, Moolenaar WH. The Na^+/H^+ exchanger is constitutively activated in P19 embryonal carcinoma cells, but not in a differentiated derivative. J Biol Chem 1987;262:9621–8.
18. Perona R, Serrano R. Increased pH and tumorigenicity of fibroblasts expressing a yeast proton pump. Nature (London) 1988;334:438–40.
19. Schwartz MA, Both G, Lechene C. Effect of cell spreading on cytoplasmic pH in normal and transformed fibroblasts. Proc Natl Acad Sci USA 1989; 86:4525–9.
20. Epel D. The role of Na^+/H^+ exchange and intracellular pH changes in fertilization. In: Grinstein S, ed. Na^+/H^+ exchange. Boca Raton, FL: CRC Press, 1988:57–75.
21. Boron WF, de Weer P. Intracellular pH transients in squid giant axons caused by CO_2, NH_3, and metabolic inhibitors. J Gen Physiol 1976;67: 91–112.
22. Roos A, Boron WF. Intracellular pH. Physiol Rev 1981;61:296–434.
23. Molecular Probes product information for BCECF-AM and BCECF. Apr. 1991 edition. Molecular Probes, Inc., Eugene, OR.
24. Bright GR, Fisher GW, Rogowska J, Taylor DL. Fluorescence ratio imaging microscopy: temporal and spatial measurements of cytoplasmic pH. J Cell Biol 1987;104:1019–33.
25. Baltz JM, Biggers JD, Lechene C. Apparent absence of Na^+/H^+ antiport activity in the two-cell mouse embryo. Dev Biol 1990;138:421–9.
26. Baltz JM, Biggers JD, Lechene C. Relief from alkaline-load in two-cell stage mouse embryos by bicarbonate/chloride exchange. J Biol Chem 1991;266: 17212–7.
27. Thomas RC. The role of bicarbonate, chloride and sodium ions in the regulation of intracellular pH in snail neurones. J Physiol 1977;273:317–38.
28. Boyarsky G, Ganz MB, Sterzel RB, Boron WF. pH regulation in single glomerular mesangial cells, I. Acid extrusion in absence and presence of HCO_3^-. Am J Physiol 1988;255:C844–56.
29. Boron WF, Boulpaep EL. Intracellular pH regulation in the renal proximal tubule of the salamander: Na^+/H^+ exchange. J Gen Physiol 1983;81:29–52.
30. Aickin CC. Movement of acid equivalents across the mammalian smooth muscle membrane. In: Bock G, Marsh J, eds. Proton passage across cell membranes. CIBA Foundation Symposium 139, Feb. 9–11, 1988. London: Wiley and Sons, 1988:23–46.
31. Bidani A, Brown SES, Heming TA, Gurich R, Dubose TD. Cytoplasmic pH in pulmonary macrophages: recovery from acid-loads is Na^+ independent and NEM sensitive. Am J Physiol 1989;257:C65–76.
32. Lubman RL, Danto SI, Crandall ED. Evidence for active H^+ secretion by rat alveolar epithelial cells. Am J Physiol 1989;257:L438–45.
33. Baltz JM, Biggers JD, Lechene C. Two-cell stage mouse embryos appear to lack mechanisms for alleviating intracellular acid-loads. J Biol Chem 1991; 266:6052–7.

34. Muallem S, Burnham C, Blissard D, Berglindh T, Sachs G. Electrolyte transport across the basolateral membrane of the parietal cells. J Biol Chem 1985;260:6641–53.
35. Kurtz I, Golchini K. Na^+-independent Cl^--HCO_3^- exchange in Madin-Darby canine kidney cells: role in intracellular pH regulation. J Biol Chem 1987;262:4516–20.
36. Reinertsen KV, Tonnessen TI, Jacobsen J, Sandvig K. Role of chloride/bicarbonate antiport in the control of cytoplasmic pH: cell line differences in activity and regulation of antiport. J Biol Chem 1988;263:11117–25.
37. Manejwala FM, Cragoe EJ, Schultz RM. Blastocoel expansion in the preimplantation mouse embryo: role of extracellular sodium and chloride and possible apical routes of their entry. Dev Biol 1989;133:210–20.
38. Borland RM, Hazra S, Biggers JD, Lechene C. The elemental composition of the environments of the gametes and preimplantation embryo during the initiation of pregnancy. Biol Reprod 1977;16:147–57.
39. Maas DHA, Storey BT, Mastroianni L. Hydrogen ion and carbon dioxide content of the oviductal fluid of the rhesus monkey (*Macaca mulatta*). Fertil Steril 1977;28:981–5.

Part III

Genetics of Embryo Development

9

Activation of the Embryonic Genome: Comparisons Between Mouse and Bovine Development

ANDREW J. WATSON, AILEEN HOGAN, ANN HAHNEL,
AND GILBERT A. SCHULTZ

Eutherian mammalian preimplantation development includes an essential transitional period during which the embryo proceeds from an initial interval of maternal control (established during oogenesis) to embryonic control resulting from the activation of transcriptional activity from the embryonic genome (1–3). As for most aspects of early mammalian development, the events surrounding the activation of the embryonic genome are most readily examined in the mouse. There is, however, evidence to suggest that the mouse may display a pattern of preimplantation development (certainly in the timing of the major events) that is unique. It is, therefore, essential to broaden the analysis of preimplantation development to include other mammalian species. A comparison of several species in Table 9.1 shows that there is considerable variation in the timing, cell number, and developmental stage at which the early embryos of these groups undergo the major morphogenetic events of preimplantation development (4, 5). As shown in Table 9.2, these variations also reflect contrasts in the timing of the molecular and biochemical events of preimplantation development.

In the mouse, transcriptional activity from the embryonic genome is initiated by the 2-cell stage (1, 3). During the transitional period, much of the maternal polyadenylated mRNA is degraded (6), including the maternal stores of such mRNAs as those that encode the actin and histone proteins (7, 8). While this reduction is occurring, embryonic transcription is initiated, with all classes of RNA being synthesized by the late 2-cell stage (9, 10). Mouse embryos cultured from the 1-cell stage in the presence of the transcriptional inhibitor of RNA polymerase II, α-amanitin, do not progress beyond the 2-cell stage. This demonstrates that new transcription from the embryonic genome is required for develop-

TABLE 9.1. Comparative developmental parameters for several mammalian species.

Species	Compaction		Cavitation			Implantation	
	Days	Cell no.	Days	Cell no.	Blast (mm)	Days	Stage
Mouse	3	8–16	3.5	32	0.15	4.5	BL
Hamster	2.5	8–16	3	16–24	0.14	3.5	BL
Rabbit	2	32–64	3	128	3–4	6.5	BL
Sheep	3	32+	4.5	64	2×10^2	15–16	ES
Cow	4–5	32+	6	64–128	2×10^2	17–34	ES
Pig	3.5	16	3.5–5	32	1×10^3	14–15	ES
Rhesus monkey	4–6	16+	6–7	64+	0.25	8–9	BL
Human	4–6	16	4.5–5	32–64	0.25	6–7	BL

Note: Days = of preimplantation development; cell no. = embryo cell number at time of initiation of the event; blast = final blastocyst size; BL = blastocyst; ES = early somite.
Source: Data from McLaren (4) and Papaioannou, Ebert (5).

TABLE 9.2. Stage of transition from maternal to embryonic control in embryos of several mammalian species.

	Characteristic feature (embryo cell number)			
Species	Developmental arrest due to culture in α-amanitin	Initiation of RNA synthesis from embryonic genome	Change in pattern of protein synthesis	Culture block
Mouse	2	1–2	1–2	2
Hamster	2	2	2	2
Cow	8–16	8	8–16	8–16
Sheep	8–16	NR	8–16	8–16
Pig	4	4	4	4–8
Human	4–8	4	4–8	4–8
Rabbit	8–16	1–2	2–16	NR

Note: NR = not reported in the literature.
Source: Data from Telford, Watson, Schultz (3).

ment beyond this point (11). Changes in protein synthetic patterns are also observed during this transitional period (12). Some of these changes are associated with translation of new mRNAs since their appearance can be blocked by α-amanitin (13, 14).

One additional factor that is associated with activation of the embryonic genome is the *2-cell block* observed when 1-cell embryos from most random-bred strains of mice are cultured in vitro (15–17). The precise origin of this phenomenon is unclear, but the fact that it arises coincidentally with the transitional period indicates that these embryos may be particularly sensitive to the culture conditions during this phase of preimplantation development. In other rodent species (rat, hamster, and gerbil), there is a distinct change in the protein synthetic profiles between the 1-cell and 4-cell stages (18), and in the best-studied case, the hamster, it is clear that the 2-cell stage represents the period in which the embryonic genome becomes active (19, 20). Therefore, the transition from maternal to embryonic control of development arises very early during preimplantation development in the rodents.

The dramatic loss of polyadenylated mRNA observed between 1- and 2-cell-stage mouse embryos does not occur in the rabbit. Instead, there is a gradual increase in the total RNA content during the early cleavage divisions, followed by a dramatic increase at the blastocyst stage (21). Culture of 1- or 2-cell rabbit embryos in the presence of α-amanitin results in developmental arrest at the 8- or 16-cell stage (22), and there is evidence that radioactive uridine is incorporated into RNA prior to the 16-cell stage (23-25). Studies investigating the protein synthetic patterns have revealed differences between the 2-cell and 16-cell stages (26). It is also possible to culture rabbit zygotes successfully to the blastocyst stage without any block (27).

In human embryos, cleavage can continue until the 4-cell stage in the presence of α-amanitin (28). Autoradiographic studies detecting [^{3}H]-uridine incorporation report nucleoplasm labeling (an indicator of pre-mRNA synthesis) first arising at the 4-cell stage, followed by nucleolar labeling (rRNA synthesis) at the 6- to 8-cell stage (29-31). Changes in protein synthetic patterns, as well as a culture block, occur at the 4- to 8-cell stage (28).

In the cow and sheep, the activation of the embryonic genome is delayed in comparison to the rodents. Initial culture studies showed that both sheep and cow embryos exhibited a culture block at the 8- to 16-cell stage if the culture conditions did not include coculture with granulosa or oviductal epithelial cells (32, 33).

Recently, however, improvements to the culture conditions have shown that bovine embryos can develop to the morula/blastocyst stages in a chemically defined protein-free medium (34). The incorporation of [^{3}H]-uridine into early cow embryos (detected by autoradiography) first occurs at the 8-cell stage in both the nucleoplasm and nucleoli, suggesting that both pre-mRNA and rRNA synthesis are simultaneously initiated at this stage (35-37). Changes in protein synthetic patterns for both sheep and cow embryos also arise during the 8- to 16-cell stages (38, 39). In addition, when 1- to 4-cell sheep embryos (39) or early in vitro cultured cow embryos (40) are cultured in the presence of α-amanitin, a developmental arrest is observed at the 8-cell stage, suggesting that transcription from a newly activated zygotic genome is required for further development. Pig embryos differ from cow and sheep embryos and, in fact, resemble human embryos in [^{3}H]-uridine incorporation (41, 42), changes in protein synthetic patterns (43), and the timing of the culture block (44), all demonstrating that the transition to embryonic control of development arises at the 4-cell stage. A summary of the major features of the activation of the embryonic genome in these mammalian species is shown in Table 9.2.

A number of genes are activated at the time of the transition from maternal to embryonic control. In the mouse, analysis of transcripts in a random cDNA library from late 2-cell-stage embryos (when the embryonic genome is already activated) indicates that many of the products are not represented by mRNAs in ovulated oocytes (45). The gene products from this new transcriptional activity that are of particular interest are those that influence metabolism, rate of cell proliferation, embryo cell number, and differentiation. Gene products that exert this influence over early development include the members of the growth factor families.

The application of in vitro maturation and in vitro fertilization methods to cow oocytes collected from slaughterhouse ovaries, followed by oviductal cell coculture of the ensuing embryos, has made it possible to produce preimplantation cow embryos of all stages in numbers comparable to those for the mouse (46–48). The employment of these embryos in mRNA phenotyping, using the polymerase chain reaction (49–51), represents a powerful system for the characterization of gene expression patterns during preimplantation development. We have initiated a comparative study characterizing the patterns of expression of specific genes in mouse and cow preimplantation embryos. Emphasis has been given to the analysis of small nuclear RNAs (snRNAs) because of their role in posttranscriptional processing to yield mature functional mRNAs and also to several growth factor ligand and receptor mRNAs because of their putative roles in early embryonic growth and differentiation. This type of analysis will clarify the qualitative differences in transcriptional patterns between two distinct mammalian species.

Materials and Methods

Bovine Embryo Culture System

The methods employed for the production of cow preimplantation embryos from *cumulus-oocyte complexes* (COCs) collected from slaughterhouse ovaries were adapted from other protocols (52–54). Both *grade I COCs* (oocytes surrounded by a minimum of 4–5 layers of cumulus and also displaying a homogeneous pigmentation) and *grade II COCs* (oocytes completely surrounded by 2–3 layers of cumulus cells) were placed into maturation medium composed of TCM-199 (Gibco, Grand Island, NY) and 10% (v/v) fetal bovine serum (FBS, Gibco), supplemented with sodium pyruvate (0.5 mM, Sigma, St. Louis, MO), human chorionic gonadotropin (1.15 U/mL, APL^R, Ayerst Laboratories, Montreal, Canada), follicle stimulating hormone (35 µg/mL, Follitropin^R, Vetrapharm, Inc., London, Canada) and 17β-estradiol (1 µg/mL; Sigma) for 26 h. The in vitro maturation, and all subsequent culture, was conducted at 39°C in a humidified atmosphere containing 5% CO_2 in air. Matured oocytes were fertilized in vitro following maturation using a modified "swim up" procedure (55). Zygote and embryo culture was conducted on oviductal epithelial cells and was maintained for a maximum of 8 days to allow for formation of blastocysts. The embryos were moved to fresh oviductal epithelial cell cultures every 48 h to maximize their developmental potential (48). Embryo stages from 1-cell to hatched blastocysts were collected for detailed analysis.

Mouse Embryo Collection

Preimplantation mouse embryos were collected using standard procedures (7) from superovulated, 7-week old, random-bred Swiss albino CD1 female mice (Charles River Breeding Laboratories, St. Constant, Quebec, Canada) mated to CD1 males.

In Situ Hybridization

Both mouse and cow preimplantation embryos were processed for in situ hybridization by fixation for 1 h in 4% paraformaldehyde (Polysciences, Inc., Warrington, PA) in PHEM buffer (56) at 4°C, followed by embedding in diethylene glycol distearate (57) (DGD, Polysciences) using conical-shaped BEEM embedding capsules (58) (J.B. EM Services, Inc., Pointe-Claire, Quebec, Canada). Sections of 1–2 μm were cut using a Reichart OMU-1 ultramicrotome and mounted on acid-washed poly-L-lysine (M_r 150,000; Sigma) -coated glass slides. Just prior to hybridization, the embedding medium was removed by treatment with 1-butanol (3 h), followed by rehydration of the sections through a descending ethanol series. The sections were washed in 4 × SSC (59), and prehybridized at 50°C for 3 h in 0.3 M NaCl, 1 mM EDTA, 10 mM Tris-HCl (pH 8.0), 1× Denhardt's solution (57), 50% deionized formamide (Fluka Chemical Corp., Ronkonkoma, NY), 100 mM dithiothreitol (Sigma), and 1 mg/mL E. coli tRNA.

The U1 and U2 antisense and sense riboprobes were produced by in vitro transcription using a standard procedure (60, 61) and diluted in the prehybridization solution at a concentration of 3–5 ng of RNA (1–2 × 10^6 cpm) per 20-μL aliquot per slide. The slides were coverslipped, sealed with rubber cement, and incubated at 50°C for 18 h. Following hybridization, the slides were transferred through a series of low-salt washes and then processed for autoradiography with nuclear track emulsion NTB-2 (Eastman Kodak Co., Rochester, NY) (61). All photomicroscopy was conducted with a Zeiss photomicroscope II equipped with Nomarski differential interference contrast optics using Ilford HP5 135 film.

mRNA Phenotyping by Reverse Transcription Polymerase Chain Reaction (RT-PCR)

Total RNA was extracted from groups of 100–300 mouse oocytes or embryos and from groups of 30–100 cow oocytes or embryos with the addition of 5–10 µg of E. coli ribosomal RNA (Boehringer-Mannheim, Montreal, Canada) as described (62). Following reverse transcription into cDNA with an oligo-dT primer, the preparations were divided into aliquots and subjected to the *polymerase chain reaction* (PCR) in the presence of sequence-specific oligodeoxynucleotide primer pairs (62, 63). The products of the PCR reaction were resolved electrophoretically on 2% agarose gels and stained with ethidium bromide to view the PCR fragments. The identity of the PCR fragments was confirmed by either digestion with the appropriate diagnostic restriction enzyme, by blotting and hybridization with a cDNA, or by cloning and sequencing of the amplified PCR product.

Results and Discussion

Localization of snRNAs During Mouse and Cow Preimplantation Development

The cellular machinery responsible for processing new primary transcripts into functional mRNAs must play an important role in the transition from maternal to embryonic control of development. Therefore, the gene expression patterns for the major *small nuclear RNAs* (snRNAs; U1, U2, U4, U5, and U6) were investigated by in situ hybridization of a developmental series of preimplantation mouse embryos (60, 61, 64, 65). The general pattern that emerges from these experiments is one where the snRNAs are initially confined to the *germinal vesicle* (GV). Following GV breakdown and the onset of meiotic maturation, the snRNAs become distributed throughout the ooplasm but subsequently localize to the pronuclei of the fertilized egg. From this point throughout preimplantation development, the snRNAs remain confined to the interphase nuclei of each blastomere and both the trophectoderm and inner cell mass of each blastocyst (60, 61, 64) (Figs. 9.1A, 9.1B, 9.1E, and 9.1F). In the mouse, there is no obvious decrease in snRNA message levels as observed for other maternal mRNAs during the 1- to 2-cell transition (60, 61).

In the sea urchin, where the activation of the embryonic genome is delayed, U1 snRNA is first localized within the GV, but unlike the mouse, it becomes undetectable within the nuclei of the early embryo and only reappears (apparently due to new synthesis) in the micromere nuclei of the 16-cell embryo (66). This observation implies that the high levels of snRNAs observed in the mouse could be due to the short period of maternal control.

9. Activation of the Embryonic Genome   121

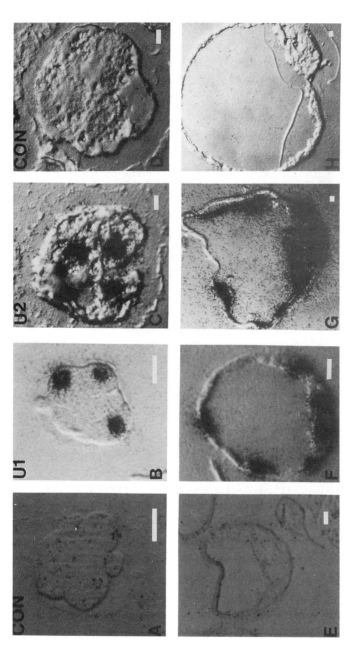

FIGURE 9.1. Localization of U1 snRNA in mouse 8-cell embryos and blastocysts and U2 snRNA in cow 8-cell embryos and blastocysts by in situ hybridization and autoradiography. U1 antisense signals are shown in *B* and *F*, while the corresponding U1 sense control signals are shown in *A* and *E*. U2 antisense signals are shown in *C* and *G*, while the corresponding U2 sense control signals are shown in *D* and *H*. The U1 and U2 antisense signals produced a distinctive cluster of silver grains overlying the nuclei of both 8-cell mouse (*B*) and cow (*C*) embryos, respectively. In the blastocyst stage for both mouse (*F*) and cow (*G*), the same U1 and U2 antisense signal was observed overlying the nuclei of both trophectoderm and inner cell mass cell types. The sense controls for each of these embryonic stages did not reveal a concentrated clustering of silver grains anywhere over the embryo section (*A*: 8-cell mouse, *D*: 8-cell cow, *E*: blastocyst mouse, and *H*: blastocyst cow). Exposure times were 6 days for both antisense and sense incubated sections. The scale bars are equal to 20 μm.

Cow preimplantation embryos are under maternal control until the third or fourth cleavage division and thus provide the opportunity to test this possibility. A developmental series of cow preimplantation embryos were examined by in situ hybridization to localize the expression of U2 snRNA. As in the mouse, U2 snRNAs are first confined to the GV but become distributed throughout the ooplasm upon its breakdown and subsequently localize to the pronuclei and blastomeric nuclei of all additional preimplantation embryo stages (67) (Figs. 9.1C, 9.1D, 9.1G, and 9.1H). There is no obvious decrease in U2 message during preimplantation development at the level of resolution of this technique. This result indicates that the eutherian mammals may maintain a maternal snRNA pool throughout early cleavage. The pulse of transcriptional activity stemming from the activation of the embryonic genome must require this pool of maternal snRNAs to expedite the rapid transition from maternal to embryonic control.

Expression of Growth Factor Ligand and Receptor Genes During Mouse and Cow Preimplantation Development

A combination of RT-PCR and immunofluorescence methods has previously been used to demonstrate that a large number of growth factor genes are expressed during the preimplantation period of mouse development. These include *transforming growth factor alpha* (TGFα) (49, 50), *transforming growth factor beta 1* (TGFβ$_1$) (49, 50), *platelet-derived growth factor A* (PDGF-A) (49, 50), *Kaposi's sarcoma-type growth factor* (kFGF) (50), and *insulin-like growth factor II* (IGF-II) (50). TGFα and PDGF-A transcripts are present in oocytes as well as in cleavage and blastocyst stages and, therefore, are derived by transcription from both the maternal and embryonic genomes. Transcripts for TGFβ$_1$, kFGF, and IGF-II are not detectable until the 2-cell stage and, therefore, must be products of transcription from the activated embryonic genome.

More recent evidence suggests that *transforming growth factor beta 2* (TGFβ$_2$) (68, 69), *transforming growth factor beta 3* (TGFβ$_3$), (68), *interleukin 3* (IL-3) (70), and *interleukin 6* (IL-6) (71) genes are also expressed by the blastocyst during early mouse development. A number of growth factor receptor genes are also expressed upon the activation of transcription from the mouse embryonic genome. These include the insulin receptor (72–74), IGF-I and IGF-II receptors (51, 75, 76), *epidermal growth factor* (EGF) receptor (77, 78), *platelet-derived growth factor alpha* (PDGFα) receptor (71), and *colony stimulating factor 1* (CSF-I) receptor (71). The activation of growth factor ligand and receptor genes is selective, however, since the transcripts for several factors, including EGF, *nerve growth factor* (NGF), *granulocyte colony stimulating factor* (G-CSF), insulin, and IGF-I, have not been detected up

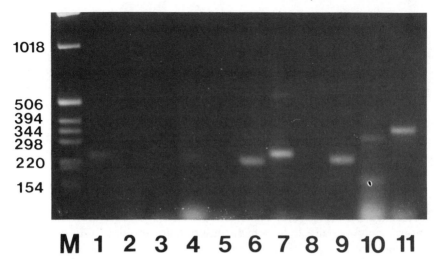

FIGURE 9.2. Detection of growth factor transcripts from preimplantation cow embryos by RT-PCR. Total RNA from 67 16-cell cow embryos was reverse transcribed, divided into aliquots, and amplified by PCR using oligonucleotide primers specific for various cDNAs. The PCR products and their sizes in base pairs (bp) are as follows: 1, actin (243); 2, EGF (247); 3, TGFβ$_2$ (273); 4, TGFα (239); 5, NGF (401); 6, PDGF-A (227); 7, IGF-II (256); 8, bFGF (282); 9, PDGFα receptor (235); 10, insulin receptor (324); 11, IGF-I receptor (354); and M: DNA markers.

to the blastocyst stage even when the sensitive RT-PCR methods are utilized (51).

We have examined whether similar patterns of expression of growth factor ligand and receptor genes occur in preimplantation cow embryos (79). An example of an mRNA phenotyping experiment on RNA extracted from a pool of 16-cell cow embryos is shown in Fig. 9.2. In this experiment, total RNA was reverse transcribed using oligo-dT as primer. The reverse transcription product was then divided into 11 aliquots that were subjected to PCR with primers specific for 11 different growth factor cDNA sequences. Amplification products of the expected size (predicted by primer design) were obtained in the RT-PCR reactions for actin, TGFβ$_2$, TGFα, PDGF-A, IGF-II, PDGFα receptor, insulin receptor, and IGF-I receptor (Fig. 9.2). PCR products were not detected for EGF, NGF, and bFGF (Fig. 9.2). The identities for all PCR products were verified either by digestion with an appropriate diagnostic restriction enzyme or by Southern blot analysis of PCR fragments and hybridization with a radiolabeled cDNA probe (data not presented).

Through RT-PCR assays of several RNA preparations of pools of cow embryos from the 1-cell through the blastocyst stages, a number of findings about growth factor gene expression have emerged (79). As in the mouse, TGFα transcripts are present throughout preimplantation development and are products of both the maternal genome (up to the 8-to 16-cell stage) and the embryonic genome (8- to 16-cell stage to the blastocyst). The same is true for TGFβ$_2$, PDGF-A, and the PDGFα receptor. Within the insulin gene family of peptides and receptors, transcripts for IGF-II ligand and insulin, IGF-I, and IGF-II receptors are detectable at all stages from the 1-cell zygote to the blastocyst. In this regard, they differ from the mouse since they are represented by both maternal and embryonic transcripts, whereas in the mouse these genes are only expressed following embryonic genome activation.

The expression of the *basic fibroblast growth factor* (bFGF) gene in the cow embryo appears to be maternal in nature since we were unable to detect transcripts beyond the 8- to 16-cell stage (Fig. 9.2). As in the mouse, we were unable to detect transcripts for insulin, EGF, or NGF at any stage of cow preimplantation development. On some occasions, we have detected transcripts for the IGF-I ligand in early cow embryos. This transcript does not appear until the peri-implantation period in mouse embryos. In summary, although many of the same growth factor and receptor genes are expressed during both murine and bovine preimplantation development, there are distinct differences in the stages in which their transcripts are first detectable and, therefore, in their maternal versus embryonic origin.

Considerable evidence has accumulated to demonstrate that growth factor and receptor genes play an important role in early mouse development. The addition of growth factors from the insulin family of peptides, the epidermal growth factor family, and TGFβ$_1$ (at physiological levels) to the culture medium of preimplantation embryos results in a broad range of effects by stimulating metabolism, rate of cell division, cell number in the blastocyst, and zona hatching (51, 80–82). Also, there is evidence supporting a role for these molecules during early bovine development. When TGFβ and bFGF are added to bovine embryo cultures, 39% of the embryos are able to develop through the 8-cell block compared to 0% for the control group (83).

Since growth factors play an important role in facilitating embryo development, experiments of this type should assist in the development of culture conditions that will optimally support the production of preimplantation embryos from domestic species. More sophisticated approaches, such as gene targeting by homologous recombination in *embryonic stem* (ES) cells, have been applied to unraveling the role of IGF-II during mouse early development (84). The heterozygous progeny from IGF-II inactivated mice are smaller than their wild-type littermates (84). These experimental approaches can ultimately be applied to any gene (or gene product) that is regulated during the preimplantation period. Continued comparative analysis of both cow and mouse preimplantation development should provide detailed functional information that will facilitate our understanding of the important molecular events that underlie mammalian preimplantation development.

Acknowledgments. We thank Dr. Steen Willadsen and Alta Genetics, Inc., for providing their facilities for the production of the bovine embryos. We also thank Dr. Klaus Wiemer and Mr. Voyteck Polanski for their assistance in the embryo culture procedures. The work referred to from the authors' laboratory was supported by Grant MT-4854 to G.A.S. from the Medical Research Council of Canada (MRC) and Grant HD-23511 from the National Institutes of Health (NIH). A.J.W. is the recipient of a postdoctoral fellowship from MRC and a research allowance from the Alberta Heritage Foundation for Medical Research (AHFMR).

References

1. Schultz GA. Utilization of genetic information in the preimplantation mouse embryo. In: Rossant J, Pedersen R, eds. Experimental approaches to mammalian embryonic development. Cambridge, UK: Cambridge University Press, 1986:239–65.
2. Schultz RM. Molecular aspects of mammalian oocyte growth and maturation. In: Rossant J, Pedersen R, eds. Experimental approaches to mammalian embryonic development. Cambridge, UK: Cambridge University Press, 1986:195–237.
3. Telford NA, Watson AJ, Schultz GA. Transition from maternal to embryonic control in early mammalian development: a comparison of several species. Mol Reprod Dev 1990;26:90–100.
4. McLaren A. The embryo. In: Austin CR, Short RV, eds. Reproduction in mammals, 2. Embryonic and fetal development. Cambridge, UK: Cambridge University Press, 1982:1–26.
5. Papaioannou VE, Ebert KM. Comparative aspects of embryo manipulation in mammals. In: Rossant J, Pedersen R, eds. Experimental approaches to mammalian embryonic development. Cambridge, UK: Cambridge University Press, 1986:67–96.
6. Piko L, Clegg KB. Quantitative changes in total RNA, total poly(A) and ribosomes in early mouse embryos. Dev Biol 1982;89:362–78.
7. Giebelhaus DH, Heikkila JJ, Schultz GA. Changes in the quantity of histone and actin messenger RNA during development of preimplantation mouse embryos. Dev Biol 1983;98:148–54.
8. Graves RA, Marzluff WF, Giebelhaus DH, Schultz GA. Quantitative and qualitative changes in histone gene expression during early mouse embryo development. Proc Natl Acad Sci USA 1985;82:5685–9.
9. Clegg KB, Piko L. Quantitative aspects of RNA synthesis and polyadenylation in 1-cell and 2-cell mouse embryos. J Embryol Exp Morphol 1983;74:169–82.
10. Clegg KB, Piko L. Poly(A) length, cytoplasmic polyadenylation and synthesis of poly(A)$^+$ RNA in early mouse embryos. Dev Biol 1983;95:331–41.

11. Golbus MS, Calarco PG, Epstein CJ. The effects of inhibitors of RNA synthesis (α-amanitin and actinomycin D) on preimplantation mouse embryogenesis. J Exp Zool 1973;186:207–16.
12. Van Blerkom J, Brockway GO. Qualitative patterns of protein synthesis in the preimplantation mouse embryo, I. Normal pregnancy. Dev Biol 1975; 44:148–57.
13. Flach G, Johnson MH, Braude PR, Taylor RAS, Bolton VN. The transition from maternal to embryonic control in the 2-cell mouse embryo. EMBO J 1982;1:681–6.
14. Bolton VN, Oades PJ, Johnson MH. The relationship between cleavage, DNA replication and gene expression in the mouse 2-cell embryo. J Embryol Exp Morphol 1984;79:139–63.
15. Muggleton-Harris AL, Whittingham DG, Wilson L. Cytoplasmic control of preimplantation development in vitro in the mouse. Nature (London) 1982; 299:460–2.
16. Pratt HPM. Isolation, culture and manipulation of preimplantation mouse embryos. In: Monk M, ed. Techniques in mammalian development. Oxford: IRL Press, 1987:13–42.
17. Pratt HPM, Muggleton-Harris AL. Cycling cytoplasmic factors that promote mitosis in cultured 2-cell mouse embryos. Development 1988;104:115–20.
18. Norris ML, Barton SC, Surani MAH. A qualitative comparison of protein synthesis in the preimplantation embryos of four rodent species (mouse, rat, hamster, gerbil). Gamete Res 1985;12:313–6.
19. Seshagiri PB, Bavister BD, Williamson JL, Aiken JM. Qualitative comparison of protein production at different stages of hamster preimplantation embryo development. Cell Differ Dev 1990;31:161–8.
20. Seshagiri PB, Aiken JM, Williamson JL, Bavister BD. The time of onset of embryonic genome activation in golden hamsters is at the early 2-cell stage. J Cell Biol 1990;111:357a.
21. Manes C. Nucleic acid synthesis in preimplantation rabbit embryos, I. Quantitative aspects, relationship to early morphogenesis and protein synthesis. J Exp Zool 1969;172:303–10.
22. Manes C. The participation of the embryonic genome during early cleavage in the rabbit. Dev Biol 1973;32:453–9.
23. Manes C. Nucleic acid synthesis in preimplantation rabbit embryos, II. Delayed synthesis of ribosomal RNA. J Exp Zool 1971;176:87–96.
24. Manes C. Nucleic acid synthesis in preimplantation rabbit embryos, III. A "dark period" immediately following fertilization and the early predominance of low molecular weight RNA synthesis. J Exp Zool 1977;201:247–58.
25. Schultz GA. Characterization of polyribosomes containing newly synthesized messenger RNA in preimplantation rabbit embryos. Exp Cell Res 1973; 82:168–74.
26. Van Blerkom J, Manes C. Development of preimplantation rabbit embryos in vivo and in vitro, II. A comparison of qualitative aspects of protein synthesis. Dev Biol 1974;40:40–51.
27. Seidel GE, Bowen RA, Kane MT. In vitro fertilization, culture and transfer of rabbit ova. Fertil Steril 1976;27:861–70.

28. Braude P, Bolton V, Moore S. Human gene expression first occurs between the four- and eight-cell stages of preimplantation development. Nature (London) 1988;332:459–61.
29. Tesarik J, Kopecny V, Plachot M, Mandelbaum J. Activation of nucleolar and extra nucleolar RNA synthesis and changes in the ribosomal content of human embryos developing in vitro. J Reprod Fertil 1986;78:463–70.
30. Tesarik J, Kopecny V, Plachot M, Mandelbaum J, Dalage C, Flechon JE. Nucleologenesis in the human embryo developing in vitro: ultrastructural and autoradiographic analysis. Dev Biol 1986;115:193–203.
31. Tesarik J, Kopecny V, Plachot M, Mandelbaum J. High resolution autoradiographic localization of DNA-containing sites and RNA synthesis in developing nucleoli of human preimplantation embryos: a new concept of embryonic nucleologenesis. Development 1987;101:777–91.
32. Camous S, Heyman Y, Meziou W, Menezo Y. Cleavage beyond the block stage and survival after transfer of early bovine embryos cultured with trophoblastic vesicles. J Reprod Fertil 1984;72:479–85.
33. Gandolfi F, Moor RM. Stimulation of early embryonic development in the sheep by co-culture with oviduct epithelial cells. J Reprod Fertil 1987; 81:23–8.
34. Pinyopummintr T, Bavister BD. In vitro-matured/in vitro-fertilized bovine embryos can develop into morulae/blastocysts in chemically defined, protein-free culture media. Biol Reprod 1991;45:736–42.
35. Camous S, Kopecny V, Flechon JE. Autoradiographic detection of the earliest stage of ^{3}H-uridine incorporation in the cow embryo. Biol Cell 1986; 58:195–200.
36. King WA, Niar A, Chartrain I, Betteridge KJ, Guay P. Nucleolus organizer regions and nucleoli in preattachment bovine embryos. J Reprod Fertil 1988; 82:87–95.
37. Kopecny V, Flechon JE, Camous S, Fulka J. Nucleologenesis and the onset of transcription in the eight-cell bovine embryo: fine structural autoradiographic study. Mol Reprod Dev 1989;1:79–90.
38. Frei RE, Schultz GA, Church RB. Qualitative and quantitative changes in protein synthesis occur at the 8–16 cell stage of embryogenesis in the cow. J Reprod Fertil 1989;86:637–41.
39. Crosby IM, Gandolfi F, Moor RM. Control of protein synthesis during early cleavage of sheep embryos. J Reprod Fertil 1988;82:769–75.
40. Barnes FL, First NL. Embryonic transcription in in vitro cultured bovine embryos. Mol Reprod Dev 1991;29:117–23.
41. Kopecny V, Flechon JE, Tomanek M, Camous S, Kanka J. Ultrastructural analysis of (^{3}H)-uridine incorporation in early embryos of pig and cow [Abstract]. 9th Nucleolar Workshop, Cracow, 1985:31.
42. Freitag M, Dopke HH, Niemann H, Elsaesser F. ^{3}H-uridine incorporation in early porcine embryos. Mol Reprod Dev 1991;29:124–8.
43. Jarrell VL, Day BN, Prather RS. The transition from maternal to zygotic control of development occurs during the 4-cell stage in the domestic pig *Sus scrofa*: quantitative and qualitative aspects of protein synthesis. Biol Reprod 1991;44:62–8.

44. Davis DL. Culture and storage of pig embryos. J Reprod Fertil Suppl 1985; 33:115–24.
45. Taylor KD, Piko L. Patterns of mRNA prevalence and expression of B1 and B2 transcripts in early mouse embryos. Development 1987;101:877–92.
46. Fukui Y. Effects of sera and steroid hormones on development of bovine oocytes matured and fertilized in vitro and co-cultured with bovine oviduct epithelial cells. J Anim Sci 1989;67:1318–23.
47. Eyestone WH, First NL. Co-culture of early cattle embryos to the blastocyst stage with oviductal tissue or in conditioned medium. J Reprod Fertil 1989; 85:715–20.
48. Wiemer KE, Watson AS, Polanski V, McKenna A, Fick GH, Schultz GA. Effects of maturation and co-culture treatments on the developmental capacity of early bovine embryos. Mol Reprod Dev 1991;30:330–8.
49. Rappolee DA, Brenner CA, Schultz R, Mark D, Werb Z. Developmental expression of PDGF, TGF-α, and TGF-β genes in preimplantation mouse embryos. Science 1988;241:1823–5.
50. Rappolee DA, Wang A, Mark D, Werb Z. Novel method for studying mRNA phenotypes in single or small numbers of cells. J Cell Biochem 1989;39:1–11.
51. Rappolee DA, Sturm K, Schultz GA, Pedersen RA, Werb Z. The expression of growth factor ligands and receptors in preimplantation mouse embryos. In: Heyner S, Wiley LM, eds. Early development and paracrine relationships. UCLA Symposia on Molecular and Cellular Biology, New York Series; vol. 117. New York: Alan R. Liss, 1990:11–26.
52. Fukui Y, Ono H. Effects of sera, hormones and granulosa cells added to culture medium for in vitro maturation, fertilization, cleavage and development of bovine embryos. J Reprod Fertil 1989;86:501–6.
53. Sirard MA, Parrish JJ, Ware CB, Leibfried-Rutledge ML, First NL. The culture of bovine embryos to obtain developmentally competent embryos. Biol Reprod 1988;39:546–52.
54. Kim CI, Ellington JE, Foote RH. Maturation, fertilization and development of bovine oocytes in vitro using TCM-199 and a simple defined medium with co-culture. Theriogenology 1990;33:433–40.
55. Parrish JJ, Susko-Parrish JL, Leibfried-Rutledge ML, Critser ES, Eyestone WH, First NL. Bovine in vitro fertilization with frozen-thawed sperm. Theriogenology 1986;25:591–600.
56. Schliwa M, Van Blerkom J. Structural interaction of cytoskeletal components. J Cell Biol 1981;90:222–35.
57. Valdimarsson G, Huebner E. Diethylene glycol distearate as an embedding medium for immunofluorescence microscopy. Biochem Cell Biol 1989; 67:242–5.
58. Watson AJ, Kidder GM. Immunofluorescence assessment of the timing of appearance and cellular distribution of the Na/K-ATPase during mouse embryogenesis. Dev Biol 1988;126:80–90.
59. Sambrook J, Fritsch EF, Maniatis T. Molecular cloning: a laboratory manual. 2nd ed. Cold Spring Harbor, NY: Cold Spring Harbor Laboratory Press, 1982:447.

60. Dean WL, Seufert AC, Schultz GA, et al. The small nuclear RNAs for pre-mRNA splicing are coordinately regulated during oocyte maturation and early embryogenesis in the mouse. Development 1989;106:325–34.
61. Lobo SM, Marzluff WF, Seufert AC, et al. Localization and expression of U1 RNA in early mouse embryo development. Dev Biol 1988;127:349–61.
62. Hahnel A, Rappolee DA, Millan JL, et al. Two alkaline phosphatase genes are expressed during early development in the mouse embryo. Development 1990;110:555–64.
63. Telford NA, Hogan A, Franz C, Schultz GA. Expression of genes for insulin and insulin-like growth factors and receptors in early preimplantation mouse embryos and embryonal carcinoma cells. Mol Reprod Dev 1990;27:81–92.
64. Prather R, Simerly C, Schatten G, et al. U3 snRNPs and nucleolar development during oocyte maturation, fertilization and early embryogenesis in the mouse: U3 snRNA and snRNPs are not regulated coordinate with other snRNAs and snRNPs. Dev Biol 1990;138:247–55.
65. Dean WL, Schultz GA. Relocalization of small ribonucleoprotein particles (snRNPs) during the first cell cycle of mouse embryo development is independent of RNA synthesis, DNA synthesis and cytokinesis. Cell Differ Dev 1990;31:43–51.
66. Nash MA, Kozak S, Angerer L, et al. Sea urchin maternal and embryonic U1 RNAs are spatially segregated. J Cell Biol 1987;104:1133–42.
67. Watson AJ, Wiemer KE, Arcellana-Panlilio M, Schultz GA. U2 small nuclear RNA localization and expression during bovine preimplantation development. Mol Reprod Dev 1992;31:231–40.
68. Kelly D, Campbell WJ, Travis J, Rizzino A. Regulation and function of transforming growth factor beta (TGF-β) during early mammalian development. J Cell Biol 1990;111:347a.
69. Slager HG, Lawson KA, Van Den Eijnden-Van Raaij AJM, DeLaat SW, Mummery CL. Differential localization of TGF-β2 in mouse preimplantation and early postimplantation development. Dev Biol 1991;145:205–18.
70. Murray R, Choy-pik C, Lee F. Hemopoietic growth factor expression in pre- and post-implantation mouse embryos. J Cell Biochem 1990;14E:93.
71. Rappolee DA, Sturm KS, Schultz GA, et al. Expression and function of growth factor ligands and receptors in preimplantation mouse embryos. In: Schomberg DW, ed. Growth factors in reproduction. New York: Springer-Verlag, 1991:207–18.
72. Schultz GA, Dean WL, Telford NA, Rappolee DA, Werb Z, Pedersen RA. Changes in RNA and protein synthesis during development of the pre-implantation mouse embryo. In: Heyner S, Wiley LM, eds. Early embryo development and paracrine relationships. New York: Alan R. Liss, 1990: 27–46.
73. Rosenblum HY, Mattson BM, Heyner S. Stage-specific insulin binding in mouse preimplantation embryos. Dev Biol 1986;116:261–3.
74. Harvey MB, Kaye PL. Visualization of insulin receptors on mouse pre-embryos. Reprod Fertil Dev 1991;3:9–15.
75. Mattson BA, Rosenblum HY, Smith RM, Heyner S. Autoradiographic evidence for insulin and insulin-like growth factor binding to early mouse embryos. Diabetes 1988;37:585–9.

76. Harvey MB, Kaye PL. IGF-II receptors are first expressed at the 2-cell stage of mouse development. Development 1991;111:1057–60.
77. Wood SA, Kaye PL. Effects of epidermal growth factor on preimplantation mouse embryos. J Reprod Fertil 1989;85:575–82.
78. Paria BC, Dey SK. Preimplantation embryo development in vitro: cooperative interactions among embryos and role of growth factors. Proc Natl Acad Sci USA 1990;87:4756–60.
79. Watson AJ, Hogan A, Hahnel A, Wiemer KE, Schultz GA. Expression of growth factor ligand and receptor genes in the preimplantation bovine embryo. Mol Reprod Dev 1992;31:87–95.
80. Heyner S, Rao LV, Jarett L, Smith RM. Preimplantation mouse embryos internalize maternal insulin via receptor-mediated endocytosis: pattern of uptake and functional correlations. Dev Biol 1989;134:48–58.
81. Heyner S, Smith RM, Schultz GA. Temporally regulated expression of insulin and insulin-like growth factors and their receptors in early mammalian development. Bioessays 1989;11:171–6.
82. Werb Z. Expression of EGF and TGF-α genes in early mammalian development. Mol Reprod Dev 1990;27:10–15.
83. Larson RC, Ignotz GG, Currie WB. Defined medium containing TGF-β and bFGF permits development of embryos beyond the "8-cell block." J Reprod Fertil Abstr Ser 1990;4.
84. DeChiara TM, Efstratiadis A, Robertson EJ. A growth deficiency phenotype in heterozygous mice carrying an insulin-like growth factor II gene disrupted by targeting. Nature (London) 1990;345:78-80.

10

Mutations Affecting Early Development in the Mouse

TERRY MAGNUSON, SHYAM K. SHARAN,
AND BERNADETTE HOLDENER-KENNY

Mutations are important for defining and dissecting complex pathways of normal development and for relating biological function to protein structure. Most of the known mutations in mouse have occurred spontaneously and were identified because of visible phenotypic effects associated with the mutated gene in the heterozygous state. More recently, an intense effort has been under way to induce random or site-directed germ-line mutations. For example, the specific-locus method has been used to generate mutations in localized regions of the genome (1). Irradiated animals are mated to a test stock homozygous for a number of visible markers. The resulting array of mutations, many of which are deletions, are detected because of visible phenotypes produced in F_1 offspring. The deletions are useful for studying the marker locus as well as the surrounding chromosomal region. The dilute-short ear-deletion complex of chromosome 9 (2, 3) and the albino-deletion complex of chromosome 7 (4–7) represent two examples of this approach.

For a fine structure analysis, chemical germ-cell mutagenesis has been used to saturate specific chromosomal regions with point mutations. This type of analysis allows one to correlate function with individual loci. Two notable examples include the *t*-complex of chromosome 17 and the *c*-region of chromosome 7 (8–12).

More recently, mutations have been introduced into the germ line by injecting foreign DNA directly into the pronucleus of a fertilized egg (13) or by retroviral infection of eggs or early embryos (14–16). The inserted DNA not only can cause a mutation by disrupting or affecting the expression of an endogenous gene, but it also can serve as a tag for molecular cloning of the affected gene in a manner not possible with spontaneous mutations or with mutations induced by chemicals or radiation.

An alternative approach for producing mutations by insertion is the use of gene-trap constructs (17). These constructs consist of a promoterless *lacZ* gene placed directly 3' to a splice-acceptor site. Integration of this construct into introns of genes in the correct orientation produces a functional fusion protein when appropriate splicing takes place. Alternatively, an enhancer-trap construct has been used. This construct consists of a *lacZ* gene fused in-frame to a minimal promoter. Expression of *lacZ* depends on *cis*-acting regulatory proteins. In both cases, *lacZ* expression serves as an indicator for integration into or near an endogenous gene. Clonal lines of *embryonic stem* (ES) cells carrying either reporter construct can be used to detect *lacZ* expression in chimeric embryos without generating transgenic mouse lines. Consequently, a large number of insertions can be analyzed for developmentally regulated expression of the reporter gene that then serves as a tag for molecular cloning of the insertion site.

In addition to random mutagenesis, gene targeting is now a possibility in the mouse (18). The ability to target specific genes involves standard genetic engineering techniques to introduce desired mutations into a cloned DNA sequence of interest. The targeting vector can be designed to produce a null mutation (19–21) or, alternatively, more subtle alterations can be made that affect gene function (22, 23). The mutation is then introduced into the endogenous locus in ES cells via homologous recombination. Clonal lines carrying the mutation of interest are established and then subsequently injected into a host to create germ-line chimeras. When heterozygous siblings are bred, animals homozygous for the desired mutation can be obtained. Thus, the potential now exists to generate mice of any desired genotype. Examples of targeted genes that have produced interesting phenotypes when bred to homozygosity include β_2-microglobulin (24, 25), *En*-2 (26), *hox*-1.5 (27), immunoglobulin μ-chain gene (28), insulin-like growth factor II (29), *int*-1 (30), c-*myb* (31), c-*src* (32), and *Wnt*-1 (33).

The work from our laboratory has concentrated on random mutagenesis induced by radiation. We have focused our attention on the genetic system known as the *albino-deletion complex*. This complex represents more than 37 overlapping chromosomal deletions that remove varying amounts of chromosome 7 (4, 5, 7). Embryological and complementation analyses have resulted in classification of these deletion chromosomes into several groups defining at least 4 genes that affect prenatal development (7, 34–38). For example, embryos homozygous for the c^{1DThWb} or the c^{23DVT} deletion (Fig. 10.1) were reported to die sometime between implantation and midgestation (7). We have completed an extensive phenotypic analysis of homozygous embryos and find that both deletions remove a gene(s) (*msd*) needed for mesoderm production at the time of gastrulation (Fig. 10.1) (Holdener-Kenny, Magnuson, unpublished results).

Similar studies with the c^{11DSD}, $c^{5FR60Hg}$, c^{2YPSj}, $c^{4FR60Hd}$, and c^{6H} deletions showed that all 5 chromosomes remove a gene(s) (*eed*) needed for development of the embryonic ectoderm (Fig. 10.1) (34, 37, 38). This locus was defined primarily by the inability to establish ES cell lines from embryos homozygous for the deletions. Two of these deletions ($c^{4FR60Hd}$ and c^{6H}) also remove an additional gene(s) (*exed*) needed for development of the extraembryonic ectoderm (Fig. 10.1) (34, 37, 38). This was defined by a complete lack of extraembryonic structures in $c^{4FR60Hd}$ and c^{6H} homozygotes and extensive development of these structures in c^{11DSD}, $c^{5FR60Hg}$, and c^{2YPSj} homozygotes. Furthermore, the latter three deletions were capable of complementing the $c^{4FR60Hd}$ and c^{6H} chromosomes by allowing for development of the extraembryonic structures in compound heterozygotes (37, 38).

Fifteen of the albino deletions remove a gene(s) (*pid*) needed for preimplantation development. The homozygous phenotype associated with one of these deletions (c^{25H}) has been examined in detail (Fig. 10.1) (35, 36). Embryos were found to cease cell division sometime between the 2- and 6-cell stages, with death occurring 1–2 days after that time. The only obvious ultrastructural abnormality was aberrantly shaped nuclei. The fact that the homozygous embryo is affected beginning at the 2-cell stage is intriguing because this is the time at which the embryonic genome is activated. The c^{25H} homozygous embryos have yet to be examined for embryonic gene transcription, polarization, compaction, or other hallmarks of early preimplantation development.

Each of the loci described above is defined by the phenotype exhibited by deletion homozygotes. In reality, these phenotypes could be the result of deletion of one important gene or a combinatorial effect of the loss of more than one gene. A molecular analysis is needed to ascertain the identity and function of these genes. To obtain molecular markers located within the region covered by the albino deletions, we produced a partial genomic library of the distal region of a nondeleted chromosome 7 using the techniques of chromosome microdissection and microcloning (39). One of the microclones was found to define a locus, *D7Cw18*, that maps to a region of chromosome 7 removed by the c^{11DSD} deletion, but not by the $c^{5FR60Hg}$, c^{2YPSj}, $c^{4FR60Hd}$, or c^{6H} deletions. Thus, *D7Cw18* maps proximal to *c*, but distal to *hsdr*-1 (hepatocyte-specific developmental regulator) (Fig. 10.1) (40, 41). In a chromosome walk initiated from *D7Cw18*, an albino-region-specific repeat was identified (39). This repeat was found to produce different banding patterns when hybridized to c^{11DSD}, $c^{5FR60Hg}$, c^{2YPSj}, $c^{4FR60Hd}$, and c^{6H} deletion DNAs. Therefore, it was possible to define the order of the proximal breakpoints by examining the Southern blot banding patterns produced by the repeat probe when hybridized to these deletion DNAs. This relative order of proximal breakpoints was confirmed by other single-copy probes isolated from the region (see Fig. 10.1 for map) (39).

Cloning of one or more of the deletion breakpoints would access the distal side of the deletions where *eed*, *exed*, and *pid* are located. In this

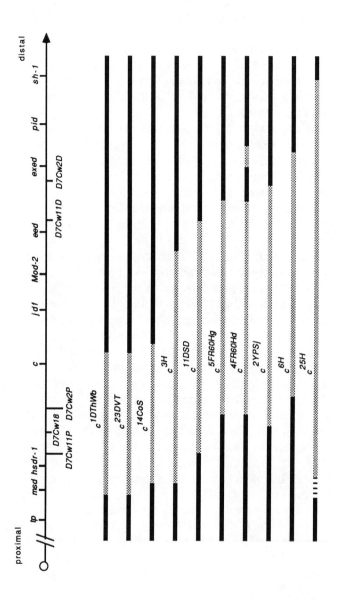

chapter, we describe the cloning of the c^{2YPSj} and the c^{11DSD} breakpoints and an ordering of the distal side of the c^{11DSD}, $c^{5FR60Hg}$, c^{2YPSj}, $c^{4FR60Hd}$, and c^{6H} deletion breakpoints relative to one another. A detailed description of this work is given in reference 42.

Materials and Methods

Deletion Mice

The c^{6H}, c^{3H}, and c^{25H} mice originated at the MRC Radiobiology Unit, Harwell, England (4). All of the remaining deletions referred to in this report originated at the Oak Ridge National Laboratory (5).

Mus spretus/Mus musculus *Interspecies Cross*

Wild-type (nondeletion) *M. spretus* males were crossed with $Df(c)/c^{ch}$ *M. musculus* females. For mapping of *D7Cw2D* or *D7Cw11D*, high-molecular weight DNA was prepared from spleen or brain of appropriate

⬅───────────────────────────────

FIGURE 10.1. Complementation map of the albino deletions cited in this chapter. Deleted regions are represented by stippled lines and nondeleted regions by solid lines. Symbols above the stippled line represent the name of the deletion chromosome. The hatched line on the distal end of the c^{25H} reflects the fact that appropriate complementation studies have not yet been done to know if this deletion removes *msd*. The solid line located within the $c^{4FR60Hd}$ reflects the fact that both *eed* and *exed* are inactivated, but *D7Cw2D* is not deleted (see text). Although the exact positions of the chromosomal breakpoints are not known, the approximate lengths of the deletions with respect to the genetic map are based on complementation analyses (4, 7, 37–39, 44). Loci defined by complementation and embryological analyses are indicated above the chromosomal line, whereas loci defined by cloned DNA probes are indicated below the chromosomal line. Marker loci include *tp* (taupe), *c* (albino), *Mod-2* (mitochondrial form of malic enzyme), and *sh-1* (shaker 1). *D7Cw18*, *D7Cw11P*, *D7Cw2P*, *D7Cw11D*, and *D7Cw2D* are loci defined by cloned fragments (39, 42). The latter four are cloned deletion breakpoints. The following loci are defined by complementation and embryological analyses using the various albino deletions: *msd* (mesoderm deficient, formerly designated implantation survival region [7 and Holdener-Kenny, Magnuson, unpublished results]), *hsdr-1* (hepatocyte-specific developmental regulator, formerly designated perinatal survival region [4, 7, 40, 41]), *jdf* (juvenile development and fertility, formerly designated juvenile survival region [7]), *eed* (embryonic ectoderm development, formerly designated embryonic ectoderm region [37, 38]), *exed* (extraembryonic ectoderm development, formerly designated extraembryonic ectoderm region [37, 38]), and *pid* (preimplantation development, formerly designated preimplantation survival region [4, 7, 35, 36]). The actual phenotype resulting from the deletion of any of these loci may be due to the removal of more than one gene.

F_1 progeny, along with their $Df(c)/c^{ch}$ dams and *M. spretus* sires, as previously described (43).

Genomic Clones

λ14RIc2 represents a 14-kb *Eco*RI fragment isolated from a c^{2YPSj}/c^{3H} subgenomic library (42). pv0.4AP represents a 0.4-kb *Pst* I/*Asp* 718 fragment subcloned into bluescript from a larger insert obtained from a jump/walk initiated from *D7Cw18*. λE10.8c11 was isolated from a c^{11DSD}/c^{3H} subgenomic library screened with pv0.4AP. The library was prepared from size-selected DNA (10- to 12-kb range) as described for λ14RIc2. pA4.2c11 represents a 4.2-kb *Asp* 718 fragment subcloned into pBS from λ10.8RIc11.

Results

Cloning of the c^{2YPSj}-Deletion-Breakpoint-Fusion Fragment

The albino-specific-repeat probe detected a 14-kb fragment when hybridized to c^{2YPSj} DNA. This fragment was not detected with any of the other deletion DNAs tested. Furthermore, this aberrant fragment was not the result of a single *Eco*RI polymorphism associated with the c^{2YPSj} chromosome since several other restriction enzymes (*Hind*III, *Xba*I, *Pst*I, *Sac*I) also produced an altered fragment. Given the relative order of the proximal breakpoints (39), it should not be possible to detect a fragment present in c^{2YPSj} DNA that is not also present in $c^{4FR60Hd}$, $c^{5FR60Hg}$, c^{6H}, or c^{ch} DNA (Fig. 10.1). Thus, we hypothesized that the altered fragment contained the c^{2YPSj}-deletion-breakpoint-fusion fragment.

A subgenomic library generated from size-selected c^{2YPSj}/c^{3H}-*Eco*RI-digested DNA was screened with the albino-specific-repeat probe. One independent isolate containing a 14-kb *Eco*RI insert, λ14RIc2, was analyzed further. End-specific riboprobes were hybridized to nondeleted c^{ch}/c^{ch} DNA and to c^{3H}/c^{3H} and c^{2YPSj}/c^{3H} deletion DNA. The proximal end was found to be present in the nondeleted c^{ch} chromosome as well as the c^{2YPSj}-deletion chromosome, but absent from the c^{3H}-deletion chromosome. In contrast, the distal end of the clone was present in all three chromosomes. These results are consistent with λ14RIc2 spanning the c^{2YPSj}-deletion-breakpoint-fusion fragment. Thus, we propose that the sequence included in the T7 riboprobe is homologous to a genomic locus (*D7Cw2P*) that lies on the proximal side of the c^{2YPSj} breakpoint (Fig. 10.1), whereas the sequence included in the T3 riboprobe is homologous to a genomic locus (*D7Cw2D*) located on the distal side of the c^{2YPSj} distal breakpoint (Fig. 10.1).

Ordering of the Distal Breakpoints Relative to D7Cw2D

Distal end-specific riboprobe prepared from λ14RIc2 detects an EcoRI-restriction-fragment-length polymorphism in DNA of different genetic backgrounds. This polymorphism was used to order the distal breakpoints relative to D7Cw2D. A 16-kb fragment was detected with wild-type c^{ch}/c^{ch} DNA, whereas a 9-kb fragment was detected with c^{14CoS}/c^{14CoS} DNA. When hybridized to c^{11DSD}/c^{14CoS}-double-heterozygous DNA, the c^{14CoS}-associated 9-kb fragment was present, as well as a 16-kb fragment derived from the c^{11DSD} chromosome. The presence of the c^{11DSD}-associated fragment indicates that D7Cw2D lies distal to the c^{11DSD} deletion. In a similar manner, the position of D7Cw2D relative to the $c^{5FR60Hg}$, $c^{4FR60Hd}$, and c^{6H} distal breakpoints was determined. From the data generated, it was concluded that the c^{11DSD}, $c^{5FR60Hg}$, and $c^{4FR60Hd}$ distal breakpoints lie proximal to the c^{2YPSj} distal breakpoint, which lies proximal to the c^{6H} distal breakpoint (Fig. 10.1).

Cloning of the c^{11DSD}-Breakpoint-Fusion Fragment

To determine the proximal-to-distal order of c^{11DSD}, $c^{5FR60Hg}$, and $c^{4FR60Hd}$ distal breakpoints relative to one another, experiments were undertaken to clone the c^{11DSD}-breakpoint-fusion fragment. A combination of chromosome walking and jumping initiated from D7Cw18 (deleted from the c^{11DSD} chromosome, see Fig. 10.1) was used to isolate clone pv0.4AP. This clone hybridizes to nondeleted c^{ch} DNA and to c^{11DSD}/c^{14CoS}-deletion DNA, but not to c^{14CoS}/c^{14CoS} DNA. These data are consistent with the genomic locus D7Cw11P, defined by pv0.4AP, being proximal to the c^{11DSD} deletion. Furthermore, pv0.4AP detects an aberrant 10.8-kb EcoRI fragment in c^{11DSD} DNA as compared to a 7.4-kb c^{ch} fragment. To determine if the aberrant fragment contains the deletion breakpoint, a subgenomic library was generated from size-selected c^{11DSD}/c^{3H}-EcoRI-digested DNA. One positive clone, λ10.8c11, containing a 10.8-kb insert was isolated using pv0.4AP as a probe. T3 riboprobe prepared from pA4.2c11 (a 4.2-kb Asp 718 subclone from one end of λ10.8c11) hybridized to nondeleted c^{ch} DNA and to c^{11DSD}, as well as to c^{14CoS}/c^{14CoS}-deletion DNAs. These results are consistent with the prediction that a clone (pA4.2c11) containing DNA from the distal side of the deletion should hybridize to all three chromosomes, whereas the proximal end (pv0.4AP) should hybridize only to c^{ch} and c^{11DSD} DNA. Thus, it can be concluded that pv0.4AP defines a genomic locus (D7Cw11P) that lies on the proximal side of the c^{11DSD} breakpoint (Fig. 10.1). In contrast, the pA4.2c11 sequence generating the T3 riboprobe is homologous to a genomic locus (D7Cw11D) that lies on the distal side of the c^{11DSD} distal breakpoint (Fig. 10.1).

Ordering of the Distal Breakpoints Relative to D7Cw11D

To map the distal breakpoints relative to *D7Cw11D*, we utilized restriction-fragment-length polymorphism detected in *M. musculus* and *M. spretus* interspecific crosses. pA4.2c11 was hybridized to *Taq*I-digested DNA from an F_1 hybrid that carried an *M. spretus* nondeletion chromosome and either an *M. musculus* $c^{5FR60Hg}$-, $c^{4FR60Hd}$-, c^{6H}-, or c^{2YPSj}-deletion chromosome. Only an *M. spretus* fragment was detected in the F_1 DNA. Thus, the *D7Cw11D* locus is removed by all of the deletions. From these data, it can be concluded that the proximal-to-distal breakpoint order of the distal side of the deletions is $c^{11DSD} \rightarrow c^{5FR60Hg}$ or $c^{4FR60Hd} \rightarrow c^{2YPSj} \rightarrow c^{6H}$.

Discussion

Based on the distal breakpoint order predicted by our earlier genetic and embryological analyses (37, 38), *D7Cw2D* and *D7Cw11D* should map to positions that lie between *eed* and *exed*. To confirm these predictions, the distal breakpoints of the c^{11DSD}, $c^{5FR60Hg}$, $c^{4FR60Hd}$, and c^{6H} deletions were mapped relative to *D7Cw2D* and *D7Cw11D*. *D7Cw11D* was deleted from $c^{5FR60Hg}$, $c^{4FR60Hd}$, c^{2YPSj}, and c^{6H}, whereas *D7Cw2D* was deleted only from c^{6H}. These results are in agreement with the prediction that *D7Cw2D* and *D7Cw11D* map between *eed* and *exed* (Fig. 10.1); they also establish the relative proximal-to-distal order of the distal breakpoints as being $c^{11DSD} \rightarrow c^{5FR60Hg}$ and $c^{4FR60Hd} \rightarrow c^{2YPSj} \rightarrow c^{6H}$.

Ordering of the c^{11DSD}, $c^{5FR60Hg}$, c^{2YPSj}, and c^{6H} distal breakpoints established the molecular limits of the *eed* and *exed* genes. The region of chromosome 7 containing the *eed* gene is bounded on the proximal side by the c^{3H} distal breakpoint (44) and on the distal side by the c^{11DSD} distal breakpoint (42). The region containing the *exed* gene is delimited by the c^{2YPSj} deletion on the proximal side and the c^{6H} deletion on the distal side (42). We have cloned two of the breakpoints (c^{11DSD} and c^{2YPSj}) delimiting the regions containing *eed* and *exed*. Our immediate work will concentrate on the isolation of the c^{6H}- and c^{3H}-breakpoint-fusion fragments, as well as establishment of a long-range physical map of the regions containing *eed* and *exed*. The cloning of the c^{6H} breakpoint will also access the region containing *pid*. This work will result in the construction of a detailed genetic and physical map that delineates the minimal region within which the genes of interest may be found. The flanking markers demarcating the boundaries of these regions will be used as the starting point for a YAC-based chromosome walk.

Placement of the $c^{4FR60Hd}$ distal breakpoint proximal to that of c^{2YPSj} is in direct contrast to the order suggested by our genetic data (37, 38). Embryos homozygous for the c^{2YPSj} deletion show the embryonic-ectoderm defect, but not the extraembryonic-ectoderm defect. In contrast, the

$c^{4FR60Hd}$ homozygotes show a phenotype consistent with both *eed* and *exed* being deleted or inactivated. Furthermore, c^{2YPSj} can complement the extraembryonic defect by providing the wild-type copy of *exed* in $c^{2YPSj}/c^{4FR60Hd}$ compound heterozygotes. All of the genetic and embryological data are consistent with the $c^{4FR60Hd}$ distal breakpoint lying distal to the c^{2YPSj} distal breakpoint. Yet, our molecular data indicate that the $c^{4FR60Hd}$ breakpoint lies proximal to that of c^{2YPSj}. One possible explanation for this discrepancy is that the $c^{4FR60Hd}$ deletion is discontinuous, thereby "skipping" and not deleting the region containing *D7Cw2D*. Precedent for radiation-induced, noncontinuous, or skipping, mutations can be found within the dilute-short ear-deletion complex of mouse chromosome 9 (2, 45). At least 5 of the dilute-short ear deletions appear to skip and inactivate genes on both sides of an active functional group. Although one can postulate several reasons for the inactivation of *exed* by the $c^{4FR60Hd}$ deletion—such as deletion, translocation, inversion, or position effect—the molecular basis for the inactivation will only be resolved when the breakpoint-fusion fragment has been cloned and a physical map of the region is completed.

Several strategies can be followed for identifying and recovering candidate genes based on their location between flanking markers. These methods are aimed at estimating the total number of genes within a particular region, analyzing the defined region for structural features characteristic of genes, and then looking for potential transcription units among the genomic DNA from this region. For the purpose of determining the total number of genes present in specific genomic regions, chemical saturation mutagenesis screens can be done, provided suitable closely linked markers are available. Such a screen is now under way for the albino region, and to date, 4 independent, complementing, lethal mutations, as well as 2 new mutations of the shaker 1 locus and 2 mutations at a new locus called *fitness 1*, have been recovered from an initial screen of 972 mutagenized gametes (11, 12). After screening more than 3000 gametes, mutations that do not complement some of those produced in the initial screen have been found, as well as additional lethals that have not yet been tested for complementation (12). The ability to determine whether saturation mutagenesis is being reached is dependent on the recovery of repeat mutations; therefore, these results are encouraging.

Many constitutively expressed and some regulated coding sequences have been found to be marked at their 5' ends by a high density of hypomethylated CpG residues (46). For this reason, it will be important to establish the position of these islands within the defined intervals using physical mapping techniques. In this manner, the location of potential transcription units can be predicted. Interspecies crosshybridization of "zoo" blots would also be informative for identifying genes within an interval, as a significant number of unique or low-copy sequences conserved between species represent genes (47).

Transcribed sequences could be identified by direct screening of cDNA libraries or by Northern blots. Although cDNA libraries for appropriate stages of early mouse development exist, the relative abundance of the mRNA in question may be a limiting factor. An alternative approach is exon trapping using a retroviral/SV40 shuttle vector for identifying transcription units (48). Because genomic sequences of nonviral origin are correctly spliced during the retrovirus life cycle, sequences that reconstitute an exon-intron-exon motif can be recovered. Finally, some progress has been made in transfecting YAC clones into mammalian cells (49, 50), which raises the possibility of identifying a YAC clone containing the appropriate wild-type copy of the gene by its ability to complement the mutated version with mice created from transfected ES cells.

The cloning of the *T*, or brachyury gene, on mouse chromosome 17 represents an elegant example of the use of a combination of these approaches to clone a mouse developmental gene first identified as a spontaneous mutation (51). The ultimate proof that this candidate gene is in fact *T* will be obtained by producing homozygous mutant mice that carry the wild-type transgene. When such studies are accomplished, they will complete the cycle of using positional cloning to identify unequivocally developmentally important genes.

Acknowledgments. The work reported here was supported in part by grants to T.M. from the National Institutes of Health (Grant HD-2446), March of Dimes (Grant 1-1180), the Ohio Edison Animal Biotechnology Center, and the Pew Scholars Program in the Biomedical Sciences.

References

1. Russell WL. X-ray-induced mutations in mice. Cold Spring Harbor Symposium Quant Biol 1951;16:327–36.
2. Rinchik EM, Russell LB, Copeland NG, Jenkins NA. Molecular genetic analysis of the *Dilute-Short Ear* (*D-SE*) region of the mouse. Genetics 1986; 112:321–42.
3. Strobel MC, Seperack PK, Copeland NG, Jenkins NA. Molecular analysis of two mouse dilute locus deletion mutations: spontaneous dilute lethal[20J] and radiation-induced dilute prenatal lethal Aa2 alleles. Mol Cell Biol 1990; 10:501–9.
4. Gluecksohn-Waelsch S, Schiffman MB, Thorndike J, Cori CF. Complementation studies of lethal alleles in the mouse causing deficiencies of glucose-6-phosphatase, tyrosine aminotransferase and serine dehydratase. Proc Natl Acad Sci USA 1974;71:825–9.
5. Russell LB, Russell WL, Kelly EM. Analysis of the albino-locus region of the mouse, I. Origin and viability of whole body and fractional mutants. Genetics 1979;91:127–39.
6. Russell LB, Raymer GD. Analysis of the albino-locus region of the mouse, III. Time of death of prenatal lethals. Genetics 1979;92:205–13.

7. Russell LB, Montgomery CS, Raymer GD. Analysis of the albino-locus region of the mouse, IV. Characterization of 34 deficiencies. Genetics 1982; 100:427–53.
8. Shedlovsky A, Guenet J-L, Johnson LL, Dove WF. Induction of recessive lethal mutations in the *T/t-H-2* region of the mouse genome by a point mutagen. Genet Res 1986;47:135–42.
9. Shedlovsky A, King TR, Dove WF. Saturation germ line mutagenesis of murine *t* region including a lethal allele at the quaking locus. Proc Natl Acad Sci USA 1988;85:180–4.
10. King TR, Dove WF, Herrmann B, Moser AR, Shedlovsky A. Mapping to molecular resolution in the *T* to *H-2* region of the mouse genome with a nested set of meiotic recombinants. Proc Natl Acad Sci USA 1989;86:222–6.
11. Rinchik EM, Carpenter DA, Selby PB. A strategy for fine-structure functional analysis of a 6- to 11-centimorgan region of mouse chromosome 7 by high-efficiency mutagenesis. Proc Natl Acad Sci USA 1990;87:896–900.
12. Rinchik EM. Chemical mutagenesis and fine-structure functional analysis of the mouse genome. Trend Genet 1991;7:15–21.
13. Westphal H, Gruss P. Molecular genetics of development studied in the transgenic mouse. Annu Rev Cell Biol 1989;5:181–96.
14. Gridley T, Soriano P, Janeisch R. Insertional mutagenesis in mice. Trends Genet 1987;3:162–6.
15. Lock LF, Keshet E, Gilbert DJ, Jenkins NA, Copeland NG. Studies of the mechanism of spontaneous germline ecotropic provirus acquisition in mice. EMBO J 1988;7:4169–77.
16. Copeland NG, Lock LF, Spence SE, et al. Spontaneous germ-line ecotropic murine leukemia virus infection: implications for retroviral insertional mutagenesis and germ-line gene transfer. Prog Nucleic Acid Res Mol Biol 1989;36:221–34.
17. Gossler A, Joyner AL, Rossant J, Skarnes WC. Mouse embryonic stem cells and reporter constructs to detect developmentally regulated genes. Science 1989;244:463–5.
18. Capecchi MR. The new mouse genetics: altering the genome by gene targeting. Trends Genet 1989;5:70–6.
19. Thomas KR, Capecchi MR. Site-directed mutagenesis by gene targeting in mouse embryo-derived stem cells. Cell 1987;51:503–12.
20. Mansour SL, Thomas KR, Capecchi MR. Disruption of the proto-oncogene *int-2* in mouse embryo-derived stem cells: a general strategy for targeting mutations to non-selectable genes. Nature (London) 1988;336:348–52.
21. Doetschman T, Maeda N, Smithies O. Targeted mutation of the *Hprt* gene in mouse embryonic stem cells. Proc Natl Acad Sci USA 1988;85:8583–7.
22. Hasty P, Ramirez-Solis R, Krumlauf R, Bradley A. Introduction of a subtle mutation into the *Hox-2.6* locus in embryonic stem cells. Nature (London) 1991;350:243–6.
23. Valancius V, Smithies O. Testing an "in-out" targeting procedure for making subtle genomic modifications in mouse embryonic stem cells. Mol Cell Biol 1991;11:1402–8.
24. Zijlstra M, Bix M, Simister NE, Loring JM, Raulet DH, Jaenisch R. β2-microglobulin deficient mice lack CD4-8+ cytolytic T cells. Nature (London) 1990;344:742–6.

25. Koller BH, Marrack P, Kappler JW, Smithies O. Normal development of mice deficient in β2M, MHC class I proteins, and CD8+ T cells. Science 1990;248:1227–30.
26. Joyner AL, Herrup K, Auerbach BA, Davis CA, Rossant J. Subtle cerebellar phenotype in mice homozygous for a targeted deletion of the *En-2* homeobox. Science 1991;251:1239–43.
27. Chisaka O, Capecchi MR. Regionally restricted developmental defects resulting from targeted disruption of the mouse homeobox gene *hox-1.5*. Nature (London) 1991;350:473–9.
28. Kitamura D, Roses J, Kuhn R, Rajewsky K. AB cell-deficient mouse by targeted disruption of the membrane exon of the immunoglobulin µ chain gene. Nature (London) 1991;350:423–6.
29. DeChiara TM, Efstratiadis A, Robertson EJ. A growth-deficiency phenotype in heterozygous mice carrying an insulin-like growth factor II gene disrupted by targeting. Nature (London) 1990;345:78–80.
30. Thomas KR, Capecchi MR. Targeted disruption of the murine *int-1* protooncogene resulting in severe abnormalities in midbrain and cerebellar development. Nature (London) 1990;346:847–50.
31. Mucenski ML, McLain K, Kier AB, et al. A functional c-*myb* gene is required for normal murine fetal hepatic hematopoiesis. Cell 1991;65:677–89.
32. Soriano P, Montgomery C, Geske R, Bradley A. Targeted disruption of the c-*src* proto-oncogene leads to osteopetrosis in mice. Cell 1991;64:693–702.
33. McMahon AP, Bradley A. The *Wnt*-1 (*int*-1) proto-oncogene is required for development of a large region of the mouse brain. Cell 1990;62:1073–85.
34. Lewis SE, Turchin HA, Gluecksohn-Waelsch S. The developmental analysis of an embryological lethal (c^{6H}) in the mouse. J Embryol Exp Morphol 1976;36:363–71.
35. Lewis S. Developmental analysis of lethal effects of homozygosity for the c^{25H} deletion in the mouse. Dev Biol 1978;65:553–7.
36. Nadijcka MD, Hillman N, Gluecksohn-Waelsch S. Ultrastructural studies of lethal c^{25H}/c^{25H} mouse embryos. J Embryol Exp Morphol 1979;52:1–11.
37. Niswander L, Yee D, Rinchik EM, Russell LB, Magnuson T. The albino deletion complex and early postimplantation survival in the mouse. Development 1988;102:45–53.
38. Niswander L, Yee D, Rinchik EM, Russell LB, Magnuson T. The albino-deletion complex in the mouse defines genes necessary for development of embryonic and extraembryonic mesoderm. Development 1989;105:175–82.
39. Niswander L, Kelsey G, Schedl A, et al. Molecular mapping of albino deletions associated with early embryonic lethality in the mouse. Genomics 1991;9:162–9.
40. Gluecksohn-Waelsch S. Regulatory genes in development. Trend Genet 1987;3:123–7.
41. McKnight SL, Lane MD, Gluecksohn-Waelsch S. Is CCAAT/enhancer-binding protein a central regulator of energy metabolism? Genes Dev 1989;3:2021–4.
42. Sharan SK, Hodener-Kenny B, Ruppert S, et al. The albino-deletion complex of the mouse: molecular mapping of deletion breakpoints that define regions necessary for development of the embryonic and extraembryonic ectoderm. Genetics 1991;129:825–32.

43. Johnson DK, Hand RE, Rinchik EM. Molecular mapping within the mouse albino-deletion complex. Proc Natl Acad Sci USA 1989;8862–6.
44. Eicher EM, Lewis SE, Turchin HA, Gluecksohn-Waelsch S. Absence of mitochondrial malic enzyme in mice carrying two complementing lethal albino alleles. Genet Res 1978;32:1–7.
45. Russell LB. Definition of functional units in a small chromosomal segment of the mouse and its use in interpreting the nature of radiation-induced mutations. Mutat Res 1971;11:107–23.
46. Bird AP. CpG islands as gene markers in the vertebrate nucleus. Trends Genet 1987;3:342–7.
47. Monaco AP, Neve RL, Colletti-Feener C, Bertelson CJ, Kurnit DM, Kunkel LM. Isolation of candidate cDNAs for portions of the Duchenne muscular dystrophy gene. Nature (London) 1986;323:646–50.
48. Duyk GM, Kim S, Myers RM, Cox DR. Exon trapping: a genetic screen to identify transcribed sequences in cloned mammalian genomic DNA. Proc Natl Acad Sci USA 1991;87:8995–9.
49. Eliceiri B, Labella T, Hagino Y, et al. Stable integration and expression in mouse cells of yeast artificial chromosomes harboring human genes. Proc Natl Acad Sci USA 1991;88:2179–83.
50. Pachnis V, Pevny L, Rothstein R, Costantini F. Transfer of a yeast artificial chromosome carrying human DNA from *Saccharomyces cerevisiae* into mammalian cells. Proc Natl Acad Sci USA 1990;87:5109–13.
51. Herrman BG, Labeit S, Poustka A, King TR, Lehrach H. Cloning of the *T* gene required in mesoderm formation in the mouse. Nature (London) 1990; 343:617–22.

11

Parental Imprinting in Mammalian Development

ANNE C. FERGUSON-SMITH AND M. AZIM SURANI

The basic laws of Mendelian genetics tell us that an equal genetic contribution is inherited from each parent and that these inherited genes function equally in the offspring. Studies in the mouse, however, have shown that for some genes this is not the case, and the finding that parental genomes were not functionally equivalent has established the study of parental imprinting as one of the most exciting areas of genetic regulation at this time.

Parental imprinting is a parental origin-specific epigenetic marking process that renders homologous chromosomes functionally nonequivalent. This non-Mendelian mode of regulation results in the expression of certain genes being dependent on their inheritance through either the egg or the sperm. In the mouse there are genes crucial for normal development that are regulated in this way, as exemplified by the inability of parthenogenetic/gynogenetic and androgenetic conceptuses to develop further than midgestation (1, 2). Not only does this epigenetic marking process distinguish between maternally and paternally inherited chromosomes, but these epigenetic modifications also must be stably inherited over many somatic generations. Finally, these modifications must be removed in the germline where totipotency is restored.

Thus, the expression of an imprinted gene is dependent on the germline through which it has passed and, indeed, the dosage of that gene can be doubled or lost completely if there is a uniparental duplication or deficiency of the corresponding chromosomal region in which it resides (Fig. 11.1). Although such duplications/deficiencies are responsible for many mutations ascribed to imprinting and will be the subject of this chapter, more subtle molecular aberrations affecting the actual mechanism of the process would also give rise to altered dosage of an imprinted gene. The mechanism of parental imprinting, currently unknown, is the subject of intense scrutiny by many investigators and will not be addressed in any depth here.

11. Parental Imprinting in Mammalian Development   145

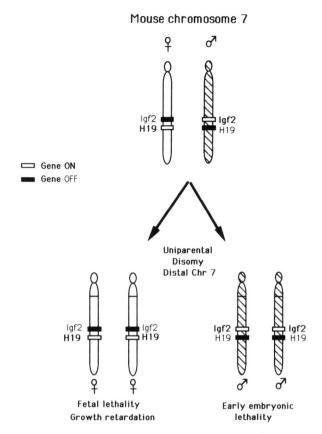

FIGURE 11.1. Dependency of imprinted gene expression on its parent of origin. Mouse chromosomes 7 inherited either through egg (circle with cross) or sperm (circle with arrow) illustrate the differential activity of the two reciprocally imprinted genes IGF-II (Igf2) and H19. IGF-II is paternally expressed and maternally repressed by imprinting. In contrast, H19 is expressed only from the maternally inherited chromosome. Dosage of both of these imprinted genes is disrupted by uniparental disomy of this region, which may explain the lethalities observed. In actuality, the genes are more distal on the chromosome than indicated in this schematic, and it is not known whether IGF-II is proximal or distal to H19. The position of a translocation breakpoint (T9H) used to generate the distal disomy is indicated by a line below the centromere, as described in reference 9.

Identification of Imprinted Loci

Genetic complementation tests have identified at least 8 subchromosomal regions on 5 mouse chromosomes that harbor imprinted genes (3). Reciprocal and Robertsonian translocations that generate unbalanced gametes containing chromosomal regions in which both copies are derived from one parent (by meiotic nondisjunction) can be crossed to produce genetically balanced zygotes. Resulting embryos are uniparentally disomic for certain chromosomal regions, and a mutant phenotype may result if there is a requirement for both a maternal and paternal complement of that region (Fig. 11.1) (4). Such studies have identified mouse chromosomes 2, 6, 7, 11, and 17 as harboring imprinted genes. However, the regions involved in these studies span many megabases of DNA. In order to identify the imprinted genes themselves, higher-resolution molecular studies must be undertaken.

A prerequisite for the identification of imprinted genes is an experimental system in which either both copies of a particular chromosome region are derived from one parent, as just described, or in which parent-specific gene activity can be distinguished in some other way. Four different approaches have proved fruitful with 3 endogenous imprinted genes identified to date. The first approach made use of the only known mutation in the mouse exhibiting parental origin effects in its pattern of inheritance. This locus, *T maternal effect* (Tme), is associated with a deletion that maps to the imprinted region on chromosome 17. This deletion is lethal at around day 17 of gestation, but only when maternally inherited (5, 6). Inheritance of the mutation through the male germline has no major effect. A reverse genetic approach was taken to physically map a portion of the deletion and analyze the transcription of 4 genes mapping to that region including 2—insulin-like growth factor II receptor (IGF-II-R) and superoxide dismutase-2 (Sod-2)—located within the deletion. In this way, Barlow and her colleagues were able to show the absence of IGF-II-R expression when the deletion was inherited maternally; transcription was normal when the deletion was inherited from the male (7). The 3 other genes were expressed normally from both chromosomes, except when they were deleted. Thus, the paternal allele of the IGF-II-R gene is repressed, and the maternal allele expressed. It is not known whether the absence of activity of this gene alone in Tme embryos is responsible for the lethality observed.

Three different approaches have been taken to identify 2 imprinted genes mapping to the distal imprinted portion of chromosome 7 (Fig. 11.1). Embryos carrying maternal disomy of distal chromosome 7 die at around day 16 of gestation and are approximately 50% smaller than nondisomic littermates. Embryos paternally disomic for distal chromosome 7 die earlier in gestation than the maternal counterparts and have never been identified (8). However, the maternal disomy embryos isolated prior to their death provide an excellent source of material in which to analyze the transcription products of candidate imprinted genes in the absence of any paternally inherited alleles in that region. The reciprocal translocation used to generate uniparental disomic embryos is t(7,15) 9H (8), hereafter referred to as T9H.

One gene that maps to this region of chromosome 7 is the IGF-II gene that encodes a growth factor implicated in embryonic growth. Northern analysis has shown that the growth-retarded maternal disomy embryos have 2 repressed alleles of IGF-II (Fig. 11.1). Thus, IGF-II is expressed from the paternally inherited chromosome, with the gene being repressed by imprinting on the maternally inherited chromosome (9). This is consistent with the growth retardation observed after inactivation of the paternally inherited IGF-II allele by gene targeting (10, 11). Mutation of the maternal IGF-II allele results in normal-sized mice. This third mode of identification of an imprinted gene, by targeted mutagenesis, is not practical for a systematic search for imprinted genes. However, the absence of parental origin effects in mutant phenotypes seen with other genes mutated in this way has ruled out imprinting in the regulation of several candidate developmentally regulated genes, notably *int-2*, also located on distal chromosome 7 in the mouse (12 and Cappecchi, personal communication).

Even though the IGF-II "knockout" embryos and maternal disomy 7 embryos share growth retardation phenotypes, an important distinction is the absence of any lethality in the former. This implies the presence of at least one other imprinted gene in the distal chromosome 7 imprinted region that may be responsible for the nonviability of disomic conceptuses. This is indeed the case. Tilghman and coworkers have shown that H19, a gene tightly linked to IGF-II, is imprinted (13). The approach taken to identify H19 as an imprinted gene utilized strain-specific RNA polymorphisms to distinguish the activity of maternally and paternally inherited alleles after appropriate reciprocal crosses. Using an RNAase protection assay, a single species-specific fragment could be identified for each of several reciprocal interspecific crosses. This indicated that only one of the two H19 alleles was active. In these cases, the gene was always only expressed from the maternally inherited chromosome; reciprocally, to the neighboring IGF-II gene. Maternal disomy 7 embryos, therefore, are expressing a double dose of H19 (Fig. 11.1). The function of the H19 gene product is unknown, but the absence of an open reading frame and lack of polysomal association suggests the absence of a protein product (14). Like IGF-II, the gene is expressed at very high abundance during embryogenesis in both extraembryonic and embryonic tissues, especially, but not exclusively, in tissues of mesodermal origin. Indeed, during development the sites of expression of both genes are remarkably similar.

The only site of adult H19 expression is in skeletal muscle where the gene maintains its imprinted state (13). In contrast, the sites of expression of IGF-II in the adult are the choroid plexus and leptomeninges of the brain; interestingly, the gene is not imprinted in these tissues (11). It is not known if the double dose of H19 is responsible for the demise of maternal disomy 7 embryos or if its absence (along with or in addition to excess IGF-II expression) may be responsible for earlier lethality of paternal disomy 7 embryos. The presence of other imprinted genes in this region is also not excluded.

Developmental Roles of Imprinted Genes

The cumulative effects of all imprinted genes have been studied (summarized in Table 11.1). In pronuclear transfer experiments generating gynogenetic (bimaternal) and androgenetic (bipaternal) zygotes, development was shown to fail. A proportion of *gynogenetic* (GG) embryos—and *parthenogenetic* (PG) embryos derived from egg activation—can survive to day 10.5 of gestation and reach up to the 25-somites stage, but they are growth retarded. In addition, the extraembryonic components are poorly developed with failure of the trophoblast to proliferate normally. (There are no phenotypic differences between components derived gynogenetically or parthenogenetically.) In contrast, fewer *androgenetic* (AG) embryos implant and give rise to a severely retarded embryonic component developing at best to the 4- to 6-somite stage with well-developed trophoblast tissue. These studies prove the requirement for both maternal and paternal chromosomes in normal development and implicate reciprocal roles for the parental genomes in the development of embryonic and extraembryonic tissues (1, 2).

Further analyses to determine the developmental potential of embryonic cells with different phenotypes can be achieved by making chimeras with PG/GG or AG embryos and *normal* (N) embryos. Such chimeras again show reciprocity in many of their phenotypic properties (Table 11.1). Chimeras made from PG and normal fertilized embryos (PG→N) will survive to term, but are smaller than control N→N chimeras by up to 50%. Furthermore, analysis of the tissue distribution of PG cells in these animals shows a marked selection against these cells in some tissues of mesodermal origin, notably skeletal muscle, where commencing at day 13–15, there is a progressive loss of PG cells such that at birth they are usually absent. This is in contrast to the brain and germline that retain a relatively high proportion of PG cells at birth (15, 16). There were no differences in the properties of PG/GG cells in chimeras when they were derived from different strains (17).

AG→N chimeras, however, show survival that is inversely proportional to the contribution of AG cells incorporated into the embryo and will therefore survive to term only if there is contribution of less than 20%. A contribution of around 50% or greater results in lethality prior to day 15. These chimeric embryos exhibit a marked growth enhancement, elongation of the anteroposterior axis, and often an enlarged heart. In addition, the tissue distribution of AG cells seems reciprocal to that seen in PG→N chimeras and shows preferential retention of such cells in some tissues of mesodermal origin, such as skeletal muscle and heart and in progenitors of the skeletal system. Indeed, one of the AG→N chimeras surviving to term exhibited severe skeletal anomalies associated with axial elongation and an abnormal proliferation of chondrocytes (18). AG cells rarely contribute to the brain and other neuroectodermal components, in

11. Parental Imprinting in Mammalian Development   149

TABLE 11.1. Summary of the characteristics of parthenogenetic (PG), androgenetic (AG), and maternal and paternal disomy for distal chromosome 7 (MatDi7 and PatDi7, respectively) embryos.

	PG	AG	MatDi7	PatDi7
Phenotype	Growth-retarded embryo Poor trophoblast Midgestation lethal	Very poor embryo Healthy trophoblast Lethal before 4–6 somite stage	Growth-retarded embryo Retarded placenta Lethal at day 16 of gestation	Unknown phenotype Early embryonic lethal
Phenotype in chimeras	Viable fertile but growth-retarded mice	Growth-enhanced embryo with elongated A-P axis Neonates rare Mesoderm anomalies	Complete apparent rescue Viable full-sized mice	Growth enhanced Adults unknown
Lineage selection in chimeras	PG cells in brain and germline, but low in mesodermal tissues—absent in muscle by birth	High in mesoderm tissues, notably muscle and cartilage Low in brain	No tissue restrictions	No tissue restrictions
Predicted expression of imprinted genes	IGF-II −/− H19 +/+ IGF-II-R +/+	IGF-II +/+ H19 −/− IGF-II-R −/−	IGF-II −/− H19 +/+ IGF-II-R +/−	IGF-II +/+ H19 −/− IGF-II-R +/−

Note: The predicted expression of the 3 imprinted genes in these embryos is shown (+ = expressed allele; − = repressed allele), and phenotypes observed in chimeras derived from all 4 cell types in the context of normal cells are also given. Chimeras made with PG and AG cells show loss or retention of these cells in particular lineages (see text); however, this cell selection is not seen in MatDi7 or PatDi7 chimeras.

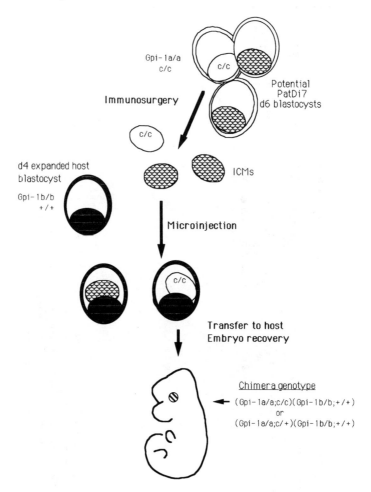

FIGURE 11.2. Generation of chimeras. Experimental blastocysts are isolated from translocation heterozygote intercrosses (T9H +/++ × T9Hc/+c). Many unbalanced gametes are produced in these crosses by nondisjunction, and balanced zygotes carrying the paternal disomy of distal chromosome 7 (PatDi7) occur at a low frequency. A higher frequency of balanced, nondisomic embryos is also expected (8). All experimental blastocysts are Gpi-1a/a, and albino males (c/c, mapping to distal chromosome 7) are used in the intercrosses to identify progeny PatDi7 cells in the chimeras, although most chimeras will contain nondisomic c/+ cells. The trophectoderm is removed by immunosurgery to release inner cell masses (ICMs) and allow better incorporation of experimental cells into the embryonic components. ICMs are then microinjected into normal host blastocysts (N), wild-type for albino and type Gpi-1b/b. The manipulated blastocysts are transferred to pseudopregnant recipients and the embryos recovered at day 15–17 of gestation. AG→N and PG→N chimeras were also made in this way, except a second marker for experimental cells is not needed as only AG or PG ICMs are microinjected.

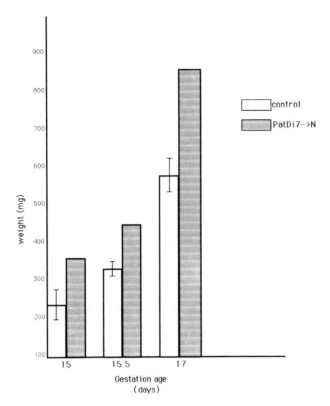

FIGURE 11.3. Histogram illustrating the growth enhancement of 3 PatDi7→N chimeras compared with pooled data from manipulated and nonmanipulated controls, as described in reference 9.

contrast to PG cells (18). The tissue distributions of AG and PG cells in chimeras suggest roles for imprinted genes in the proliferation and development of particular lineages, and it is likely that the 3 aforementioned imprinted genes play a role in these phenotypic effects as they are all expressed in the lineages where these restrictions on experimental cells occur.

In order to examine the effects of a subset of imprinted genes on development, those on distal chromosome 7, chimeras were constructed between normal and putative *paternal disomy 7* (PatDi7) embryos. Nonchimeric PatDi7 embryos are lethal prior to day 11 of gestation and cannot be identified. Chimeras were made by blastocyst injection of *inner cell masses* (ICMs) from putative PatDi7 embryos (expected at a very low frequency [8]) into normal host blastocysts (Fig. 11.2). All PatDi7→N chimeras exhibited growth enhancement (Fig. 11.3) similar to that seen in AG→N chimeras. By contrast, it is particularly interesting to note that the nonchimeric maternal disomy 7 embryos are growth retarded prior to their demise (9). However, the other anomalies associated with AG→N chimeras, notably in skeletal lineages, were not seen. This was also true of MatDi7→N chimeras in which the presence of normal cells led to the apparent total rescue of both the lethality and growth retardation phenotypes. Again, these chimeras did not show the same tissue restrictions as PG→N chimeras, most notably in skeletal muscle (9) (Table 11.1). Thus, these lineage effects in chimeras containing whole genome duplications are probably due to the cumulative effects of several imprinted genes, including those on chromosome 7, or effects of imprinted gene/ genes elsewhere. It is also possible that there is some redundancy of gene activity in the developing embryo such that absence of activity from one gene may be compensated for by others in the case of a subgenomic disomy.

The developmental anomalies described prove a role for the imprinted gene IGF-II in the regulation of embryonic growth and suggest interactive roles of several imprinted genes in the appropriate proliferation of sclerotome derivatives and the formation of normal mesodermal and neuroectodermal structures. The mitogenic role of the embryonic growth factor IGF-II seems to be mediated through the insulin-like growth factor I receptor (IGF-I-R). The IGF-I-R is also located on distal chromosome 7 and is not imprinted (Ferguson-Smith, unpublished; Glaser, Ferguson-Smith, Barton, Surani, Ohlsson, manuscript in preparation). However, IGF-II can also act as a ligand for the IGF-II-R, whose best-described role is as a receptor for mannose-6-phosphate. This latter interaction results in lysosomal targeting. The binding of IGF-II to IGF-II-R involves a separate binding site, and it is not known whether the complex gets targeted to lysosomes (19, 20). Nonetheless, even though the biological significance of interaction between IGF-II and IGF-II-R is unclear at present, it is intriguing that they are imprinted reciprocally, the paternal allele being expressed in the former and the maternal allele in the latter.

Haig and Moore have postulated a competition theory to explain the evolutionary constraints resulting in the reciprocal imprinting of these two genes (21, 22). This theory is based on the hypothesis that like its role as a mannose-6-phosphate receptor, the IGF-II-R targets IGF-II to lysosomes for degradation, and a balance of growth factor/receptor is maintained by the paternal chromosome synthesizing the growth factor to make large embryos (in his interest) at the expense of the female's reproductive strength. To counteract this, the maternal chromosome synthesizes a receptor to prevent excess growth, therefore causing less drain on her resources during this pregnancy (and also future pregnancies, most likely with a different male). Interestingly, Tme embryos, lacking IGF-II-R, are larger than controls, consistent with excess IGF-II even though both a maternal and paternal allele are present. The strength of this hypothesis will be tested when the interactive role(s) of IGF-II and IGF-II-R during development is better understood.

Imprinting Disorders in Humans

In general, the phenotypes observed in the presence of uniparental whole or partial genome duplications exhibit patterns of growth and proliferative inhibition in the cases of maternal disomy and patterns of growth and proliferative excess in the presence of 2 paternal complements. This has implications in the study of certain tumor disorders in humans that exhibit parental origin effects in their pattern of inheritance, most often seen in tumors with an imbalance in parental chromosomes (23, 24). The most extreme example is the complete hydatidiform mole, where proliferation and hyperplasticity of the extraembryonic cytotrophoblast, often in the absence of any embryonic component, is due to 2 paternal sets of chromosomes and no maternal chromosomes (25, 26). This is comparable to the androgenetic mouse conceptuses described previously.

Though not exclusively a tumor syndrome, *Beckwith-Wiedemann syndrome* (BWS) is a fetal overgrowth syndrome associated with paternal duplication/maternal deficiency of human chromosome 11p15.5 (27). Indeed, in cases of BWS in trisomy 11 individuals, the additional chromosome 11 material is always paternal in origin (28). Loss of heterozygosity, also seen in BWS patients, is associated with loss of maternal chromosomal material (paternal disomy) and includes the maternal IGF-II allele (28). It is not surprising that human chromosome 11p shares syntenic homology with distal mouse chromosome 7. BWS individuals exhibit neonatal gigantism, macroglossia, and neonatal hypoglycemia resulting in mental retardation if untreated. It has been suggested that overactivity of the IGF-II gene may be responsible for some of these phenotypes (29). Rhabdomyosarcoma and Wilms' tumor (which also occur in 12.5% of BWS patients) are associated with maternal deficiencies of the same region, and overexpression of IGF-II has been reported in Wilms' tumor (30, 31).

Two completely different syndromes, *Prader-Willi syndrome* (PWS) and *Angelman syndrome* (AS), involve sequences on the same region of human chromosome 15q11-13, which also shares syntenic homology with the more proximal portion of distal mouse chromosome 7. The outcome, PWS or AS, depends on the parental origin of a deletion or uniparental disomy for 15q13-15; therefore, both a maternal copy and a paternal copy of this region are necessary for normal human development (32–34). PWS patients have maternal excess/paternal deficiency of 15q and exhibit developmental delay, growth retardation, hypogonadism, cognitive disabilities, and subsequent hyperphagia with obesity. In contrast, AS is associated with severe mental retardation, absence of speech, hyperactivity, ataxia, seizures, and abnormal EEG. The AS patients have paternal excess/maternal deficiency of chromosome 15q. Mouse loci homologous to these 15q deleted regions have been cloned and assayed in maternal disomy 7 mouse embryos for perturbations in gene expression, but as yet, none have been found. However, we believe that this MatDi7 assay system has great potential for the study of human-mouse imprinting homologies.

Conclusion

The molecular basis for the phenotypes observed in androgenetic and parthenogenetic embryos is slowly beginning to be unraveled with the identification of the first three endogenous imprinted genes in the mouse. It will be of interest to determine whether these genes are imprinted in other organisms, which should provide information about the evolution of this process. One of the most important questions that can now be addressed using these genes is, What are the epigenetic mechanisms responsible for parental origin-specific repression of imprinted genes? One of the best-described modes of epigenetic modification is CpG methylation, which is usually correlated with gene inactivity (35, 36). Comparative methylation differences in maternal and paternal chromosomes are currently under investigation by several groups. Again, the MatDi7 assay system is very useful in these comparative molecular studies. In addition, the ability to answer questions of how and when the epigenetic modification(s) become initiated, established and maintained, and then erased during gametogenesis will provide deeper insight into questions of establishment and loss of totipotency during mammalian embryogenesis. The stability of the imprints during development, tumor progression, and in the culture of embryonic stem cells, for example, can also be addressed. These studies will assist in our understanding of non-Mendelian patterns of inheritance as they relate to human disease. In addition, they will provide fundamental information about the development dynamics of epigenetic inheritance and the roles of epigenetic modification, not only in the transcriptional regulation of the parentally imprinted genes, but also as a general mechanism affecting the activity of other genes.

References

1. Surani MAH, Barton SC, Norris ML. Development of reconstituted mouse eggs suggests imprinting of the genome during gametogenesis. Nature (London) 1984;308:548–50.
2. Solter D. Differential imprinting and expression of maternal and paternal genomes. Annu Rev Genet 1988;22:127–46.
3. Cattanach BM. Parental origin effects in mice. J Embryol Exp Morphol 1986;97:137–50.
4. Searle AG, Beechey CV. Noncomplementation phenomena and their bearing on nondisjunctional events. In: Dellarco VL, et al., eds. Aneuploidy. New York: Plenum Press, 1985:363–76.

5. Johnston DR. Further observations on the hairpin-tail (Thp) mutation in the mouse. Genet Res 1974;24:207–13.
6. Winking H, Silver L. Characterisation of a recombinant mouse t-haplotype that expresses a dominant lethal maternal effect. Genetics 1984;108:1013–20.
7. Barlow DP, Stoger R, Herrman BG, Saito K, Schweifer N. The mouse insulin-like growth factor type-2 receptor is imprinted and closely linked to the Tme locus. Nature (London) 1991;349:84–7.
8. Searle AG, Beechey CV. Genome imprinting phenomena on mouse chromosome 7. Genet Res 1990;56:237–44.
9. Ferguson-Smith AC, Cattanach BM, Barton SC, Beechey CV, Surani MA. Embryological and molecular investigations of parental imprinting on mouse chromosome 7. Nature (London) 1991;351:667–70.
10. DeChiara TM, Efstratiadis A, Robertson EJ. A growth deficiency phenotype in heterozygous mice carrying an insulin-like growth factor II gene disrupted by targetting. Nature (London) 1990;345:78–80.
11. DeChiara TM, Robertson EJ, Efstratiadis A. Parental imprinting of the mouse insulin-like growth factor II gene. Cell 1991;64:849–59.
12. Mansour SL, Thomas KP, Deng CX, Cappecchi MR. Introduction of a lacZ reporter gene into the mouse int-2 locus by homologous recombination. Proc Natl Acad Sci USA 1990;87:7688–92.
13. Bartolemei MS, Zemel S, Tilghman SM. Parental imprinting of the mouse H19 gene. Nature (London) 1991;351:153–5.
14. Brannan CI, Dees EC, Ingram RS, Tilghman S. The product of the H19 gene may function as an RNA. Mol Cell Biol 1990;10:28–36.
15. Fundele RH, Norris ML, Barton SC, Reik W, Surani MA. Systematic elimination of parthenogenetic cells in mouse chimaeras. Development 1989;106:28–35.
16. Fundele RH, Norris ML, Barton SC, et al. Temporal and spatial selection against parthenogenetic cells during development of fetal chimaeras. Development 1990;108:203–11.
17. Fundele R, Howlett SK, Kothary R, Norris ML, Mills WE, Surani MA. Developmental potential of parthenogenetic cells: role of genotype-specific modifiers. Development 1991;113:941–6.
18. Barton SC, Ferguson-Smith AC, Fundele RH, Surani MA. Influence of paternally imprinted genes on development. Development 1991;113:679–88.
19. Heyner S, Smith R, Schultz GA. Temporally regulated expression of insulin and insulin-like growth factors and their receptors in early mammalian development. Bioessays 1989;11:171–6.
20. Morgan DO, Edman JC, Standring DN, et al. Insulin-like growth factor II receptor as a multifunctional binding protein. Nature (London) 1987;329:301–7.
21. Haig D, Graham C. Genomic imprinting and the strange case of the insulin-like growth factor II receptor. Cell 1991;64:1045–6.
22. Moore T, Haig D. Genomic imprinting in mammalian development: a parental tug-of-war. Trends Genet 1991;7:45–9.
23. Ferguson-Smith AC, Reik W, Surani MA. Genomic imprinting and cancer. Cancer Surveys 1990;9:487–503.
24. Reik W. Genomic imprinting and genetic disorders in man. Trends Genet 1989;5:331–6.

25. Kajii T, Ohama K. Androgenetic origin of hydatidiform moles. Nature (London) 1977;268:633–4.
26. Bagshawe KD, Lawler SD. Unmasking moles. Br J Obstet Gynaecol 1982;89:255–9.
27. Hall J. Genomic imprinting: review and relevance to human diseases. Am J Hum Genet 1990;46:857–73.
28. Henry I, Bonaiti-Pellie C, Chehensse V, et al. Uniparental paternal disomy in a genetic cancer-predisposing syndrome. Nature (London) 1991;351:665–7.
29. Little M, Van Heyningen V, Hastie N. Dads and disomy and disease. Nature (London) 1991;351:609–10.
30. Reeve AE, Eccles MR, Wilkins RJ, Bell GI, Millow LJ. Expression of insulin-like growth factor II transcripts in Wilms' tumour. Nature (London) 1985;317:258–60.
31. Scott J, Cowell J, Robertson ME, Priestley LM, et al. Insulin-like growth factor II gene expression in Wilms' tumour and embryonic tissues. Nature (London) 1985;317:260–2.
32. Knoll JHM, Nicholls RD, Magenis RE, Graham JM, Lalande M, Latt SA. Angelman and Prader-Willi syndrome share a common chromosome 15 deletion but differ in parental origin of the deletion. Am J Med Genet 1989;32:285–90.
33. Malcolm S, Nichols M, Clayton-Smith J, et al. Angelman syndrome can result from uniparental disomy. Am J Hum Genet 1990;47:A227.
34. Nicholls RD, Knoll JHM, Butler MG, Karam S, Lalande M. Genetic imprinting suggested by maternal disomy heterodisomy in non-deletion Prader-Willi syndrome. Nature (London) 1989;342:281–5.
35. Reik W, Collick A, Norris ML, Barton SC, Surani MAH. Genomic imprinting determines methylation of parental alleles in transgenic mice. Nature (London) 1987;238:248–51.
36. Yen PH, Patel P, Chinault AC, Mohandas T, Shapiro LJ. Differential methylation of hypoxanthine phosphoribosyltransferase genes on active and inactive X-chromosome. Proc Natl Acad Sci USA 1984;81:1759–63.

12

Developmental Potential of Mouse Embryonic Stem Cells

JANET ROSSANT, ELIZABETH MERENTES-DIAZ, ELEN GOCZA,
ESZTER IVANYI, AND ANDRAS NAGY

Embryonic stem (ES) cells are pluripotent cell lines that have been derived from the *inner cell mass* (ICM) of the mouse blastocyst in tissue culture. When maintained on feeder layers or in the presence of leukemia inhibitory factor (1, 2), ES cells retain an undifferentiated phenotype, but when culture conditions are altered or the cells are transplanted into ectopic sites in adult animals, the cells can differentiate into a wide variety of different cell types (3, 4). Furthermore, ES cells can contribute to normal development of both somatic and germline tissues when injected into blastocysts (5). The ability to take a cell line from culture and introduce its genotype into the mouse gene pool has opened up whole new areas of genetic investigation in mammals (reviewed in 6, 7). Any kind of genetic manipulation that can be performed on somatic cell lines can also be performed on ES cells, with the advantage that the effects of such manipulation can be assessed in the intact animal.

In the 10 years since ES cells were first described (8, 9), the number of different uses for them has increased rapidly year by year. Perhaps the most exciting and widely applicable has been the development of gene targeting, in which homologous recombination is used to mutate a specific gene in ES cells in culture and ES chimeras are used to transmit that mutation into the germline (reviewed in 10). However, the ability of ES cells to contribute to a wide range of somatic cells in chimeras also allows study of the potential of genetically manipulated cells prior to germline transmission. For example, the effects of dominant-acting genetic alterations, which may be incompatible with survival to term, can be studied in ES chimeras (11). Also, ES cells can be used as recipients of *lacZ* reporter constructs designed to identify new developmentally regulated genes by their pattern of expression in ES chimeras (12). In all these kinds of chimera study, the validity of the approach depends on ES cells behaving like normal ICM cells and contributing widely and uniformly to all developing embryonic cell lineages.

The literature reveals very little detailed study of the behavior of ES cells in chimeras; in many studies, coat color contribution alone is reported. Beddington and Robertson (13) report contribution to both embryonic and extraembryonic lineages in midgestation embryos, and two groups have shown that ES cells expressing *E. coli* β-galactosidase can contribute widely to the developing embryo proper (14, 15). We have shown that ES cells can form an entire fetus if aggregated with developmentally compromised tetraploid embryos, demonstrating that these cells have full developmental potential (16). However, none of these completely ES-derived fetuses survived beyond term. Here, we describe further studies with both tetraploid and diploid ES aggregation chimeras that demonstrate that lethality is a function of extent of ES contribution, but not a result of lack of potential to form specific tissues.

Perinatal Lethality of ES-Tetraploid (ES-T) Chimeras

In an attempt to generate mice in which all tissues, including the germline, were derived from ES cells, we previously reported that aggregation of small numbers of ES cells with pairs of tetraploid embryos could generate newborn mice that were completely ES derived (16). Tetraploid embryos alone rarely proceed beyond early postimplantation stages (17). In ES-T chimeras, tetraploid cells rarely contributed to embryonic tissues, but contributed extensively to the extraembryonic lineages. The production of entirely ES-derived fetuses at term by this means provided the first demonstration that ES cells had full embryonic potential. However, none of the ES fetuses produced survived more than 2 days beyond term. The pathological cause of death was unclear; no obvious morphological defects were found. The studies previously published were performed with two ES cell lines, D3 (3) and α (16), making it possible that the lethality might be a consequence of the particular cell lines used. Since then we have repeated the experiment with four different cell lines and have found essentially similar results (Table 12.1). Cell line AB1 (18) was derived from the same strain, 129J, as the originally tested D3; β7 and β1 (unpublished) were derived from C57BL/6; and A3 was derived from a 129 × T-MβG1 (19) cross, showing that the lethality is neither cell line specific nor genotype specific.

The number of ES-derived fetuses that survived to term was quite low, but not significantly different for any of the cell lines. Importantly, A11, which was a retrovirally infected and G418-selected subclone of D3 (Kang, Chambers, Hozumi, Nagy, unpublished observations), was capable of producing ES-derived progeny, showing that retroviral infection and G418 selection are compatible with retention of full ES potential.

TABLE 12.1. Development of ES-T chimeras.

ES cell line	No. transferred	No. resorptions	No. dead fetuses	No. normal fetuses
D3	124	46	21	14
A11	56	29	3	2
AB-1	81	44	5	5
A3	34	0	8	4
	12	0	1	9*
β7	54	Gave birth		1
β7-31	43	Gave birth		1
β1	28	4	0	1

* Embryos were dissected on day 12.5.
Note: All aggregations were performed with F2 (for 129/Sv-derived ES cells) or CD-1 (preselected GPI-AA phenotypes for C57BL/6-derived ES cells) tetraploid embryos and transferred to CD-1 recipients. Except where noted, delivery was by cesarean section at day 19 of gestation. All normal live fetuses delivered were completely ES derived by GPI analysis, and all failed to oxygenate their lungs and died shortly after delivery.

In practical terms, this means that the ES-T chimera approach can be used to provide access to totally ES-derived fetal stem cells of various sorts. For example, we have recently shown that ES-derived fetal liver cells from ES-T chimeras can be used to repopulate the adult hemopoietic system (20), providing a means of genetically manipulating hemopoietic stem cells without germline transmission.

Effects of Increasing ES Cell Number on ES-Diploid (ES-D) Aggregation Chimeras

The most obvious conclusion from the ES-T results is that extensive ES contribution to newborn mice is incompatible with postnatal survival. If this were the case, one would predict that increasing the extent of ES contribution to normal chimeras made with diploid embryos would result in the same problems. Most studies on ES-D chimeras have utilized blastocyst injection, where only a limited number of cells can be injected and where ES contributions greater than 70% or so tend to be rare. ES cells can be aggregated with pairs of diploid 8-cell embryos, as reported previously for embryonal carcinoma cells (21), and such aggregations allow one to vary the ES cell number added in a systematic way.

We performed a series of aggregation chimera experiments in which 5–10, 15–25, or 50 ES cells were aggregated with (C57 × CBA)F2 8-cell embryos. Successful aggregates were transferred to pseudopregnant recipients and allowed to develop to 19 days of gestation, when pups were delivered by cesarean section. As controls, 129/Sv-CP ICMs were aggregated with pairs of F2 embryos and allowed to develop in the same

TABLE 12.2. Development to term of ES-D and ICM-D chimeras.

No. ES cells	No. transferred	No. implanted (% transferred)	No. resorbed (%)	No. retarded fetuses (%)	No. normal fetuses (%)	No. chimeras (%)
5–10	70	53(76)	14(20)	4(6)	35(50)	19(27)
15–25	124	98(79)	52(42)	7(6)	39(31)	25(20)
ICM	48	30(63)	12(25)		18(38)	17(35)

Note: Table shows pooled data of experiments with D3, A11, and AB-1 ES cell lines aggregated with diploid 8-cell-stage F2 embryos. ICM-D aggregates were made from 129/Sv ICM and F2 8-cell-stage embryos. All newborns were delivered by cesarean section at day 19 of pregnancy.

manner. Aggregation with 50 ES cells prevented development to term, with all embryos dying and resorbing in the early postimplantation period. Aggregation of lower numbers of ES cells did allow some development to term (Table 12.2). Fifteen to 25 ES cells resulted in a lower proportion of offspring born from the embryos transferred, but the percentage of chimeras among the newborns was slightly higher than with 5–10 cells. The end result was that the rate of chimera production per starting number of aggregates was similar for the two groups. Chimera production was, however, higher in the ICM-D aggregates (Table 12.2).

When the extent of ES or ICM contribution to resulting newborns was assessed by GPI isozyme analysis, a clear relationship between number of ES cells aggregated and extent of contribution was seen; many more chimeras were apparently 100% ES derived when 15–25 rather than 5–10 cells were aggregated (Fig. 12.1). ICM chimeras showed a balanced distribution of ICM contribution, and there were very few nonchimeric newborns, unlike the situation with either ES group. Thus, ES cells can contribute as extensively as ICM cells to chimeras, but they do this with less consistency. However, unlike ICM cells, extensive ES contribution to newborn mice militates against neonatal survival. When data from all ES-D experiments were pooled, it became clear that the chance of neonatal mortality was correlated with the extent of ES contribution (Fig. 12.2). To date, no embryo with ES contribution over 70% by GPI has survived.

Why Do ES-Derived Mice Die?

The results with ES-D chimeras confirm and extend the results with ES-T chimeras by showing that although ES cells can produce newborn mice that are extensively ES-derived, such extensive ES contribution is usually incompatible with survival. Thus, ES cells resemble ICM cells in their pluripotency, but differ in their effects on neonatal survival. They also differ from ICM cells in showing considerable variation in ability to

12. Developmental Potential of Mouse Embryonic Stem Cells   161

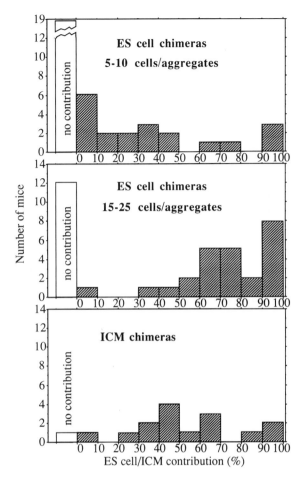

FIGURE 12.1. Distribution of contributions from ES/ICM component to chimeras at birth. Histograms of percentage ES or ICM contributions observed in chimeras made with 5–10 or 15–25 ES cells or 15 ICM cells.

produce chimeras (Fig. 12.1) and in final birth weight of chimeric offspring (not shown). We have previously discussed several reasons why extensive ES chimeras should suffer high perinatal mortality. First, the ES cells could carry an infectious agent whose effect is manifest at birth. Second, the population of cells used could contain a high proportion of aneuploid cells, so that any extensive ES chimera is likely to contain a proportion of such cells high enough to cause mortality. Third, continued growth in tissue culture could affect the maintenance of genomic imprinting and thus cause disturbances in normal development.

The first two reasons seem unlikely to provide a full explanation for the effects observed. Perinatal lethality has been observed with five different

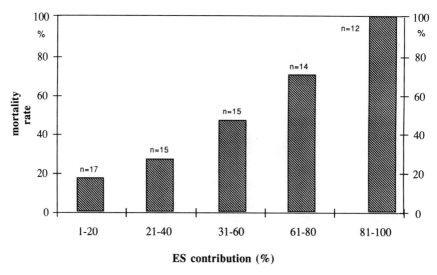

FIGURE 12.2. Perinatal mortality of ES chimeras as a function of ES contribution.

cell lines maintained in two different labs, making a common infectious agent unlikely. A high degree of aneuploidy in all cell lines also seems unlikely, and the production of the same phenotype with clonal sublines argues against, but cannot eliminate, this possibility. The suggestion that genomic imprinting could be affected is an interesting one, but hard to assess. A number of genetic and nuclear transplantation experiments has shown that normal mammalian development requires both male- and female-derived genomes (reviewed in 22). Direct evidence that this effect depends on differential gene expression from the two genomes is provided by the recent description of three genes that show such parental-specific gene expression (23–25). At the moment, it is not clear to what extent chromosomal memory of parent of origin can be retained in cell lines in culture, although the abnormal development of androgenetic ES cells in chimeras suggests that at least part of the imprint is retained (26). One of the features of parental imprinting is that the effects can be tissue specific, depending on the expression of the particular imprinted gene (27–30). Such tissue-specific effects can be revealed by studying the distribution of cells derived from uniparental embryos in chimeras.

We reasoned that if imprinting had been disturbed at certain genes during the passage of ES cells in culture, one might observe a consistent reduction in the ability of the cells to contribute to tissues where those genes were normally active or inactive. To this end, we performed a careful, tissue-by-tissue study of the contribution of two ES cell lines, the 129J-derived AB1 and the C57BL/6-derived β1, to diploid aggregation chimeras. The results showed that both cell lines were capable of contributing to all tissues examined and that there was generally a good correlation among contributions to different tissues. When a series of 26 chimeras was analyzed, the progeny of ES cells were consistently underrepresented in certain tissues. However, the tissues with low ES contribution were different for the two cell lines examined, providing no evidence for a common defect. Strikingly, when the results were compared with those from ICM chimeras of matched genotypes, the exact same tissues were shown to have low contributions from ICM cells of the relevant genotype. Thus, the skewed distributions in certain tissues were due to genotype-specific differences and were not ES specific. At the level of analysis performed, we could not find any evidence to suggest that ES cells differ from ICM cells in their contributions to specific embryonic lineages in chimeras, and the reason for the postnatal survival problems of ES-derived fetuses remains obscure.

Conclusions

These results show that ES cells are fully capable of contributing to all tissues in a chimera and behave identically to ICM cells in this regard. Unlike ICM cells, extensive ES contribution to chimeras is, however, not normally compatible with postnatal survival. The reasons for this failure are still unclear, but there is no evidence for any tissue specificity to the defect. The fact that ES cells and ICM cells of the same genotype behave identically in their tissue distribution within any chimera validates the use of ES cells as models of embryonic development. ES-T and ES-D aggregation chimeras can thus provide a rapid means of screening the effect of a specific genetic manipulation on fetal development. Further, the derivation of completely ES-derived fetuses from ES-T chimeras provides access to genetically altered fetal stem cell populations without germline transmission.

Acknowledgments. This work was supported by grants from the Medical Research Council of Canada, the National Institutes of Health, Bristol Myers/Squibb, and the Hungarian Science Foundation (OTKA). J.R. is a Terry Fox Cancer Research Scientist of the NCIC and an International Scholar of the Howard Hughes Medical Institute.

References

1. Williams RL, Hilton DJ, Pease S, et al. Myeloid leukaemia inhibitory factor maintains the developmental potential of embryonic stem cells. Nature (London) 1988;336:684-7.

2. Smith AG, Heath JK, Donaldson DD, et al. Inhibition of pluripotential embryonic stem cell differentiation by purified peptides. Nature (London) 1988;336:688–90.
3. Doetschman TG, Eistetter M, Katz M, Schmidt W, Kemler R, In vitro development of blastocyst-derived embryonic stem cell lines: formation of visceral yolk sac, blood islands and myocardium. J Embryol Exp Morphol 1985;87:27–45.
4. Evans M, Kaufman MH. Pluripotential cells grown directly from normal mouse embryos. Cancer Surveys 1983;2:186–207.
5. Bradley A, Evans M, Kaufman MH. Formation of germline chimaeras from embryo-derived teratocarcinoma cell lines. Nature (London) 1984;309:255–6.
6. Robertson EJ. Using embryonic stem cells to introduce mutations into the mouse germ line. Biol Reprod 1991;44:238–45.
7. Rossant J. Manipulating the mouse genome: implications for neurobiology. Neuron 1990;4:323–34.
8. Martin GR. Isolation of a pluripotent cell line from early mouse embryos cultured in medium conditioned by teratocarcinoma stem cells. Proc Natl Acad Sci USA 1981;78:7634–8.
9. Evans M, Kaufman MH. Establishment in culture of pluripotential cells from mouse embryos. Nature (London) 1981;292:154–5.
10. Capecchi MR. Altering the genome by homologous recombination. Science 1989;244:1288–92.
11. Boulter CA, Aguzzi A, Williams RL, Wagner EF, Evans MJ, Beddington R. Expression of v-src induces aberrant development and twinning in chimaeric mice. Development 1991;111:357–66.
12. Gossler A, Joyner AL, Rossant J, Skarnes WJ. Mouse embryonic stem cells and reporter constructs to detect developmentally regulated genes. Science 1989;244:463–5.
13. Beddington RSP, Robertson EJ. An assessment of the developmental potential of embryonic stem cells in the midgestation mouse embryo. Development 1989;105:733–7.
14. Lallemand Y, Brulet P. An in situ assessment of the routes and extents of colonisation of the mouse embryo by embryonic stem cells and their descendants. Development 1990;110:1241–8.
15. Suemori H, Kadodawa Y, Goto K, Araki I, Kondoh H, Nakatsuji N. A mouse embryonic stem cell line showing pluripotency of differentiation in early embryos and ubiquitous β-galactosidase expression. Cell Differ Dev 1990;29:181–6.
16. Nagy A, Gocza E, Diaz EM, et al. Embryonic stem cells alone are able to support fetal development in the mouse. Development 1990;110:815–22.
17. Kaufman MH, Webb S. Postimplantation development of tetraploid mouse embryos produced by electrofusion. Development 1990;110:1121–32.
18. McMahon AP, Bradley A. The Wnt-1 (int-1) proto-oncogene is required for development of a large region of the mouse brain. Cell 1990;62:1073–85.
19. Varmuza S, Prideaux V, Kothary R, Rossant J. Polytene chromosomes in mouse trophoblast giant cells. Development 1988;102:127–34.
20. Forrester LM, Bernstein A, Rossant J, Nagy A. Long-term reconstitution of the mouse hematopoietic system by embryonic stem cell-derived fetal lines. Proc Natl Acad Sci USA 1991;88:7514–7.

21. Stewart CL. Aggregation between teratocarcinoma cells and preimplantation mouse embryos. J Embryol Exp Morphol 1980;58:289–302.
22. Solter D. Differential imprinting and expression of maternal and paternal genomes. Annu Rev Genet 1991;22:127–46.
23. Dechiara TM, Robertson EJ, Efstratiadis A. Parental imprinting of the mouse insulin-like growth factor II gene. Cell 1991;64:849–60.
24. Barlow DP, Stoger R, Herrmann BG, Saito K, Schweifer N. The mouse insulin-like growth factor type-2 receptor is imprinted and closely linked to the Tme locus. Nature (London) 1991;349:84–7.
25. Bartolomei MS, Zemel S, Tilghman SM. Parental imprinting of the mouse H19 gene. Nature (London) 1991;351:153–5.
26. Mann JR, Gadi I, Harbison ML, Abbondazo SJ, Stewart CL. Androgenetic mouse embryonic stem cells are pluripotent and cause skeletal defects in chimeras; implications for genetic imprinting. Cell 1990;62:251–60.
27. Nagy A, Sass M, Markkula M. Systematic non-uniform distribution of parthenogenetic cells in adult mouse chimaeras. Development 1989;106:321–4.
28. Clarke H, Varmuza S, Prideaux V, Rossant J. The development potential of parthenogenetically-derived cells in chimeric mouse embryos: implications for action of imprinted genes. Development 1988;104:175–82.
29. Fundele RH, Norris ML, Barton SC, et al. Temporal and spatial selection against parthenogenetic cells during development of fetal chimeras. Development 1990;108:203–11.
30. Thomson JA, Solter D. The development fate of androgenetic, parthogenetic, and gynogenetic cells in chimeric gastrulating mouse embryos. Genes Dev 1988;2:1344–51.

Part IV

Differentiation of the Embryo

13
Blastocyst Development and Growth: Role of Inositol and Citrate

M.T. KANE AND M.M. FAHY

Information on the control of development and growth in the preimplantation mammalian embryo is of major interest both because of its intrinsic biological importance and its relevance to human and animal fertility. Yet in spite of this situation, information about even such a basic area as preimplantation embryo growth patterns is confined to a very limited number of species, and there appears to be little information for any species on the mechanisms controlling growth in preimplantation embryos.

Growth Patterns in Preimplantation Embryos

In all species studied so far, it is clear that during the cleavage process from 1-cell to early morula, there is little or no increase in protein content and, thus, no true growth of the embryo. In species in which there is a large increase in protein content during the preimplantation period, it occurs at the blastocyst stage. Data for a number of species are summarized in Table 13.1. While the mouse embryo has long been the standard model for the study of preimplantation development, it is not a suitable model for the study of preimplantation embryo growth because there is very little increase in protein content over the entire preimplantation period. The 1-cell mouse embryo contains about 20–25 ng of protein, and the 4-day blastocyst with about 90 cells has only about 20–40 ng just before implantation (1, 2). This is the basic reason why it is possible to culture 1-cell mouse embryos to blastocyst stages in a simple salt solution supplemented with a macromolecule and a suitable energy source, but without amino acids, vitamins, or trace elements. The rat embryo has a similar growth pattern.

In striking contrast, the 1-cell rabbit embryo has a protein content of about 160 ng, and the 6-day blastocyst with about 80,000 cells (3) has about 26,000-ng protein (Morgan, Kane, unpublished observations) of which about 25% is contained in the embryonic cells and almost all the rest in the embryonic coverings (zona, neozona, mucin coat, etc.). The protein content of the rabbit blastocyst is closely correlated with the blastocyst surface area. This is not unexpected if one considers that the late rabbit blastocyst is largely just a fluid-filled sphere. This great increase in protein content of the rabbit blastocyst before implantation is the reason why the growth of rabbit embryos to blastocysts requires amino acids, vitamins (4), and as yet undefined growth factors (5). It also makes the rabbit an extremely interesting model for studying the control of preimplantation growth. Pig embryos exhibit similar increases in protein content over the first 6 days of development (6). Our own recent work in this area with rabbit embryos has concentrated on two main aspects relevant to blastocyst growth: (i) the role of inositol and the phosphatidylinositol second-messenger system and (ii) the nature of the embryotrophic low-molecular weight contaminants bound to serum albumin.

Inositol and Preimplantation Embryos

Roles of Inositol in Cell Function

There is tremendous current research interest in the roles played by *myo*-inositol, inositol-containing phospholipids, and inositol phosphates in the control of cell function. First defined as a B vitamin by the work of Woolley in 1940 (7), *myo*-inositol was later shown by Eagle to be essential for growth of a number of human cell lines (8). There is extensive evidence that inositol itself functions as a controller of cell volume and cell osmolarity in certain cell types (9–11). A second very well documented function is the use of glycosylphosphoinositides by cells as anchors for attaching proteins to the cell membrane (12, 13). There is also evidence that an inositol phosphate glycan may act as the intracellular second messenger for insulin (14, 15).

However, most interest is focused on the role of the *phosphatidylinositol* (PtdIns) cycle as a second-messenger system for a number of hormones and growth factors in a wide range of animal cells (16–20). In the PtdIns cycle, *myo*-inositol is combined with phosphatidic acid to form PtdIns. PtdIns is phosphorylated successively to PtdIns(4)P and PtdIns(4,5)P_2. Stimulation of cell receptors by a hormone or other agonist activates the enzyme phosphoinositidase C (phospholipase C) via the intermediary action of G-proteins. Phosphoinositidase C breaks down PtdIns(4,5)P_2 to two diverging second messengers—inositol *1,4,5-trisphosphate* (Ins(1,4,5)P_3) and *diacylglycerol* (DAG). The Ins(1,4,5)P_3 raises cell Ca^{++} levels, and the DAG activates protein kinase C that phosphorylates a range of cell

13. Blastocyst Development and Growth: Role of Inositol and Citrate

TABLE 13.1. Cell numbers and protein content of mammalian embryos at different preimplantation stages.

Mouse			Rat			Rabbit			Pig		
Timing (h)	Cell no.	Protein (ng/embryo)	Timing (h)	Cell no.	Protein (ng/embryo)	Timing (h)	Cell no.	Protein (ng/embryo)	Timing (h)	Cell no.	Protein (ng/embryo)
12	1	24.5 ± 6.5	12	1	36.6 ± 4.6	18–23	1	160 ± 10	24	1	273 ± 16
36	2	26.2 ± 6.6	36	2	32.5 ± 4.3	22–24	2	169 ± 20	72	8	334 ± 19
60	8	22.0 ± 5.5	84	8	32.8 ± 5.2	45–46 (early morula)	16	259 ± 50	96 (morula)		491 ± 21
84 (morula)	25.9 ± 4.8	20.9 ± 5.0	104 (morula)	14.6 ± 4.2	29.0 ± 4.1	70–72 (late morula)	141	419 ± 86	120 (blast)		620 ± 24
84 (blast)	34.5 ± 4.0	24.8 ± 6.2	104 (blast)	29.6 ± 6.6	27.7 ± 4.5	118–125 (blast)	1112	470 ± 50	144 (blast)		1115 ± 82
96 (blast)	85.9 ± 16.0	40.8 ± 8.4	120 (blast)	46.0 ± 5.1	26.8 ± 5.6	122 (blast)	9000	2950 ± 260	168 (blast)		1499 ± 79
						144 (blast)	80,259	25,030 ± 3040	192 (blast)		2163 ± 225

Note: Values for protein content and cell number are mean values ± SEM. The data for rabbit embryo protein are for intact embryos and thus include the embryonic coverings laid down in the oviduct and uterus (mucin coat and neozona).

Source: Data are extracted from Schiffner and Spielmann (2) for the mouse and rat, from Morgan and Kane (J Reprod, in press) for the rabbit embryo protein content, from Daniel (3) for rabbit embryo cell numbers, and from Wright et al. (6) for the pig.

proteins, including cell enzymes. Ins(1,4,5)P$_3$ may be further phosphorylated to Ins(1,3,4,5)P$_4$ that also acts as a Ca^{++}-mobilizing agent. The cycle is completed by the breakdown of Ins(1,4,5)P$_3$ and Ins(1,3,4,5)P$_4$ by a series of phosphatase enzymes to inositol monophosphates and, finally, free inositol that can then be recycled.

Effects of Inositol on Blastocyst Growth and Development In Vitro

Over 20 years ago, Daniel showed a stimulatory effect of inositol on rabbit blastocyst expansion (21). Recently, we examined the vitamin requirements for development of rabbit morulae to blastocysts in vitro and found that, of the 11 water-soluble vitamins studied, the omission of inositol had the most serious effect on blastocyst formation and expansion (22). In the absence of inositol, while blastocyst formation and the beginning of blastocyst expansion did occur, these blastocysts rapidly shrank to their original unexpanded diameter (22, 23). Later work in our laboratory using [^{3}H]thymidine incorporation into acid-precipitable material to monitor blastocyst cell division showed that omission of inositol from the culture medium drastically reduced cell proliferation as well as blastocyst expansion (Fig. 13.1). This work indicates clearly that inositol is essential for rabbit blastocyst growth.

The importance of inositol in blastocyst development is not confined to the rabbit. Addition of inositol to the culture medium stimulates zona shedding or hatching of hamster blastocysts in vitro (24). Also, while preimplantation mouse embryos do not require exogenous inositol for development to viable blastocysts, mouse blastocysts do actively take up inositol and incorporate it into phosphoinositides (discussed in next section), suggesting that endogenous inositol may play a role in mouse blastocyst formation.

Information on inositol levels in the female reproductive tract is extremely limited. Gregoire et al. in 1962 reported a concentration of 2.6 mg/100 mL (144 µM) in rabbit oviductal fluid and levels in uterine fluid approximately 3 times that value (25). Lewin et al. (26) found a concentration of 54 µg/g (equivalent to about 299 µm) in rat proestrous uterine fluid. There are now sensitive modern methods available for the measurement of free inositol (27). It would be interesting to have comparative information on inositol concentrations in reproductive tract fluids of other species, particularly humans and primates.

Inositol Uptake by Preimplantation Embryos

There are three possible sources of inositol for synthesis of phosphoinositides, and the like: (i) Cells may recycle inositol from the breakdown

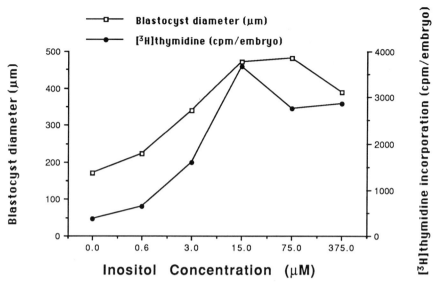

FIGURE 13.1. Effects of *myo*-inositol concentration on mean blastocyst expansion and incorporation of [^{3}H]thymidine by rabbit blastocysts cultured from the early morula stage over a period of 5 days. [^{3}H]thymidine (5 µCi/mL) was added for 4 h at the end of culture, and the uptake of label into acid-precipitable material was measured. Data points are mean values based on 64–67 embryos per inositol level. Data from Fahy and Kane, unpublished observations.

of inositol phosphates, (ii) cells may metabolize glucose to inositol, and (iii) cells may take up inositol from the surrounding fluid. Mouse embryos can convert glucose to inositol (28). Uptake of inositol by preimplantation embryos is stage dependent in both mouse and rabbit. In the mouse embryo (Fig. 13.2), there is about a 12-fold increase in uptake from 1-cell to 2-cell, a slight decline at the 8-cell stage, and a further 6-fold increase at the blastocyst stage as compared with the 2-cell stage (Kane, Norris, Harrison, unpublished observations). These changes take place in embryos that have a virtually unchanging protein content (Table 13.1). Uptake by the mouse blastocyst is temperature and sodium dependent, inhibited by glucose, and appears to occur mainly by a saturable mechanism. There is no information available on the uptake mechanism for mouse embryo stages other than the blastocyst.

In the rabbit embryo, there is a gradual increase in uptake from 1-cell to morula, but uptake then increases dramatically with blastocyst expansion and increase in protein content (Conlon, Kane, unpublished observations). Preliminary information indicates that uptake in cleavage-stage rabbit embryos is sodium dependent, but uptake in 6-day blastocysts is not. Also in the 6-day blastocyst, about two-thirds of [^{3}H]inositol

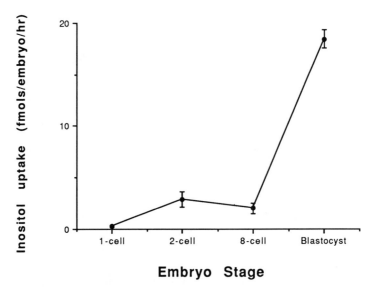

FIGURE 13.2. Effect of embryo stage on uptake of *myo*-inositol by mouse embryos. Embryos were cultured in 250-μCi [^{3}H]-inositol/mL for 2 h and then washed free of label and total uptake measured. Data points are mean values ± SEM based on 210–278 embryos per stage. Data from Kane, Norris, and Harrison, unpublished observations.

incorporated after 2 h is found in blastocoelic fluid. Uptake of inositol by cells generally takes place both by a Na-dependent saturable mechanism and a nonsaturable lipophilic mechanism (29, 30). In the Na^+-dependent mechanism, Na^+ and inositol appear to be cotransported across the membrane, and this mechanism appears to dominate in most cell types at physiological levels of inositol.

Incorporation of Inositol into Phosphinositides and Inositol Phosphates by Preimplantation Embryos

We have examined the incorporation of [^{3}H]inositol into phosphoinositides and inositol phosphates by rabbit blastocysts cultured from the morula stage (Fahy, Kane, unpublished observations). Embryos were cultured in 5 μCi of *myo*-[2-^{3}H]inositol/mL for 3 days. The phosphoinositides were precipitated with perchloric acid and separated by thin-layer chromatography along with phosphoinositide standards. The labeled phosphoinositides were identified based on their comigration with the standards. The results indicated that the rabbit blastocysts incorporated [^{3}H]inositol into PtdIns, PtdIns(4)P, and PtdIns(4,5)P_2, with the vast bulk of the label going into PtdIns. The supernatant from the perchloric acid precipitation was used for the separation of inositol phosphates by HPLC

Mono Q anion exchange chromatography. The HPLC analysis indicated that [^{3}H]inositol was being incorporated into inositol mono-, bis-, tris-, and tetrakis-phosphates. The presence of the second messengers Ins(1,4,5)P$_3$ and Ins(1,3,4,5)P$_4$ was identified based on their elution at the same position as labeled standards. These results indicate that most of the components of the PtdIns cycle are present in rabbit blastocysts.

We have preliminary evidence in mouse blastocysts that [^{3}H]inositol is incorporated at high rates into PtdIns and to a much lesser extent into PtdInsP$_1$ (Kane, Norris, Harrison, unpublished observations). However, there was no detectable incorporation into PtdInsP$_2$. The failure to detect any incorporation into PtdInsP$_2$, the immediate precursor of the second messengers diacylglycerol and Ins(1,4,5)P$_3$, suggests that the PtdIns cycle is not fully active in these blastocysts. This could be due to a number of reasons. First, in these uptake studies, blastocysts were cultured in the absence of amino acids, vitamins (other than inositol), trace elements, and growth factors. Full activation of the cycle could require the presence of peptide growth factors. Second, it is also possible that at the blastocyst stage, mouse embryos are merely storing phosphoinositides for later use at the time of implantation.

Effects of Lithium on Rabbit Blastocysts

Lithium disrupts the PtdIns cycle by inhibiting the breakdown of inositol monophosphates (31), thus preventing inositol recycling and causing an accumulation of inositol phosphates in cells. Treatment of *Xenopus* embryos with lithium disrupts differentiation by interfering with the PtdIns cycle and starving the cells of recycled inositol (32). These effects were reversible on microinjection of the embryo cells with *myo-* but not *epi*-inositol. We have found that long-term treatment (5 days) of rabbit embryos in culture with concentrations of 5–20mM LiCl inhibits blastocyst expansion and cell proliferation (Fahy, Kane, unpublished observations). Short-term treatment (1 h) with similar levels of lithium causes a significant accumulation of inositol monophosphates, indicating that the growth-inhibiting effect of lithium on blastocyst growth could be due to disruption of the PtdIns cycle. However, it was not possible to reverse the inhibitory effect of lithium on growth by the simultaneous addition of very high concentrations of *myo*-inositol (up to 9mM) to the culture medium. Either the inhibitory effect of lithium is due to factors other than its depletion of inositol for the PtdIns cycle or the raising of external inositol levels is not reflected in sufficiently elevated inward inositol transport to counteract the inositol depletion.

Possible Roles for the PtdIns Cycle in Blastocysts

Our data indicate that much of the machinery of the PtdIns cycle second-messenger system is in place in rabbit blastocysts. It is interesting to speculate on possible roles. One obvious possibility is that the system normally responds to peptide growth factors present in the oviduct and uterus. Recent work on the importance of peptide growth factors, such as insulin and *insulin-like growth factors* (IGFs)—for example, TGFs, PDGF, and EGF—and their receptors in mouse embryos lends support to this view (33–39). Thus, the PtdIns cycle in embryos could form part of a classical hormone second-messenger endocrine system regulating overall growth and development of the embryo.

A second possibility is that in blastocysts, the PtdIns cycle could be involved in local signaling from inner cell mass to trophoectoderm, or vice versa, and thus form part of a paracrine signaling system. The evidence that preimplantation mouse embryos both have receptors for IGFs (37) and also express IGF messenger RNA (35) is interesting tentative support for this possibility.

Embryotrophic Effect of Bovine Serum Albumin (BSA) Preparations

Variability in the Embryotrophic Effect of BSA

The standard culture medium, which we first developed for rabbit embryo culture (4, 40) and which is variously known with limited alterations as either BSMII (38–40) or modified F10 medium (4, 5), was very successful in supporting culture of 1-cell rabbit embryos through to viable blastocysts (43, 44). This medium basically consisted of the salts of Brinster's mouse medium (45) and the amino acids, vitamins, and trace elements of Ham's F10 medium (46) supplemented with a high level of BSA (either 15 mg/mL in early work or, more recently, 5 mg/mL). However, it rapidly became clear that this medium only supported the start of blastocyst growth and expansion. Blastocyst growth in the medium was much more limited and variable than growth in utero (5). While part of this poor growth could be related to suboptimal levels of inositol in the medium (23), it was evident that much of the variation in growth was related to the use of BSA as a protein supplement. Replacement of the BSA by uterine proteins or whole uterine flushings improved growth (42, 47).

Extraction and Purification of a Low-Molecular Weight Embryotrophic Contaminant from BSA Preparations

Commercial batches of BSA varied greatly in their ability to support blastocyst growth (48, 49). This suggested very strongly that part of the embryotrophic activity of BSA was due to a contaminant whose concentration varied from batch to batch. A number of known contaminants of BSA that might possess embryotrophic activity were eliminated as playing a major role; for example, fatty acids (48) and proteases (50). Finally, extraction of the BSA with formic acid and molecular filtration of the extract through a membrane filter with a cutoff of M 10,000 provided decisive proof that the embryotrophic activity was largely due to a low-molecular weight contaminant (51). We then purified the active factor by chromatography using in sequence G-10 Sephadex, QAE-Sephadex A-25 anion exchange, and HPLC reverse-phase columns (52).

Identification of the BSA Acid-Extractable Embryotrophic Factor as Citrate

Mass spectrometry carried out on this purified material by Dr. Steve Howell at the School of Pharmacy in London indicated that this material had a molecular weight of 192 and showed the presence of carboxylic acid groups. Chromatography of the material on a HPLC Mono Q anion exchange column along with a number of carboxylic acids showed that the embryotrophic activity eluted at the same position as citrate, which also has a molecular weight of 192, thus indicating that the embryotrophic BSA factor was citrate. Conclusive evidence of this fact was provided (50) as follows: (i) The mass spectra of the active reverse-phase material and citrate were shown to be identical; (ii) the original BSA sample was shown by enzyme assay to be heavily contaminated by citrate, and the degree of citrate contamination of various other BSA samples and fractions was correlated with their embryotrophic activity; and (iii) citrate stimulated blastocyst cell proliferation and expansion (Fig. 13.3).

The presence of citrate is clearly significant if not essential in stimulating rabbit blastocyst growth. However, the growth-promoting effect of commercial BSA preparations on preimplantation embryos is *not* solely due to their citrate content, even for rabbit embryos. Addition of citrate does not improve the medium to the extent of reproducing full in vivo growth of blastocysts in vitro. We have demonstrated the effect of citrate in the presence of charcoal-treated BSA. Replacing the charcoal-treated BSA by a synthetic polymer, such as *polyvinylalcohol* (PVA), only allows a very limited degree of blastocyst growth to take place even in the presence of citrate. The possibility that even charcoal-treated BSA may be contaminated with other growth factors—for example, peptide growth factors—is suggested by the fact that we have found that blastocyst growth in the presence of citrate varies to some degree with the batch of charcoal-treated BSA. This could, however, also be due to a toxic effect of some batches of BSA (53).

Previous data on the effects of citrate on preimplantation embryo development are limited. Brinster (54) found that while citrate did not support cleavage of 2-cell embryos, it did, even in the absence of any

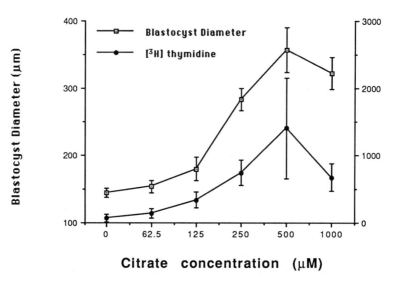

FIGURE 13.3. Effects of citrate concentration on mean blastocyst expansion and incorporation of [^{3}H]thymidine by rabbit blastocysts cultured from the early morula stage over a period of 5 days. [^{3}H]thymidine (5 µCi/mL) was added for 4 h at the end of culture, and the uptake of label into acid-precipitable material was measured. Data points are mean values ± SEM based on 22–26 embryos per treatment. Modified with permission from Gray, Morgan, and Kane (52).

other carbon sources, support the development of 8-cell mouse embryos to the blastocyst stage (55). Daniel (56) found that citrate had a very limited ability to act as the sole energy source for growth of rabbit blastocysts.

Possible Roles for Citrate

It is unlikely that the growth-promoting effect of citrate on rabbit blastocysts is due to its acting as an energy source since the basic medium contains a wide range of potential energy sources, such as glucose, pyruvate, and amino acids. There are a number of other ways in which citrate might stimulate rabbit blastocyst growth.

Citrate is an active chelator of metal ions with a high affinity for such cations as Ca^{++}, Sr^{++}, Mn^{++}, Mg^{++}, and Fe^{+++} (57). Thus, citrate could be involved in either chelating toxic metal ions or in promoting the uptake of ions such as Ca^{++} or Fe^{++} into blastocysts in chelated form. It has also been suggested that citrate, acting as a Ca^{++} chelator, may open tight junctions in epithelia, thereby enhancing the paracellular shunt pathway of solute transport (58). There is good evidence that chelating agents such as EDTA stimulate development of mouse embryos (59–61).

If citrate is taken up by blastocysts, it could be affecting fatty acid metabolism. Fatty acid synthesis is actively stimulated by citrate and isocitrate. As an allosteric activator of acetyl-CoA carboxylase, citrate plays a key role in the control of fatty acid synthesis (62, 63). Citrate promotes acetyl-CoA carboxylase activity by inducing 30-40 monomers of inactive enzyme to assemble into a fully active polymer. It is possible that fatty acid synthesis for the production of new membranes for cell proliferation is a factor that limits the rapid growth of the rabbit blastocyst in the absence of citrate.

While there are data available on the concentrations of citrate in mouse embryos (64), it is unfortunate that there appear to be no data on the concentrations of citrate in female reproductive tract fluids. Such information would provide some support for the notion that the role of exogenous citrate is a true physiological one or would indicate if it is an artifact of other deficiencies of the culture environment.

Conclusions

Study of the requirements for growth and the control of growth in pre-implantation embryos of species such as the rabbit, in which the embryo markedly increases its protein content before implantation, can be expected to shed considerable light on the processes underlying control of development and the initiation of growth in mammalian embryos in general. Evidence that growth of the rabbit blastocyst requires inositol (22) has led to work indicating that inositol may play an important role in hamster (65) and mouse (Kane, Norris, Harrison, unpublished) embryos. Furthermore, more attention should be paid to the peri-implantation and post-implantation mouse embryo (66) as models for the control of embryonic growth.

Our current knowledge on the initiation of true growth in mammalian embryos is limited. We know that a supply of amino acids and a range of vitamins (including inositol) are required, and in the rabbit blastocyst, exogenous citrate appears to be necessary. We do not know what, if any, peptide growth factors are required, although there is some suggestive evidence in this regard based on work in the mouse (33-39, 67). Also, if such growth factors are required, we are totally ignorant of their mechanisms of action in the embryo. Work in this area promises some exciting and interesting discoveries in the future.

Acknowledgments. We wish to thank the Irish Health Research Board, the Wellcome Trust, and the Irish National Board for Science and Technology for financial support of the research described in this paper.

References

1. Brinster RL. Protein content of the mouse embryo during the first five days of development. J Reprod Fertil 1967;13:413–20.
2. Schiffner J, Spielmann H. Fluorometric assay of the protein content of mouse and rat embryos during preimplantation development. J Reprod Fertil 1976; 47:145–7.
3. Daniel JC Jr. Early growth of rabbit trophoblast. Am Naturalist 1964; 98:85–97.
4. Kane MT, Foote RH. Culture of two- and four-cell rabbit embryos to the expanding blastocyst stage in synthetic media. Proc Soc Exp Biol Med 1970; 113:921–5.
5. Kane MT. In vitro growth of preimplantation rabbit embryos. In: Bavister BD, ed. The mammalian preimplantation embryo: regulation of growth and differentiation in vitro. New York: Plenum Press, 1987:193–217.
6. Wright RWJ, Grammer J, Bondioli K, Kuzan F, Menino AJ. Protein content of porcine embryos during the first nine days of development. Theriogenology 1981;15:235–9.
7. Woolley DW. The nature of the anti-alopecia factor. Science 1940;92:384–5.
8. Eagle H, Oyama VI, Levy M, Freeman AE. *myo*-Inositol as an essential growth factor for normal and malignant human cells in tissue culture. J Biol Chem 1957;226:191–207.
9. McConnell F, Goldstein L. Volume regulation in Elasmobranch red blood cells. In: Beyenbach KW, ed. Cell volume regulation. Basel: Karger, 1990: 114–31.
10. Kleinzeller A, Ziyadeh FN. Cell volume regulation in epithelia with emphasis on the role of osmolytes and the cytoskeleton. In: Beyenbach KW, ed. Cell volume regulation. Basel: Karger, 1990:59–86.
11. Strange K, Morrison R, Heilig CW, DiPietro S, Gullans SR. Upregulation of inositol transport mediates inositol accumulation in hyperosmolar brain cells. Am J Physiol 1991;260:C784–90.
12. Low MG. Biochemistry of the glycosyl-phosphatidylinositol membrane protein anchors. Biochem J 1987;244:1–13.
13. Ferguson MAJ, Williams AF. Cell-surface anchoring of proteins via glycosyl-phosphatidylinositol structures. Annu Rev Biochem 1988;57:285–320.
14. Saltiel AR, Cuatrecasas P. Insulin stimulates the generation from hepatic membranes of modulators derived from an inositol glycolipid. Proc Natl Acad Sci USA 1986;83:5793–7.
15. Saltiel AR. The role of glycosyl-phosphoinositides in hormone action. J Bioenerg Biomembr 1991;23:29–41.
16. Nishizuka Y. Perspectives on the role of protein kinase C in stimulus-response coupling. J Natl Cancer Inst 1986;76:363–70.
17. Berridge MJ. Inositol lipids and cell proliferation. Biochim Biophys Acta 1987;907:33–45.
18. Berridge MJ, Irvine RF. Inositol phosphates and cell signalling. Nature (London) 1989;341:197–204.
19. Michell RH, Drummond AH, Downes CP, eds. Inositol lipids in cell signalling. London: Academic Press, 1989.
20. Bell RM, Burns DJ. Lipid activation of protein kinase-C. J Biol Chem 1991;266:4661–4.

21. Daniel JC Jr. Vitamins and growth factors in the nutrition of rabbit blastocysts in vitro. Growth 1967;31:71–7.
22. Kane MT. The effects of water soluble vitamins on the expansion of rabbit blastocysts. J Exp Zool 1988;245:220–3.
23. Kane MT. Effects of the putative phospholipid precursors, inositol, choline, serine and ethanolamine, on formation and expansion of rabbit blastocysts in vitro. J Reprod Fertil 1989;87:275–9.
24. Kane MT, Bavister BD. Vitamin requirements for development of eight-cell hamster embryos to hatching blastocysts in vitro. Biol Reprod 1988;39:1137–43.
25. Gregoire AT, Gongsakdi D, Rakoff AE. The presence of inositol in genital tract secretions of the female rabbit. Fertil Steril 1962;13:432–5.
26. Lewin LM, Yannai Y, Melmed S, Weiss M. *myo*-Inositol in the reproductive tract of the female rat. Int J Biochem 1982;14:147–50.
27. Maslanski JA, Busa WB. A sensitive and specific mass assay for *myo*-inositol and inositol phosphates. In: Irvine RF, ed. Methods in inositide research. New York: Raven Press, 1990:113–26.
28. Flynn TJ, Hillman N. Lipid synthesis from [U-^{14}C] glucose in preimplantation mouse embryos in culture. Biol Reprod 1978;19:922–6.
29. Simmons D, Bomford J, Ng LL. *myo*-Inositol influx into human leucocytes: methods of measurement and the effect of glucose. Clin Sci 1990;78:335–41.
30. Wiesinger H. myo-Inositol transport in mouse astroglia-rich primary cultures. J Neurochem 1991;56:1698–1704.
31. Sherman WR. Inositol homeostasis, lithium and diabetes. In: Michell RH, Drummond AH, Downes CP, eds. Inositol lipids in cell signalling. London: Academic Press, 1989:39–79.
32. Busa WB, Gimlich RL. Lithium-induced teratogenesis in frog embryos prevented by a polyphosphoinositide cycle intermediate or a diacylglycerol analog. Dev Biol 1989;132:315–24.
33. Harvey MB, Kaye PL. Insulin stimulates protein synthesis in compacted mouse embryos. Endocrinology 1988;122:1182–4.
34. Harvey MB, Kaye PL. Insulin and IGF-1 are anabolic and mitogenic in preimplantation mouse embryos. Cell Differ Dev 1989;27:S31.
35. Heyner S, Smith RM, Schultz GA. Temporally regulated expression of insulin and insulin-like growth factors and their receptors in early mammalian development. Bioessays 1989;11:171–6.
36. Harvey MB, Kaye PL. Insulin increases the cell number of the inner cell mass and stimulates morphological development of mouse blastocysts in vitro. Development 1990;110:963–7.
37. Harvey MB, Kaye PL. IGF-2 receptors are first expressed at the 2-cell stage of mouse development. Development 1991;111:1057–60.
38. Rappolee DA, Brenner CA, Schultz R, Mark D, Werb Z. Developmental expression of PDGF, TGF-α, and TGF-β genes in preimplantation mouse embryos. Science 1988;241:1823–5.
39. Paria BC, Dey SK. Preimplantation embryo development in vitro: cooperative interactions among embryos and role of growth factors. Proc Natl Acad Sci USA 1990;87:4756–60.
40. Kane MT. In vitro culture of two- to four-cell rabbit embryos to expanding blastocysts in serum extracts and synthetic media [Ph.D. thesis]. Cornell University, 1969.

41. Maurer RR. Advances in rabbit embryo culture. In: Daniel JC Jr, ed. Methods in mammalian reproduction. New York: Academic Press, 1978: 259-72.
42. Fischer B. Development retardation in cultured preimplantation rabbit embryos. J Reprod Fertil 1987;79:115-23.
43. Kane MT. Energy substrates and culture of single-cell rabbit ova to blastocysts. Nature (London) 1972;238:468-9.
44. Seidel GE Jr, Bowen RA, Kane MT. In vitro fertilization, culture and transfer of rabbit ova. Fertil Steril 1976;27:861-70.
45. Brinster RL. A method for the in vitro cultivation of mouse ova from two-cell to blastocyst. Exp Cell Res 1963;32:205-8.
46. Ham RG. An improved nutrient solution for diploid Chinese hamster and human cell lines. Exp Cell Res 1963;29:515-26.
47. Maurer RR, Beier HM. Uterine proteins and development in vitro of rabbit pre-implantation embryos. J Reprod Fertil 1976;48:33-41.
48. Kane MT, Headon DR. The role of commercial bovine serum albumin preparations in the culture of one-cell rabbit embryos to blastocysts. J Reprod Fertil 1980;60:469-75.
49. Kane MT. Variability in different lots of commercial bovine serum albumin affects both cell multiplication and hatching of rabbit blastocysts in culture. J Reprod Fertil 1983;69:555-8.
50. Kane MT. A survey of the effects of proteases and glycosidases on culture of rabbit morulae to blastocysts. J Reprod Fertil 1986;78:225-30.
51. Kane MT. A low molecular weight extract of bovine serum albumin stimulates rabbit blastocyst cell division and expansion. J Reprod Fertil 1985;73:147-50.
52. Gray CW, Morgan PM, Kane MT. Purification of an embryotrophic factor from commercial bovine serum albumin and its identification as citrate. J Reprod Fertil 1991;94:471-80.
53. McKiernan SH, Bavister BD. Different lots of bovine serum albumin inhibit or stimulate in vitro development of hamster embryos. In Vitro Cell Dev Biol 1992;28A:154-6.
54. Brinster RL. Studies on the development of mouse embryos in vitro, II. The effect of energy source. J Exp Zool 1965;158:59-68.
55. Brinster RL, Thomson JL. Development of eight-cell mouse embryos in vitro. Exp Cell Res 1966;42:308-15.
56. Daniel JC Jr. The pattern of utilization of respiratory metabolic intermediates by preimplantation rabbit embryos in vitro. Exp Cell Res 1967;47:619-24.
57. Chaberek S, Martell AE. Organic sequestering agents. New York: Wiley and Sons, 1959.
58. Cho MJ, Scieszka JF, Burton PS. Citric acid as an adjuvant for transepithelial transport. Int J Pharmaceutics 1989;52:79-81.
59. Abramczuk J, Solter D, Koprowski H. The beneficial effects of EDTA on development of mouse one-cell embryos in chemically defined medium. Dev Biol 1977;61:378-83.
60. Chatot CL, Ziomek CA, Bavister BD, Lewis JL, Torres I. An improved culture medium supports development of random-bred 1-cell mouse embryos in vitro. J Reprod Fertil 1989;86:679-88.
61. Toyoda Y, Azuma S, Takeda S. Effects of chelating agents on preimplantation development of mouse embryos fertilized in vitro. In: Yoshinaga K, Mori T,

eds. Development of preimplantation embryos and their environment. New York: Alan R. Liss, 1989:171–9.
62. Goodridge AG. Regulation of the activity of acetyl coenzyme A carboxylase by palmitoyl coenzyme A and citrate. J Biol Chem 1972;247:6946–52.
63. Goodridge AG. Regulation of fatty acid synthesis in isolated hepatocytes. Evidence for a physiological role for long chain fatty acyl coenzyme A and citrate. J Biol Chem 1973;248:4318–26.
64. Barbehenn EK, Wales RG, Lowry OH. The explanation for the blockade of glycolysis in early mouse embryos. Proc Natl Acad Sci USA 1974;71:1056–64.
65. Kane MT, Bavister BD, Fahy MM. Water-soluble vitamins stimulate in vitro development of rabbit and hamster blastocysts [473]. Proc 11th Annu Congr Anim Reprod Artif Insem, 1988.
66. Rizzino A. Defining the roles of growth factors during early mammalian development. In: Bavister BD, ed. The mammalian preimplantation embryo: regulation of growth and differentiation in vitro. New York: Plenum Press, 1987:151–74.
67. Wood SA, Kaye PL. Effects of epidermal growth factor on preimplantation mouse embryos. J Reprod Fertil 1989;85:575–82.

14

Development of Human Blastocysts In Vitro

KATE HARDY

Until recently, relatively little was known about human preimplantation development, mainly due to the inaccessibility of the human embryo after normal conception. In a classic study, some descriptive anatomical and morphological information was obtained from a limited series of embryos both at preimplantation and early postimplantation stages that were fortuitously obtained over many years from uteri removed at hysterectomy (1). However, it was not until the advent of *in vitro fertilization* (IVF) for the treatment of infertility over a decade ago—whereby techniques were developed to fertilize oocytes in vitro and culture the resulting embryos to early preimplantation stages before transfer back to the oocyte donor— that the human preimplantation embryo became accessible for study. Currently, in the UK, all oocytes retrieved from the patient are inseminated (day 0), and after examination for the presence of 2 pronuclei on day 1 to confirm normal fertilization, embryos are cultured for a further 24 h or 48 h until transfer on day 2 or 3. Two or rarely 3 embryos of the best morphology and most advanced stages of development are transferred. Following ethical approval and after confirming patient's consent, it is possible to study the remaining surplus embryos, and such research has provided clues as to possible reasons for the low success rate of human IVF.

In this chapter, recent studies of human embryo development in vitro are compared with development in vivo after transfer at cleavage stages to assess the extent to which suboptimal media and culture conditions affect viability. These studies include observations on the proportion of surplus embryos developing to the blastocyst stage in vitro, changes in uptake of energy substrates during preimplantation development, and allocation of cells to the *trophectoderm* (TE) and *inner cell mass* (ICM) in human blastocysts. Finally, the implications for low IVF success rates are discussed.

Pregnancy Rates Following IVF

Despite recent improvements, the success rate of IVF and embryo transfer remains disappointingly low (overall figures for all clinics throughout the UK show that only 21% of patients who have embryos transferred become pregnant [2]). However, in larger centers, better rates are achieved. Over a recent 15-month period at the IVF Unit at Hammersmith Hospital, London, 656 patients had between 1 and, exceptionally, 3 (mean 2.2 ± 0.02) embryos transferred. Of these patients, 292 (45%) became pregnant, as determined by raised serum *human chorionic gonadotropin* (hCG) levels. In 43 (15% of patients with raised hCG) cases, the pregnancy did not progress and was defined as a biochemical pregnancy. In a further 8 patients, the pregnancies terminated as an ectopic pregnancy. In 222 (34%) patients, the stage of pregnancy was reached where the fetal heart could be detected by ultrasound. In total, 1427 embryos were transferred. In each of the 43 patients who had a biochemical pregnancy and the 8 patients who had an ectopic pregnancy, at least 1 embryo must have implanted. In addition, 346 embryos reached the fetal sac stage; therefore, a total of at least 397 embryos must have implanted, giving a minimum implantation rate of 28% per embryo transferred. Since 309 embryos developed further to the fetal heart stage, an overall fetal development rate of 22% per embryo transferred was obtained.

Preimplantation Development In Vitro

Following IVF, pronuclei are visible on the morning of day 1, 16 h postinsemination. By days 2 and 3, when embryos are normally transferred, embryos developing at the normal rate have reached the 2- to 4-cell and 8-cell stages, respectively. Compaction appears to occur mainly at the 16-cell stage (day 4), which is later than in the mouse, where it occurs at the 8-cell stage. Expanded blastocysts are seen from day 5 onwards in vitro (3), as they are in vivo (4).

Information on the precise timing of preimplantation developmental events in vitro is sparse due to a variety of reasons. There is an obvious reluctance to disturb embryos that are to be transferred more than absolutely necessary; thus, observations on their development are made only once or twice a day so that accurate timings of cleavage divisions are not available. This is further complicated by the fact that about half of human embryos arrest at all stages between 2-cell and morula. Combining the results of three recent studies, only 107/224 (48%) of normally fertilized surplus human embryos have developed to the blastocyst stage (3, 5, 6) by day 6 or 7 in vitro.

Figure 14.1 provides a schematic representation of the approximate times at which various developmental preimplantation stages in the human have been observed, with a comparison with cleavage times known for the mouse. The first 3 cell cycles in vitro are similar in length in the mouse and the human. Cell cycles lengthen in human embryos after the 8-cell stage, with blastocysts appearing a day later than in the mouse.

Cleavage-Stage Arrest

What causes cleavage-stage arrest in vitro is not known. Most culture media used for human IVF are based on those developed for mouse embryos, and indications that these may not be optimal for human embryo culture include uneven cleavage, cytoplasmic fragmentation, and a high incidence of polyploid, anucleate, and multinucleate cells. However, cleavage-stage arrest has also been observed in vivo. Buster et al. (4) report the flushing on day 5 post LH surge of 21 embryos from the uteri of women acting as embryo donors after artificial insemination. These embryos were estimated to be between 93- and 130-h postovulation, and while 24% were blastocysts, 33% were still at early cleavage stages. Thus, although some variability may be explained by the timing of fertilization, approximately one-third of the embryos flushed had arrested at early cleavage stages.

Chromosome analysis of surplus embryos after transfer has shown that a high proportion (up to 40%) are chromosomally abnormal (14). As the incidence of congenital abnormality after IVF (1.4%) is low and not higher than that after natural conception (1.8%) (2), it is possible that embryos arresting in vitro may have gross genetic defects. However, a report comparing the incidence of chromosomal abnormalities in pre-implantation embryos with that in early clinically recognized pregnancies suggests that there is little difference between the two groups, except in the case of monosomies (15). Monosomies are rarely seen in clinically recognized pregnancies, but are seen in preimplantation embryos with an incidence similar to trisomies. Thus, it is possible that monosomies (with an incidence of 10%) (15) may contribute to some of the cleavage arrest in vitro, while embryos with other chromosomal abnormalities are capable of implanting, although a large proportion of these later miscarry.

In the mouse, proteins necessary for compaction are synthesized very early in development before the embryonic genome is fully activated (16). This suggests that compaction can occur in the absence of activation of the embryonic genome. This may also be the case in the human. Tesarik et al. (17) showed that 8-cell embryos that failed to show the enhancement of transcriptional activity associated with the activation of the embryonic genome still developed into normal-appearing morulae. However, further work showed that activation of the embryonic genome

14. Development of Human Blastocysts In Vitro   187

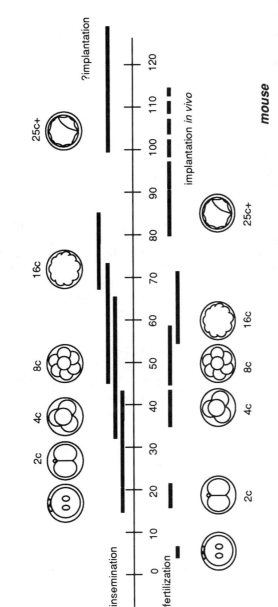

FIGURE 14.1. Comparative timing of preimplantation development of human and mouse embryos in vitro. Time scale is in hours using activation of egg as $t = 0$. For human development, the bars represent the range of times at which these stages have been observed, as described in references 3, 7, 8, and 9. For mouse development, the bars represent the range of times at which cleavage to a particular stage occurs, as described in references 10, 11, 12, and 13.

was necessary for blastocyst formation (18), particularly for the development of vectorial fluid transport mechanisms and an efficient permeability seal. Thus, it is possible that human embryos that arrest at the morula stage have done so because of the failure of activation of the embryonic genome. Further evidence for this from a limited series of human embryos showed that if activation of the genome is blocked with the transcriptional inhibitor α-amanitin, development is arrested, but only after the 4- to 8-cell stage (19).

Energy Substrate Uptake In Vitro

While extensive work has been carried out over the last 40 years into the metabolism and in vitro culture of embryos from species such as the mouse, little work has been done on the human embryo, mainly due to the paucity of surplus embryos and the lack of sensitive assay techniques. Studies on preimplantation mouse embryo metabolism have shown that during early cleavage, pyruvate, rather than glucose, is the major energy substrate (20). There is then a switch in substrate preference at the 8-cell/morula stage in favor of glucose, which becomes the predominant energy source at the blastocyst stage (21).

Culturing individual embryos in microdrops of medium such as HLT6—a modified T6 medium containing 5 mM lactate, 0.47 mM pyruvate, and 1 mM glucose (5, 6)—or Earle's (22) enables the measurement of the uptake of such substrates as pyruvate or glucose from the medium over a 24-h culture period. The amount of pyruvate and glucose in a microdrop can be analyzed after completion of the 24-h incubation using an ultramicrofluorescence assay (23). Comparison of the amounts of substrate in sample medium with control medium, in which no embryo has been incubated, allows calculation of the depletion of these substrates by the embryo.

During development to the blastocyst stage in vitro, the metabolism of the embryo, as reflected by variations in the uptake of both pyruvate and glucose, changes dramatically (Fig. 14.2). Pyruvate uptake increased significantly ($P < 0.05$) over both the third and the fourth cleavage divisions, from 24 pmoles/embryo/h at the 2- to 4-cell stage to 38 pmoles/embryo/h at the 9- to 16-cell stages, and peaked at the morula stage (42 pmoles/embryo/h). Between the morula and the blastocyst stages, there was a significant decline ($P < 0.005$) in pyruvate uptake to 34 pmoles/embryo/h. As there was no difference in pyruvate uptake between early and expanded blastocysts, the values for these stages were combined. Overall, glucose uptake was lower than pyruvate uptake throughout preimplantation development in vitro. During cleavage up to the 16-cell stage, glucose uptake was <10 pmoles/embryo/h. However, between the 16-cell and the blastocyst stages, there was a significant increase in glucose uptake, to a maximum of 24 pmoles/embryo/h. Thus, during early cleavage, the

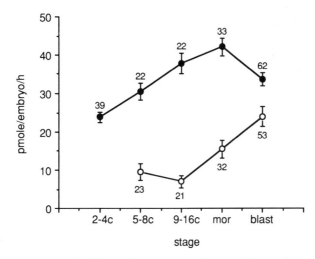

FIGURE 14.2. Pyruvate (solid circle) and glucose (open circle) uptake by normally fertilized human preimplantation embryos that developed to the blastocyst stage in vitro. Values are means ± SEM, and the number of observations is given next to each point.

embryo has a high uptake of pyruvate that peaks at the morula stage and then declines, whereas glucose uptake remains low before increasing after the 16-cell stage.

This profile of pyruvate and glucose uptake during preimplantation human development in vitro is qualitatively similar to that observed for mouse embryos (21, 23), where it was demonstrated that pyruvate is taken up preferentially during early cleavage stages, but with a switch to glucose uptake during the transition between morula and blastocyst stages. In the human, the switch from a pyruvate- to a glucose-based metabolism is not as dramatic as is seen in the mouse in that pyruvate uptake does not fall below the uptake of glucose at the blastocyst stage.

The reasons for a switch from a pyruvate-based to a glucose-based metabolism remain obscure, but it is notable that several important events occur at this stage of development that would justify an increase in glucose uptake. First, several pieces of evidence suggest that in the human, the embryonic genome is activated between the 4- and 8-cell stages of development. At this stage, ribosomal particles mature into a functional morphological form (24), uptake and incorporation of uridine into ribosomes increases (25), and development is blocked for the first time by transcriptional inhibitors (19). Accompanying these events are changes in the types of proteins that are synthesized (19). Leese (26) has noted that glucose is a precursor of many macromolecular cell constituents, such as ribose moieties for nucleic acid biosynthesis, glycerol phosphate for the formation of phospholipids and complex sugars in mucoproteins and mucopolysaccharides, and he has suggested that the switching on of the embryonic genome may impose an increased requirement for glucose.

Second, accompanying blastocyst formation is the first differentiative step during embryogenesis, the formation of the TE and ICM. Outer TE cells develop a polar epithelial morphology, and formation of zonular junctions between cells creates a tight permeability seal, allowing the development of the blastocoel cavity. The inner ICM cells, precursors of the fetus, retain a rounded stem cell morphology. It would not be surprising if cell differentiation, involving a change in cell function and morphology, should involve a change in metabolism, and, indeed, the appearance of the first differentiated cell type coincides with a decline in pyruvate uptake and a sharp increase in glucose uptake.

Third, the process of compaction and blastocoel formation imposes a heavy demand for ATP, thereby increasing the embryo's metabolic requirements (26). Finally, it is possible that the changes in substrate requirement coincide with the time at which the embryo moves from the oviduct to the uterus in vivo, as has been suggested in the mouse (27). Fertilized human embryos cannot be recovered from the uterus before the 4th or 5th day postovulation (28), which is around the time of the increase in glucose uptake (Fig. 14.3).

In this context, it is of interest that human embryos that fail to reach the blastocyst stage in vitro fail to show the increase in glucose uptake that characterizes blastocyst formation in normally developing embryos (Fig. 14.3). This strongly suggests that the increase in glucose uptake is not under temporal control, but is dependent on blastocyst formation and, therefore, probably activation of the embryonic genome. In addition, throughout preimplantation development, the pyruvate uptake rates of embryos that arrest are between ~10% and 30% lower than the rates of embryos that develop to the blastocyst stage (Fig. 14.3). This suggests that the pyruvate uptake rates of preimplantation human embryos reflect their viability and could be used to predict which embryos are most likely to implant after transfer (22).

Interestingly, in these studies in which embryos were cultured in microdrops of a simple medium supplemented with *bovine serum albumin* (BSA), 59% (45/75) developed to the blastocyst stage (5). Whether culturing embryos in small volumes of medium enhances development or whether BSA is better than serum at supporting blastocyst formation is not yet known, and further investigation is needed. However, it is notable that despite the relatively high proportion of embryos developing to the blastocyst stage in microdrops of HLT6 + BSA, this medium is unable to sustain continued development of blastocysts, with collapse and degeneration following within 24 h of expansion. Upon blastocyst formation, the embryos' metabolic requirements become more complex, with serum supplementation being essential for further culture.

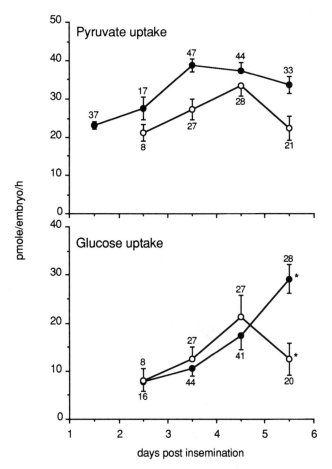

FIGURE 14.3. Pyruvate and glucose uptake by normally fertilized embryos that developed to the blastocyst stage by day 5 or 6 (solid circle) and by embryos that arrested during cleavage (open circle). Values are means ± SEM. Points marked by an asterisk are significantly different ($P \leq 0.0005$), and the number of observations is given next to each point. Modified with permission from Hardy, Hooper, Handyside, Rutherford, Winston, and Leese (5).

Cell Numbers in Human Blastocysts

Apart from a series of ultrastructural studies of human blastocysts (29, 30), there is relatively little information about cleavage rates in late preimplantation development and no information about the total numbers of cells in human blastocysts or about the first differentiative divisions that give rise to the TE and ICM. In the mouse preimplantation embryo, the division of polarized cells at the 8-cell stage generates phenotypically

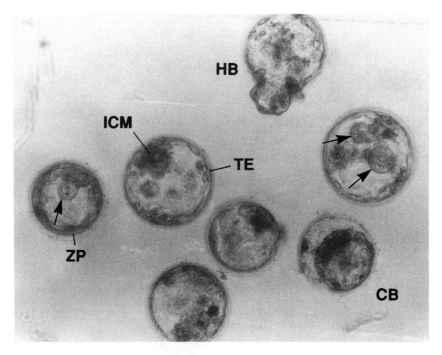

FIGURE 14.4. Phase contrast micrograph of 7 human blastocysts on day 6 post insemination. (TE = trophectoderm; ICM = inner cell mass; ZP = zona pellucida.) Note the blastocyst hatching from the zona (HB) and the collapsing blastocyst (CB), which earlier had been expanded. In some blastocysts, large cells can be seen lying in the blastocoel cavity (arrowed). These could be cleavage-stage blastomeres that arrested during development and have been excluded from blastocyst formation (130×).

distinct outer and inner cells that differentiate into the outer epithelial layer of TE and the ICM at the blastocyst stage. This first differentiative division, allocating cells to the TE and ICM, is of fundamental importance for later development. At the 32-cell stage, during blastocyst formation, the TE is responsible for blastocyst fluid accumulation and becomes specialized for implantation. Following blastocyst expansion, hatching from the zona, and implantation, the TE gives rise to components of the placenta and extraembryonic membranes, while the ICM gives rise to the fetus (31). The human blastocyst is morphologically similar to the mouse blastocyst (Fig. 14.4). There is some ultrastructural evidence that blastomeres of human 8-cell embryos undergo a similar process of surface polarization as described for mouse embryos at this stage and that inner cells have different plasma membrane characteristics than outer cells (18,

32). Beyond this, nothing is known about the stage at which blastocyst formation occurs or how many cells are allocated to the TE and ICM.

It is possible to count both TE and ICM cells in the blastocyst by differential labeling of the nuclei with polynucleotide-specific fluorochromes. The outer TE nuclei are labeled with one fluorochrome during immunosurgical lysis before fixing the embryo and labeling both the TE and ICM nuclei with a second fluorochrome (3). Since the emission spectra of the two fluorochromes are different, the labeled nuclei can be distinguished by the color of the fluorescence.

All of the normally fertilized blastocysts of good morphology examined by this technique on days 5–7 had ICMs (Table 14.1), although the range in cell numbers was considerable. The average proportion of ICM cells varied from about 33% on days 5 and 7 to 50% on day 6. Apoptotic cell death, characterized by fragmenting nuclei, was widespread and increased between days 5 and 7; 95% of the blastocysts examined had evidence of cell death in either the TE or the ICM, and 56% had evidence of cell death in both lineages. Low numbers of ICM cells on days 6 and 7 were associated with a high proportion of dead cells (Fig. 14.5) The mitotic index of both the TE and ICM was relatively high on day 5, but was reduced by day 7. Taken together, these observations on increased cell death and a reduced mitotic index by day 7 (Table 14.1) could suggest that the culture conditions for blastocysts are inadequate.

However, it should be noted that even freshly flushed mammalian embryos show evidence of cell death in vivo, as shown by ultrastructural studies (30, 33). In the mouse, apoptotic cell death can be seen predominantly in the ICM after differential labeling. Although the significance of this form of cell death in the ICM is not known, it has been suggested that it is a mechanism for the elimination of cells retaining the potential to form TE (34). In addition, over this culture period, the total cell number in normally fertilized blastocysts increased from day 5 onwards and significantly between days 6 and 7 (Table 14.2). Thus, cell division continues in vitro at least up to day 7. The hatched blastocyst with 283 cells on day 7 is particularly encouraging for these in vitro culture conditions.

Newly expanded normally fertilized blastocysts on day 5 had a total of ~60 cells, which increased to ~80 and 125 on days 6 and 7, respectively. The range of total cell numbers in normally fertilized blastocysts increased progressively from day 5 to day 7 (Fig. 14.6). Newly expanded blastocysts on day 5 ranged from 24 to 90 cells, on day 6 from 27 to 136 cells, and on day 7 from 60 to 1 hatched blastocyst with 283 cells. The majority of day 5 blastocysts had between 31 and 60 cells, while the majority of day 6 blastocysts had between 61 and 120 cells. In contrast, morphologically abnormal blastocysts on day 6 (characterized by unincorporated blastomeres, multiple cavities, no discrete ICM visible, or low numbers of mural TE cells) had only half the cells of their morphologically normal counterparts

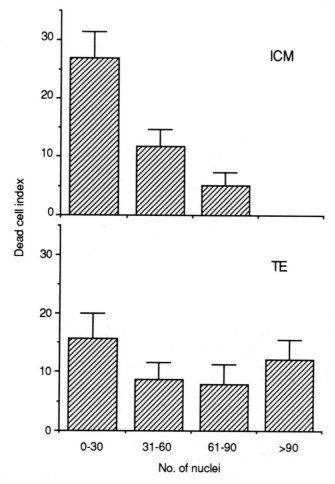

FIGURE 14.5. Comparison of the mean dead cell index ± SEM with the number of nuclei in the ICM and TE of normally fertilized human blastocysts on days 6 and 7. Modified with permission from Hardy, Handyside, and Winston (3).

and large numbers of dead cells. One morphologically abnormal blastocyst on day 5 had only 18 cells. Multiple cavities may, however, be a normal feature of primate embryo development since they have been observed in baboon and rhesus monkey blastocysts (35).

The minimum number of cells that we observed in normally fertilized, morphologically normal expanded blastocysts on days 5 and 6 was 24 and 27 cells, respectively. In a similar study in the mouse (34), the minimum number observed was 24, which suggests that cavitation occurs at the same time in both species; that is, at completion of the 5th cleavage division. An attempt has been made to estimate cell allocation at the 16-

TABLE 14.1. Numbers of TE and ICM cells in normally fertilized blastocysts grown in vitro to days 5, 6, and 7.

Morphology	TE cell number	TE mitotic index	TE dead cell index	ICM cell number	ICM mitotic index	ICM dead cell index	Total cell number
Newly expanded blastocyst day 5 n = 13	39.1 ± 4.3	4.7 ± 1.5	7.3 ± 2.0	19.3 ± 2.9	2.6 ± 1.3	10.3 ± 3.8	58.4 ± 5.9
Expanded blastocyst day 6 n = 16	44.7 ± 4.5	1.2 ± 0.5	6.6 ± 1.9	37.6 ± 4.2	0.9 ± 0.4	11.9 ± 2.4	82.3 ± 6.2
Expanded blastocyst day 7 n = 10	80.6 ± 15.2	0.4 ± 0.3	16.3 ± 2.7	45.6 ± 10.2	0.4 ± 0.3	27.8 ± 5.4	126.2 ± 21.0

Note: Values are means ± SEM; mitotic index = (no. of cells in mitosis/total no. of cells) × 100; dead cell index = (no. of dead cells/total no. of cells + no. of dead cells) × 100.

TABLE 14.2. Total numbers of nuclei in expanded and abnormal human blastocysts grown in vitro to days 5, 6, and 7.

Age and morphology	No. of embryos	Total no. of nuclei (range)	Mitotic index	Dead cell index
Day 5 newly expanded blastocysts	13	58.4 ± 5.9 (24–90)	4.3 ± 1.2	8.5 ± 2.5
Day 6 expanded blastocysts	33	83.1 ± 4.7 (27–136)	1.4 ± 0.3	10.1 ± 1.3
Day 7 expanded blastocysts	11	125.5 ± 19.0 (60–283)	0.7 ± 0.2	16.7 ± 2.9
Day 5 morphologically abnormal blastocyst	1	18	5.6	0
Day 6 morphologically abnormal blastocysts	15	38.3 ± 6.2 (3–86)	2.1 ± 0.7	27.0 ± 5.6

Note: Values are means ± SEM; mitotic index = (no. of cells in mitosis/total no. of cells) × 100; dead cell index = (no. of dead cells/total no. of cells + no. of dead cells) × 100; data presented here include total cell numbers from Table 14.1 and a further series of embryos that were simply fixed with 1 fluorochrome.

and 32-cell stages by extrapolating from TE and ICM numbers in day 5–7 blastocysts (Fig. 14.7). On this basis and ignoring cell death, at the 16-cell stage there would be approximately 5 cells allocated to the ICM and 11 to the TE and, at the 32-cell stage, there would be approximately 11 and 21 cells allocated to the ICM and TE, respectively. These numbers are similar to the numbers found in the mouse (34, 36).

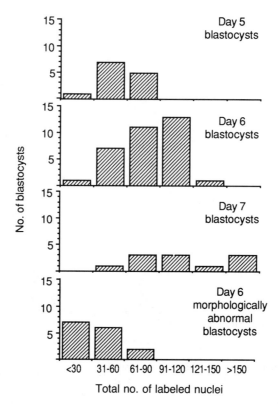

FIGURE 14.6. Distribution of total cell number in normally fertilized human blastocysts in vitro on days 5–7 and morphologically abnormal blastocysts on day 6. Modified with permission from Hardy, Handyside, and Winston (3).

Implications for Low IVF Pregnancy Rates

Although many human embryos undergo cleavage-stage arrest in vitro and only about half reach the blastocyst stage, this is still higher than the numbers implanting in vivo after transfer. If half the embryos reach the blastocyst stage in vitro and less than a third implant in vivo, this suggests that some blastocysts are developmentally incompetent. A variety of maternal or embryonic factors could be involved. From examination of numbers of TE, ICM, and total cell numbers in blastocysts, it is possible to postulate reasons for the failure of some blastocysts to implant. Morphologically abnormal blastocysts, with unincorporated blastomeres, no visible ICM, or low numbers of TE cells, have significantly fewer cells and high levels of cell death (Table 14.2). This, combined with low numbers of ICM cells in blastocysts of good morphology, raises the possibility that these blastocysts may be responsible either for the failure

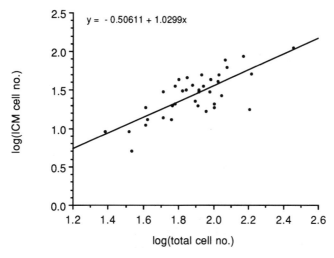

FIGURE 14.7. Logarithm (base 10) of total cell numbers of 39 day 5, 6, and 7 blastocysts plotted against the logarithm of their ICM cell numbers. The straight line is a least-squared fit to the observed data. Modified with permission from Hardy, Handyside, and Winston (3).

to implant or for biochemical pregnancies where there is an initial increase in the levels of hCG, but no fetal development. Whether these blastocysts with low cell numbers result from suboptimal conditions in vitro or inherent defects will require further investigation.

Acknowledgments. The author would like to thank Alan Handyside, Joe Conaghan, and Jaroslav Stark for useful discussion and help in preparing the manuscript; Professor Robert Winston for his generous support and encouragement; and Karin Dawson and the IVF team for their help and cooperation.

References

1. Hertig AT, Rock J, Adams EC, Mulligan WJ. On the preimplantation stages of the human ovum: a description of four normal and four abnormal specimens ranging from the second to fifth day of development. Contrib Embryol 1954;35:199–220.
2. ILA. The fifth report of the interim licensing authority for human in vitro fertilization and embryology. London: Clements House, 1990.
3. Hardy K, Handyside AH, Winston RML. The human blastocyst: cell number, death and allocation during late preimplantation development in vitro. Development 1989;107:597–604.

4. Buster JE, Bustillo M, Rodi IA, et al. Biologic and morphologic development of donated human ova recovered by nonsurgical lavage. Am J Obstet Gynecol 1985;153:211–7.
5. Hardy K, Hooper MAK, Handyside AH, Rutherford AJ, Winston RML, Leese HJ. Non-invasive measurement of glucose and pyruvate uptake by individual human oocytes and preimplantation embryos. Hum Reprod 1989; 4:188–91.
6. Hardy K, Martin KL, Leese HJ, Winston RML, Handyside AH. Human preimplantation development in vitro is not adversely affected by biopsy at the 8-cell stage. Hum Reprod 1990;5:708–14.
7. Trounson AO, Mohr LR, Wood C, Leeton JF. Effect of delayed insemination on in-vitro fertilization, culture and transfer of human embryos. J Reprod Fertil 1982;64:285–94.
8. Laufer N, Tralatzis BC, Naftolin F. In vitro fertilization: state of the art. Seminars in Reproductive Endocrinology 1984;2:197–219.
9. Mohr L, Trounson A. In vitro fertilization and embryo growth. In: Wood C, Trounson A, eds. Clinical in vitro fertilization. Berlin: Springer-Verlag, 1984: 99–115.
10. Bolton VN, Oades PJ, Johnson MH. The relationship between cleavage, DNA replication and gene expression in the mouse 2-cell embryo. J Embryol Exp Morphol 1984;79:139–63.
11. Howlett SK, Bolton VN. Sequence and regulation of morphological and molecular events during the first cell cycle of mouse embryogenesis. J Embryol Exp Morphol 1985;87:175-206.
12. Molls M, Zamboglou N, Streffer C. A comparison of the cell kinetics of preimplantation mouse embryos from two different mouse strains. Cell Tissue Kinet 1983;16:277–83.
13. Smith RKW, Johnson MH. Analysis of the third and fourth cell cycles of mouse early development. J Reprod Fertil 1986;76:393–9.
14. Plachot M, de Grouchy J, Junca A-M, et al. From oocyte to embryo: a model, deduced from in vitro fertilization, for natural selection against chromosome abnormalities. Ann Genet 1987;30:22–32.
15. Angell RR, Hillier SG, West JD, Glasier AF, Rodger MW, Baird DT. Chromosome anomalies in early human embryos. J Reprod Fertil 1988;suppl 36:73–81.
16. Levy JB, Johnson MH, Goodall H, Maro B. The timing of compaction: control of a major developmental transition in mouse early embryogenesis. J Embryol Exp Morphol 1986;95:213–37.
17. Tesarik J, Kopecny V, Plachot M, Mandelbaum J. Activation of nucleolar and extranucleolar RNA synthesis and changes in the ribosomal content of human embryos developing in vitro. J Reprod Fertil 1986;78:463–70.
18. Tesarik J. Involvement of oocyte-coded message in cell differentiation control of early human embryos. Development 1989;105:317–22.
19. Braude P, Bolton V, Moore S. Human gene expression first occurs between the four- and eight-cell stages of preimplantation development. Nature (London) 1988;332:459–61.
20. Brinster RL. Studies on the development of mouse embryos in vitro, II. The effect of energy source. J Exp Zool 1965;158:59–68.

21. Gardner DK, Leese HJ. Non-invasive measurement of nutrient uptake by single cultured preimplantation mouse embryos. Hum Reprod 1986;1:25–7.
22. Conaghan J, Martin K, Hardy K, et al. Prospects for pre-transfer assessment of human embryo viability following in vitro fertilization. Hum Reprod abstracts from the 7th world congress on IVF and assisted procreation 1991: 232.
23. Leese HJ, Barton AM. Pyruvate and glucose uptake by mouse ova and preimplantation embryos. J Reprod Fertil 1984;72:9–13.
24. Tesarik J, Kopecny V, Plachot M, Mandelbaum J. Early morphological signs of embryonic genome expression in human preimplantation development as revealed by quantitative electron microscopy. Dev Biol 1988;128:15–20.
25. Tesarik J, Kopecny V, Plachot M, Mandelbaum J, DeLage C, Flechon J. Nucleogenesis in the human embryo developing in vitro: ultrastructural and autoradiographic analysis. Dev Biol 1986;115:193–203.
26. Leese HJ. The energy metabolism of the preimplantation embryo. In: Heyner S, Wiley LM, eds. Early embryo and paracrine relationships. New York: Alan R. Liss 1990:67–78.
27. Brinster RL. Carbon dioxide production from lactate and pyruvate by the preimplantation mouse embryo. Exp Cell Res 1967;47:634–7.
28. Croxatto HB, Diaz S, Fuentealba B, Croxatto H, Carrillo D, Fabres C. Studies on the duration of egg transport in the human oviduct, I. The time interval between ovulation and egg recovery from the uterus in normal women. Fertil Steril 1972;23:447–58.
29. Lopata A, Kohlman DJ, Kellow GN. The fine structure of human blastocysts developed in culture. In: Embryonic development, part B: cellular aspects. New York: Alan R. Liss, 1982:69–85.
30. Mohr LR, Trounson AO. Comparative ultrastructure of hatched human, mouse and bovine blastocysts. J Reprod Fertil 1982;66:499–504.
31. Gardner RL, Papaioannou VE. Differentiation in the trophectoderm and inner cell mass. In: Balls M, Wild AE, eds. The early development of mammals. London: Cambridge University Press, 1975:107–32.
32. Lopata A, Kohlman D, Johnston I. The fine structure of normal and abnormal human embryos developed in culture. In: Beier HM, Lindner HR, eds. Fertilization of the human egg in vitro. Heidelberg: Springer-Verlag, 1983:189–210.
33. Enders AC, Hendrickx AG, Binkerd PE. Abnormal development of blastocysts and blastomeres in the rhesus monkey. Biol Reprod 1982; 26:353–66.
34. Handyside AH, Hunter S. Cell division and death in the mouse blastocyst before implantation. Roux's Arch Dev Biol 1986;195:519–26.
35. Enders AC, Lantz KC, Schlafke S. The morula-blastocyst transition in two old world primates: the baboon and rhesus monkey. J Med Primatol 1990; 19:725–47.
36. Handyside AH. Immunofluorescence techniques for determining the number of inner and outer blastomeres in mouse morulae. J Reprod Immunol 1981; 2:339–50.

15

Development of Na/K ATPase Activity and Blastocoel Formation

CATHERINE S. GARDINER AND ALFRED R. MENINO, JR.

The fertilized mouse egg undergoes a series of cleavage divisions that result in an embryo composed of 8 similar totipotent cells. These cells subsequently develop cytoplasmic and plasma membrane polarity with distinct apical and basolateral regions. Polarization of cells is the first step in embryonic differentiation. Cleavage divisions of these polarized cells produce two distinct cell types: polar cells on the outside of the embryo and apolar cells on the inside of the embryo. Polar cells develop into trophectoderm—the first transporting epithelium of the embryo—and subsequently will form extraembryonic membranes in the conceptus. The apolar cells develop into inner cell mass that later forms the embryo proper.

Development of the Blastocyst

Coincident with differentiation of trophectoderm and inner cell mass, the embryo forms a fluid-filled cavity or blastocoel. The blastocoel continues to expand and eventually stretches the surrounding zona pellucida, thus aiding the embryo in shedding the zona pellucida. A more detailed description of many aspects of preimplantation embryonic development may be obtained from recent reviews (1–3).

At the morula stage in mice, outer blastomeres become polarized due to asymmetric cell contacts, whereas inner blastomeres remain apolar due to symmetric cell contacts being totally surrounded by other cells (4). The mechanism of polarization appears to involve transcellular ion currents carried by Na (3, 5). Formation of the blastocoel is dependent on the formation of a permeability seal between cells on the outside of the embryo (6) and fluid accumulation between cells (7). Embryonic cells move Na from the outside of the embryo into intercellular spaces in the inside of the embryo, thus producing an ionic gradient that causes

15. Development of Na/K ATPase Activity and Blastocoel Formation 201

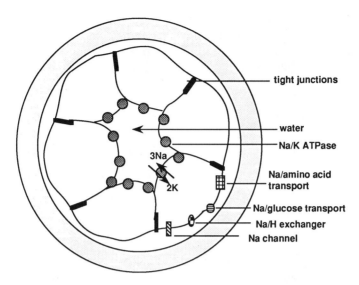

FIGURE 15.1. Diagram of blastocoel formation with important components labeled. Apical membrane structures bring Na into trophectodermal cells and the basolateral Na/K ATPase pumps Na out of cells into blastocoel, thus producing an ion gradient that leads to the osmotic accumulation of fluid.

the osmotic movement of water into intercellular spaces (8). Sodium is believed to move into trophectodermal cells through their apical surfaces due to the action of the Na/H exchanger, the Na channel (9), Na-dependent amino acid transport (10, 11), and the Na/glucose cotransporter (12). Sodium is transported out of the trophectodermal cells through their basolateral surfaces due to the action of Na/K ATPase (8). Figure 15.1 identifies the components believed to be important in blastocoel formation.

The Na/K ATPase is localized to the blastocoelic surface of embryonic cells and is believed to be involved in blastocoel formation and expansion (11). Specifically, the enzyme is localized to the basolateral domain of the plasma membrane of mural trophectoderm (13). Ouabain, a specific inhibitor of Na/K ATPase, blocks blastocoel formation and expansion; blastocysts that are collapsed by treating with cytochalasin B or D are prevented from reexpansion by treatment with ouabain; and intact blastocysts have decreased blastocoel expansion (9, 11). Wiley (14) demonstrated that Na/K ATPase is involved in cortical localization of organelles and fluid accumulation in extracellular spaces in the mouse embryo. The continued accumulation of blastocoelic fluid leads to expansion of the blastocoel and stretching and thinning of the zona pellucida. This is believed to have a role in embryonic escape from the zona pellucida, which is a prerequisite for continued growth and implantation of the embryo. It has been proposed that intracellular water droplets arise from β-oxidation of lipid by basal-lateral mitochondria; subsequently this water is moved into the blastocoel due to osmosis (15). Adenylate cyclase plays a role in blastocoel fluid accumulation, and its activity increases between the morula and blastocyst stages (16).

In addition to Na/K ATPase, other molecular components necessary for production of the transcellular ionic gradient, blastocoel fluid accumulation, and formation of the permeability seal in the embryo are beginning to be elucidated. Removal of the cell adhesion molecule, uvomorulin, blocks cell flattening and decreases blastomere polarization; however, uvomorulin is detected on all preimplantation stages of development (17). In addition, treatment of mouse embryos with antiserum to uvomorulin disrupts the normal localization of Na/K ATPase and blocks cavitation (18). The Na/glucose cotransporter is believed to be important in entry of Na into blastomeres and is preferentially localized to the apical membrane (12). Inhibition of either Na/K ATPase or the Na/glucose cotransporter causes decreased polarization of polar 4-cell/16-cell heterokaryons (3). Evidence also suggests that Na entry into the trophectoderm may occur through the Na/H exchanger and the Na channel because Na entry and blastocoel expansion can be decreased by specifically blocking these activities (9). Sodium-dependent amino acid cotransport is another pathway for Na entry into the trophectoderm, and this activity increases dramatically at the blastocyst stage (10, 11). Inhibitors of either the Cl/HCO_3 exchanger or the Cl channel do not affect blastocyst expansion or Cl uptake (9).

To maintain the osmotically accumulated fluid within the blastocoel, a permeability seal develops between cells on the outside of the embryo (6). Using immunocytochemistry, the tight junction-specific peripheral membrane protein, ZO-1, is first detected at the compacting 8-cell stage (19). Localization of ZO-1 along the boundary of apical and basolateral domains is dependent on cell adhesion and microfilaments and, therefore, is a secondary step in blastomere polarization. It is believed that tight junctions restrict paracellular movement of small molecules between the blastocoel and the outside of the embryo and aid in maintaining membrane components in their appropriate locations either in the basolateral or apical membrane. Antibodies to uvomorulin interfere with the assembly and localization of tight junctions (19).

Blastocyst formation and hatching occurs during a critical period in mammalian embryo development. During this time, the embryo is undergoing extensive cellular morphological differentiation changes that are imperative to continued development. The importance of blastogenesis in cell lineage differentiation of the embryo is suggested by the fact that differentiation of the two cell types has never been observed without the formation of a blastocoel (15). Transcriptional events necessary for compaction and hatching occur well in advance; however, the transcriptional and translational events required for blastocoel formation occur within a few hours of the start of this process (20). The complete mechanisms of blastocoel formation, cell differentiation, and the regulation of these processes in the preimplantation embryo are not yet understood.

General Aspects of Na/K ATPase

Several aspects of Na/K ATPase have been reviewed (21–24); the active form of the enzyme is composed of 2 subunits, α and β. The α-subunit is a protein of 90–120 kd and is the catalytic subunit, whereas the β-subunit is a glycoprotein of 40–60 kd whose function is just beginning to be understood. The β-subunit is necessary for enzyme function, but whether the enzyme is composed of a single or multiple α/β protomers is not yet clear. The Na/K ATPase pumps 3 Na ions out of the cell and 2 K ions into the cell on each cycle, thus producing an ionic gradient. This enzyme has several physiological functions, including maintenance of K levels for intracellular enzyme function, osmotic pressures, ionic gradients for cellular excitability, and maintenance of ionic gradients utilized for transport systems and generation of metabolic heat.

The role of the β-subunit and the functional maturation of Na/K ATPase has recently been reviewed (24). The newly synthesized α-subunit has a loose configuration and is highly sensitive to trypsinolysis. Removal of the sugar moiety from the β-subunit does not affect Na/K ATPase activity. However, the trypsin sensitivity of both subunits is increased when glycosylation of the β-subunit is prevented. It appears that the glycosylated β-subunit has a role in establishing a specific conformation of the α-subunit. It is possible that transport of functional Na/K ATPase to the membrane is controlled by the association of the α-subunit with limiting amounts of the β-subunit. In *Xenopus* oocytes, the β-subunit has a role as a receptor or stabilizer of the α-subunit (25). In addition, the presence of the β-subunit causes the α-subunit to be localized to the membrane and resistant to trypsinolysis (26). In the fully grown *Xenopus* oocyte, much less β-subunit is produced than α-subunit, and the addition of β-subunit results in an increase in the amount of trypsin-resistant α-subunits capable of cation transport (27). The assembly of α- and β-subunits occurs in the *endoplasmic reticulum* (ER), results in stabilization of the α-subunit, and is necessary for both subunits to be transported out of the ER (28). An increase in the amount of β-subunit mRNA was found to cause an increase in the amount of newly synthesized α- and β-subunits and an increase in Na/K ATPase activity (29).

The molecular genetics of Na/K ATPase has also been reviewed recently (30). There are at least 3 different isoforms of the α-subunit and 2 isoforms of the β-subunit (31, 32). The different isoforms are expressed in a tissue-specific manner and have varied cellular functions (31–33). Activity of Na/K ATPase is regulated by many factors, including steroid, thyroid, protein, and peptide hormones; developmental changes; and monovalent cations. Steroid and thyroid hormones appear to regulate Na/K ATPase by altering the amounts of mRNA and protein subunits of the enzyme. Other hormones act by altering the activity of existing Na/K ATPase pumps. In most circumstances, mRNA transcripts for the α- and β-subunit are found in equal amounts. Regulation of Na/K ATPase activity in various circumstances occurs at the transcriptional, posttranscriptional, translational, and posttranslational levels. The regulatory regions of genes encoding the α1 (34, 35), α2 (36–38), and β2 (39) isoforms have begun to be characterized and indicate the presence of a number of promoter, enhancer and, transcription factor binding sites. Future research should help us begin to understand the factors and mechanisms regulating Na/K ATPase gene transcription.

Depending on the cell type, Na/K ATPase may be evenly distributed throughout the plasma membrane, limited to the apical membrane, or restricted to the basolateral membrane, as is the case with epithelial cells (reviewed in 40). Membrane proteins are anchored in their correct membrane domain with the aid of the cytoskeleton, as is the case with fodrin maintaining the asymmetric distribution of Na/K ATPase. When Na/K ATPase molecules are trapped in the apical region of an epithelial cell membrane by tight junction formation, they are removed, and the enzyme assumes a polarized distribution in a few hours. Tight junctions are not the first step in the polarized distribution of membrane proteins, but aid in maintaining apical and basolateral domains by preventing free diffusion in the membrane. Uvomorulin is a cell adhesion molecule that plays a role in the development of membrane polarity. Inducing the expression of uvomorulin in a nonpolarized cell is sufficient to cause the cell to redistribute Na/K ATPase sites similar to a polarized epithelial cell (41). This redistribution of Na/K ATPase pumps is accompanied by changes in the membrane cytoskeleton. The adhesion molecule on glia (AMOG) is an adhesion molecule involved in neuron-astrocyte interaction, has a similar amino acid sequence to the Na/K ATPase β1 subunit, and has been identified as the β2 subunit (42). In polarized cells, newly synthesized Na/K ATPase is not sorted in the Golgi complex and is distributed to both apical and basolateral membrane domains (43). It is proposed that basolaterally localized Na/K ATPase is stabilized due to its interaction with the cytoskeleton; whereas, the apically localized Na/K ATPase is inactivated and removed (43).

Regulation of Na/K ATPase During Blastocoel Formation in the Mouse

Electrophysiological and isotope flux studies reveal the presence of Na/K ATPase activity in the mouse egg and 2-cell mouse embryo (44, 45);

however, ultrastructural cytochemistry does not detect Na/K ATPase until the morula stage (46). There is also disagreement as to the localization of Na/K ATPase at the morula stage; it has been found limited to apposed cell membranes (46) and throughout the cytoplasm (13). Watson and Kidder (13) first detected Na/K ATPase at the late morula stage using immunofluorescent localization. The amount of Na/K ATPase α-subunit present per embryo does not appear to increase during blastocoel formation in the mouse, but the amount of β-subunit may increase (47). Because this enzyme is believed to have a critical role in differentiation, it is possible that an increase in the amount of Na/K ATPase activity or the cellular location of the enzyme may be a trigger for blastocoel formation.

However, the ouabain-sensitive cation transporting activity only increases approximately 2-fold between the 2-cell and blastocyst stages (48). Comparing ouabain-inhibitable cation transport in collapsed and intact blastocysts indicates that only 20% of the Na/K ATPase activity is localized on the outside surface of the embryo at the blastocyst stage. However, there is little difference between the compacted or uncompacted 8-cell embryo in the amount of ouabain-inhibitable cation transporting activity present, indicating that ouabain is able to reach virtually all of the enzyme present in the compacted 8-cell embryo (48). This suggests that between the compacted 8-cell stage and the blastocyst stage, the amount of Na/K ATPase activity present on the outside surface of the embryo is decreased, and activity is increased on the basolateral cell surfaces. By the blastocyst stage, Na/K ATPase is localized to the blastocoel-facing membrane of blastomeres, and as such, it is a tissue-specific marker for mural trophectoderm (13). The discrepancies in the timing of appearance, amount of enzyme present, and localization of Na/K ATPase in the mouse embryo may be due to its potentially dual roles. In most, if not all, mammalian cells, Na/K ATPase is known to have the housekeeping function of maintaining cell volume. The embryo probably utilizes Na/K ATPase in this housekeeping function in addition to the role Na/K ATPase has in the transtrophectodermal transport of ions and fluid. These two roles of Na/K ATPase in the mammalian embryo may involve different enzyme isoforms, different localizations of the enzyme, and/or different regulations of enzyme activity.

Of the 3 known isoforms of the α-subunit, only mRNA encoding the $\alpha 1$ isoform has been detected in the blastocyst (49). The amount of α-subunit mRNA present in the embryo increases dramatically from the 2-cell stage to the blastocyst stage (47, 49); however, the timing of increases in the amount of β-subunit mRNA appears to be more closely correlated with blastocoel formation (49). To begin to understand the regulation of Na/K ATPase activity in the embryo, it is necessary to compare enzyme activity levels to amounts of the enzyme subunits and amounts of the mRNAs for the subunits in the embryo. This comparison is made more difficult due to the complicated metabolic changes occurring in the embryo. Total RNA and ribosomal RNA content increase dramatically during the period of blastocoel formation (50). While RNA content of the mouse embryo increases 5-fold during development from the 2-cell to the blastocyst stage, total protein content decreases slightly during the same period (51). To determine whether Na/K ATPase gene expression is regulated differently than the majority of embryonic genes, we may want to compare the profiles of total RNA and total protein content of the embryo. The amount of Na/K ATPase α-subunit (47) and the total amount of protein (51) per embryo remain relatively constant from the 2-cell to the blastocyst stage. The amount of α-subunit mRNA (47, 49) and the amount of total RNA increase dramatically during the same time. This would suggest that the α-subunit is not being regulated differently than the bulk of embryonic proteins at either the mRNA or the protein level. In contrast, the amount of β-subunit and its mRNA exhibit more of an increase during preimplantation development than total protein and RNA levels. The amount of β-subunit present in the embryo increases during the same time that the enzyme activity increases 2-fold; therefore, the amount of the β-subunit present in the embryo may aid in the regulation of Na/K ATPase activity during blastocoel formation.

Regulation of Na/K ATPase During Blastocyst Formation and Expansion in the Rabbit and Other Species of Embryo

The rabbit embryo is an example of a maximally expanding blastocyst— as is the pig, horse, cow, and sheep embryo—whereas the mouse embryo is a minimally expanding embryo (8). In the rabbit embryo, ouabain binding sites per trophoblast cell increase 7.5-fold between days 4 and 6, corresponding to the time period of maximal blastocyst expansion (52). On a per-embryo basis, Na/K ATPase increases 50-fold between days 4 and 6 in the rabbit (51). The rate of synthesis of new Na/K ATPase increases 90-fold during this same time (8, 53), and the blastocoel's diameter increases from 300 µm to 3 mm. The Na/K ATPase α-subunit content increases 22-fold, and α-subunit mRNA content increases 35-fold between days 4 and 6 in rabbit embryo development (54). This suggests that Na/K ATPase genes are expressed at a high level during blastocyst expansion and that expression of the α-subunit is at least partially regulated at the mRNA level.

It has been demonstrated that Na is also actively transported in the trophectoderm of the 7- to 10-day pig blastocyst and that the degree of Na transport is correlated with the degree of blastocyst expansion (55). Additionally, inhibition of Na/K ATPase with ouabain in the days 8 and 10 pig blastocyst at least partially inhibits Na influx (55). Total RNA content of early pig embryos has not been determined; however, total protein content increases slightly from fertilization to day 5 of development and increases dramatically from day 5 to day 9 of preimplantation development (56). We are currently studying Na/K ATPase gene expression in the pig embryo during this same period of development. Blastocyst expansion and turgidity are important for several reasons: (i) shaping the blastocyst, (ii) transport to implantation sites and spacing of embryos, and (iii) ensuring the close apposition of the blastocyst and endometrium that is necessary for implantation (57). Embryonic development through implantation is the period of highest embryonic mortality, and understanding Na/K ATPase regulation in the embryo may elucidate some mechanisms of embryonic loss.

Conclusion

Na/K ATPase has an important role in blastocoel fluid accumulation and appears to become concentrated in mural trophectoderm. Investigating the regulation of Na/K ATPase during blastocoel formation and the first cell lineage differentiation of the embryo may aid us in understanding how the signal for differentiation (i.e., polarization) is converted into a heritable cell phenotype. In some tissues, the β-subunit has a role in stabilizing the α-subunit, anchoring it in the membrane and regulating the amount of enzyme activity. This may also be the case in the preimplantation embryo. The amount of β-subunit and β-subunit mRNA increases to a greater degree than the bulk of embryonic proteins and RNAs. Whether the increase in β-subunit mRNA in the embryo is due to an increased rate of transcription is unknown. Factors regulating transcription of Na/K ATPase genes in other systems are beginning to be elucidated.

In addition to the increase in the amount of β-subunit and Na/K ATPase activity, the enzyme becomes preferentially localized to basolateral membranes at the blastocyst stage. This may cause the enzyme to be more effective at generating transcellular ion gradients. The localization of Na/K ATPase to the basolateral membrane of mural trophectodermal cells may be regulated through cell adhesion molecules, as is the case in other cell types. How uvomorulin leads to basolateral localization of Na/K ATPase in other cell types is not completely understood; however, in the embryo, uvomorulin is present long before basolateral localization of Na/K ATPase occurs.

References

1. Bavister BD, ed. The mammalian preimplantation embryo: regulation of growth and differentiation in vitro. New York: Plenum Press, 1987.
2. Prather RS, First NL. A review of early mouse embryogenesis and its applications to domestic species. J Anim Sci 1988;66:2626–35.

3. Wiley LM, Kidder GM, Watson AJ. Cell polarity and development of the first epithelium. Bioessays 1990;12:67–73.
4. Johnson MH, Ziomek CA. Cell interactions influence the fate of mouse blastomeres undergoing the transition from the 16- to the 32-cell stage. Dev Biol 1983;95:211–8.
5. Nuccitelli R, Wiley LM. Polarity of isolated blastomeres from mouse morulae: detection of transcellular ion currents. Dev Biol 1985;109:452–63.
6. McLaren A, Smith R. A functional test of tight junctions in the mouse blastocyst. Nature (London) 1977;267:351–3.
7. Wiley LM, Eglitis MA. Cell surface and cytoskeletal elements: cavitation in the mouse preimplantation embryo. Dev Biol 1981;86:493–501.
8. Benos DJ, Biggers JD, Balaban RS, Mills JW, Overstrom EW. Developmental aspects of sodium-dependent transport processes of preimplantation rabbit embryos. In: Graves JS, ed. Regulation and development of membrane transport processes. New York: Wiley and Sons, 1985:211–35.
9. Manejwala FM, Cragoe EJ Jr, Schultz RM. Blastocoel expansion in the preimplantation mouse embryo: role of extracellular sodium and chloride and possible apical routes of their entry. Dev Biol 1989;133:210–20.
10. Van Winkle LJ, Campione AL, Farrington BH. Development of system $B^{o,+}$ and a broad-scope Na^+-dependent transporter of zwitterionic amino acids in preimplantation mouse conceptuses. Biochim Biophys Acta 1990;1025:225–33.
11. Wiley LM, Lever JE, Pape C, Kidder GM. Antibodies to a renal Na^+/glucose cotransport system localize to the apical plasma membrane domain of polar mouse embryo blastomeres. Dev Biol 1991;143:149–61.
12. DiZio SM, Tasca RJ. Sodium-dependent amino acid transport in preimplantation mouse embryos: Na^+-K^+-ATPase-linked mechanism in blastocysts. Dev Biol 1977;59:198–205.
13. Watson AJ, Kidder GM. Immunofluorescence assessment of the timing of appearance and cellular distribution of Na/K-ATPase during mouse embryogenesis. Dev Biol 1988;126:80–90.
14. Wiley LM. Cavitation in the mouse preimplantation embryo: Na/K-ATPase and the origin of nascent blastocoel fluid. Dev Biol 1984;105:330–42.
15. Wiley LM. Development of the blastocyst: role of cell polarity in cavitation and cell differentiation. In: Bavister BD, ed. The mammalian preimplantation embryo: regulation of growth and differentiation in vitro. New York: Plenum Press, 1987:65–93.
16. Manejawala F, Kaji E, Schultz RM. Development of activatable adenylate cyclase in the preimplantation mouse embryo and a role for cyclic AMP in blastocoel formation. Cell 1986;46:95–103.
17. Vestweber D, Gossler A, Boller K, Kemler R. Expression and distribution of cell adhesion molecule uvomorulin in mouse preimplantation embryos. Dev Biol 1987;124:451–6.
18. Watson AJ, Damsky CH, Kidder GM. Differentiation of an epithelium: factors affecting the polarized distribution of Na^+, K^+-ATPase in mouse trophectoderm. Dev Biol 1990;141:104–14.
19. Fleming TP, McConnell J, Johnson MH, Stevenson BR. Development of tight junctions de novo in the mouse early embryo: control of assembly of the tight junction-specific protein, ZO-1. J Cell Biol 1989;108:1407–18.

20. Kidder GM, McLachlin JR. Timing of transcription and protein synthesis underlying morphogenesis in preimplantation mouse embryos. Dev Biol 1985;112:265–75.
21. Trachtenberg MC, Packey DJ, Sweeney T. In vivo functioning of the Na^+, K^+-activated ATPase. Curr Top Cell Reg 1981;19:159–217.
22. Jorgensen PL. Mechanism of the Na^+, K^+ pump, protein structure and conformations of the pure (Na^+ & K^+)-ATPase. Biochim Biophys Acta 1982;694:27–68.
23. Glynn IM. The Na^+, K^+-transporting adenosine triphosphate. In: Martonosi A, ed. The enzymes of biological membranes; vol. 3. New York: Plenum Press, 1985:35–114.
24. Geering K. Subunit assembly and functional maturation of Na, K-ATPase. J. Membr Biol 1990;115:109–21.
25. Noguchi S, Higashi K, Kawamura M. Assembly of the α-subunit of *Torpedo californica* Na^+/K^+-ATPase with its pre-existing β-subunit in *Xenopus* oocytes. Biochim Biophys Acta 1990;1023:247–53.
26. Noguchi S, Higashi K, Kawamura M. A possible role of the β-subunit of (Na,K)-ATPase in facilitating correct assembly of the α-subunit into the membrane. J Biol Chem 1990;265:15991–5.
27. Geering K, Theulaz I, Verrey F, Hauptle MT, Rossier BC. A role for the β-subunit in the expression of functional Na^+-K^+-ATPase in *Xenopus* oocytes. Am J Physiol 1989;257:C851–8.
28. Ackermann U, Geering K. Mutual dependence of Na,K-ATPase α- and β-subunits for correct posttranslational processing and intracellular transport. FEBS Lett 1990;269:105–8.
29. Lescale-Matys L, Hensley CB, Crnkovic-Markovic R, Putnam DS, McDonough AA. Low K^+ increases Na,K-ATPase abundance in LLC-PK_1/Cl_4 cells by differentially increasing β, and not α, subunit mRNA. J Biol Chem 1990;265:17935–40.
30. Lingrel JB, Orlowski J, Shull MM, Price EM. Molecular genetics of Na,K-ATPase. Prog Nucleic Acid Res Mol Bio 1990;38:37–89.
31. Martin-Vasallo P, Dackowski W, Emanuel JR, Levenson R. Identification of a putative isoform of the Na,K-ATPase β subunit. J Biol Chem 1989;264:4613–8.
32. Shyjan AW, Gottardi C, Levenson R. The Na,K-ATPase β2 subunit is expressed in rat brain and copurifies with Na,K-ATPase activity. J Biol Chem 1990;265:5166–9.
33. Shyjan AW, Levenson R. Antisera specific for the α1, α2, α3, and β subunits of the Na,K-ATPase: differential expression of α and β subunits in rat tissue membranes. Biochemistry 1989;28:4531–5.
34. Yagawa Y, Kawakami K, Nagano K. Cloning and analysis of the 5'-flanking region of rat Na^+/K^+-ATPase α1 subunit gene. Biochim Biophys Acta 1990;1049:286–92.
35. Shull MM, Pugh DG, Lingrel JB. The human Na,K-ATPase α1 gene: characterization of the 5'-flanking region and identification of a restriction fragment length polymorphism. Genomics 1990;6:451–60.
36. Kawakami K, Yagawa Y, Nagano K. Regulation of Na^+, K^+-ATPases, I. Cloning and analysis of the 5'-flanking region of the rat NKAA2 gene encoding the α2 subunit. Gene 1990;91:267–70.

37. Shull MM, Pugh DG, Lingrel JB. Characterization of the human Na,K-ATPase α2 gene and identification of intragenic restriction fragment length polymorphisms. J Biol Chem 1989;264:17532–43.
38. Sverdlov ED, Bessarab DA, Malyshev IV, et al. Family of human Na^+,K^+-ATPase genes: structure of the putative regulatory region of the α^+-gene. FEBS Lett 1989;244:481–3.
39. Kawakami K, Okamoto H, Yagawa Y, Nagano K. Regulation of Na^+, K^+-ATPases, II. Cloning and analysis of the 5'-flanking region of the rat NKAB2 gene encoding the β2 subunit. Gene 1990;91:271–4.
40. Cereijido M, Ponce A, Gonzalez-Mariscal L. Tight junctions and apical/basolateral polarity. J Membr Biol 1989;110:1–9.
41. McNeill H, Ozawa M, Kemler R, Nelson J. Novel function of the cell adhesion molecule uvomorulin as an inducer of cell surface polarity. Cell 1990;62:309–16.
42. Gloor S, Antonicek H, Sweadner K, et al. The adhesion molecule on glia (AMOG) is a homologue of the β subunit of the Na,K-ATPase. J Cell Biol 1990;110:165–74.
43. Hamilton RW, Krzeminski KA, Mays RW, Ryan TA, Wollner DA, Nelson WJ. Mechanisms for regulating cell surface distribution of Na^+,K^+-ATPase in polarized epithelial cells. Science 1991;254:847–50.
44. Powers RD, Tupper JT. Ion transport and permeability in the mouse egg. Exp Cell Res 1975;91:413–21.
45. Powers RD, Tupper JT. Developmental changes in membrane transport and permeability in the early mouse embryo. Dev Biol 1977;56:306–15.
46. Vorbrodt A, Konwinski M, Solter D, Koprowski H. Ultrastructural cytochemistry of membrane-bound phosphatases in preimplantation mouse embryos. Dev Biol 1977;55:117–34.
47. Gardiner CS, Williams JS, Menino AR. Sodium/potassium adenosine triphosphatase α- and β-subunit and α-subunit mRNA levels during mouse embryo development in vitro. Biol Reprod 1990;43:788–94.
48. Van Winkle LJ, Campione AL. Ouabain-sensitive Rb^+ uptake in mouse eggs and preimplantation conceptuses. Dev Biol 1991;146:158–66.
49. Watson AJ, Pape C, Emanuel JR, Levenson R, Kidder GM. Expression of Na,K-ATPase α and β subunit genes during preimplantation development of the mouse. Dev Genet 1990;11:41–8.
50. Piko L, Clegg KB. Quantitative changes in total RNA, total poly(A), and ribosomes in early mouse embryos. Dev Biol 1982;89:362–78.
51. Sellens MH, Stein S, Sherman MI. Protein and free amino acid content in preimplantation mouse embryos and in blastocysts under various culture conditions. J Reprod Fertil 1981;61:307–15.
52. Benos DJ. Ouabain binding to preimplantation rabbit blastocysts. Dev Biol 1981;83:69–78.
53. Overstrom EW, Benos DJ, Biggers JD. Synthesis of Na^+/K^+ ATPase by the preimplantation rabbit blastocyst. J Reprod Fertil 1989;85:283–95.
54. Gardiner CS, Grobner MA, Menino AR. Sodium/potassium adenosine triphosphatase α-subunit mRNA levels in early rabbit embryos. Biol Reprod 1990;42:539–44.

55. Overstrom EW. In vitro assessment of blastocyst differentiation. In: Bavister BD, ed. The mammalian preimplantation embryo: regulation of growth and differentiation in vitro. New York: Plenum Press, 1987:95–116.
56. Wright RW, Grammer J, Bondioli K, Kuzan R, Menino AR. Protein content of porcine embryos during the first nine days of development. Theriogenology 1981;15:235–9.
57. Biggers JD, Bell JE, Benos DJ. Mammalian blastocyst: transport functions in a developing epithelium. Am J Physiol 1988;255:C419–32.

16
Effects of Imprinting on Early Development of Mouse Embryos

R.A. Pedersen, K.S. Sturm, D.A. Rappolee, and Z. Werb

During the past decade, a major advance in understanding the development of embryonic and extraembryonic lineages of mammals has emerged from studies on the fate of parthenogenetic (or gynogenetic) embryos and their androgenetic counterparts. Diploid parthenogenotes produced by experimentally activating mouse oocytes develop to midgestation stages, then they die with a characteristic phenotype: The most advanced embryos have extensive development of the axial embryonic structures (brain and neural tube, somites) and other embryonic organs, but only rudimentary development of the trophoblast lineage (1). *Diploid androgenotes*—that is, embryos with only paternally derived chromosomes—that are manufactured by nuclear transfer (2) also die at midgestation stages, but with retarded development of the embryo proper and with normal trophoblast by gross morphological examination (1, 3). The conclusion drawn from these and other studies (reviewed in 4–6) is that maternal and paternal gametes make distinct and complementary contributions to the developing conceptus, so that normal mouse development requires both maternal and paternal haploid genomes (1, 7). This phenomenon, referred to as *genomic imprinting*, thus appears to have profound consequences for peri-implantation development in eutherian mammals (metatheria and prototheria have not been studied). Because examples of viable parthenogenesis are known among fish, amphibia, reptiles, and birds (8), it is unlikely that imprinting has the same functional impact in other vertebrate classes.

The fate of parthenogenetic or androgenetic embryos has been studied in aggregation chimeras, where the *isoparental* embryo is combined with a normal diploid embryo at early cleavage stages, then examined at the blastocyst stage or returned to the uterus of a foster mother for further development. Parthenogenetic and androgenetic embryos contribute to both the ICM and trophectoderm cell populations of such chimeras examined at the blastocyst stage, but are selectively eliminated at later stages (9, 10). Parthenogenetic cells are eliminated from the trophectoderm and primitive endoderm lineages by midgestation stages, but persist in most lineages of the embryo proper (9, 11). With further development to term, and to adult mice, parthenogenetically derived cells persist in some tissue lineages (e.g., brain, heart, kidney, spleen, and female germ cells), but are systematically eliminated from other tissues (e.g., skeletal muscle, liver, and pancreas) (12, 13). These observations indicate that parthenogenetic cells are at a strong selective disadvantage in extraembryonic and some embryonic lineages, but they contribute extensively to most embryonic tissues. However, adult mice with extensive parthenogenetic or gynogenetic contribution have reduced viability and retarded postnatal development (14, 15).

Androgenetic cells are eliminated from most embryonic lineages by midgestation stages, but persist in the trophoblast and primitive endoderm lineages of chimeric embryos (9, 11). However, androgenetic *inner cell mass* (ICM) cells and *embryonic stem* (ES) cells produced from androgenetic embryos were found to contribute to fetal tissues, especially the mesoderm lineage in chimeras (16–18, and Chapter 12, this volume).

These morphological observations provide some clues about the molecular genetic basis of imprinting in early mouse development. Some of the parthenogenetic and androgenetic deficiencies are cell autonomous because they are not corrected in chimeras by cell-cell associations between themselves and the cells derived from the normally fertilized embryo. There are also deficiencies in diffusible factors that regulate growth and differentiation of the various cell types, as suggested by Surani and coworkers (19). Recent work has shown that the expression of *insulin-like growth factor II* (IGF-II) is regulated by imprinting (20, 21). This gene is predominantly transcribed when it is inherited from the father; the maternal allele is expressed only in 2 brain tissues: the leptomeninges and choroid plexus (21). Expression of the IGF-II/mannose-6-phosphate receptor is also affected by imprinting, being transcribed only when it is maternally inherited (22). Thus, the simplest model for the molecular effects of genomic imprinting is that endogenous imprinted genes are transcribed only when they are inherited from one parent, but not from the other parent; consequently, the normally fertilized individual is functionally hemizygous for such genes. Accordingly, the perturbations arising from imprinting in the development of parthenogenotes could arise either from underexpression or from overexpression of the imprinted genes.

Genomic imprinting also appears to be involved in the etiology of several human diseases, including retinoblastoma, osteosarcoma, Angelman syndrome, and other degenerative or hyperplastic diseases (reviewed in 6). These observations strongly suggest that there are general developmental consequences of genomic imprinting among eutherian mammals, including domestic species and humans (23, 24). In the following sections, we review our recent results showing the effects of imprinting on the morphology and gene expression in developing mouse parthenogenotes. Then we consider the implications of these observations for understanding the evolution and developmental role of imprinting in mammals.

Experimental Approach and Observations

We examined the preimplantation and early postimplantation morphology of parthenogenetic mouse embryos in comparison with fertilized embryos to determine in detail the phenotype of peri-implantation developmental perturbations that arise from imprinting (25). We also studied the expression of specific genes known to be transcribed in preimplantation mouse embryos to determine whether their expression was affected by imprinting. These observations revealed unforeseen disturbances of both embryonic and extraembryonic lineages in parthenogenotes and led to the identification of imprinting effects on the transcription of a growth factor receptor gene.

Morphology

The morphology of parthenogenotes was compared to in vivo fertilized embryos of the same (C57BL/6J × CBA/J) F_1 hybrid strain. Parthenogenetic activation was accomplished by a modification of the procedure of Cuthbertson (26, 27). After culture to the 2-cell stage, parthenogenotes were transferred to the oviducts of foster mothers of the CD-1 strain; in some cases, parthenogenotes or fertilized embryos were cultured in simple medium to more advanced preimplantation stages, then were transferred to complex medium for outgrowth. For controls, fertilized embryos were removed from the oviducts at 0.5 day of gestation (approximately noon of plug day), cultured overnight to the 2-cell stage, then transferred to foster mothers. Postimplantation morphology of parthenogenotes and fertilized embyros was compared at days 7–10 after activation or fertilization. The control fertilized embryos were delayed approximately 1 day as compared to normal development in utero.

Although there was attrition during cleavage, morphogenesis of preimplantation parthenogenotes was normal by light microscopic examination. The in vitro development of parthenogenotes from the 2-cell to the blastocyst stage was reduced compared to cultured fertilized embryos. Parthenogenotes that reached the morula and blastocyst stages were morphologically similar to fertilized embryos, although we could not rule out subtle alterations in ultrastructure or cell number at these stages. Hatching from the zona pellucida was slightly impaired in cultured parthenogenetic blastocysts, but their capacity to cause extensive decidual reactions was inconsistent with any significant block to hatching in utero.

Parthenogenetic blastocysts were able to form outgrowths in vitro, but these were smaller than normal, and the ICMs were small. There were no organized regions of embryonic ectoderm, visceral embryonic endoderm, or visceral extraembryonic endoderm in the ICM-derived portion of the outgrowth. Moreover, the trophectodermal portion of the outgrowth was associated with numerous small round cells and cell fragments. By contrast, blastocysts from fertilized eggs formed egg cylinder-like structures with ectodermal and endodermal layers, and their trophoblast outgrowths had few small cells and fragments at an equivalent age (5–7 days of culture beyond the expanded blastocyst stage).

Although the foregoing observations revealed subtle manifestations of the parthenogenetic phenotype at preimplantation stages and in outgrowths, the most dramatic consequences of imprinting were evident during postimplantation development in utero. Parthenogenotes examined at 8–10 days after activation displayed a wide variety of developmental perturbations. Because most of the embryos did not conform to the characteristics of normal embryos at the same ages, a classification system derived from preliminary observations was used to score the degree of abnormality. For simplicity, only parthenogenotes dissected at 9 days are described here; however, comparable results were obtained for parthenogenotes at 8 and 10 days.

Class I embryos were smaller than normal, but had an anterior-posterior axis and possessed brain, somites, and paraxial mesoderm, as well as embryonic gut, heart, and blood, reflecting the formation and persistence of all 3 primary germ layers and their organization into organ rudiments. Extraembryonic mesoderm was present both in the yolk sac and in the allantois. The ectoplacental cone was small; both diploid and giant trophectoderm cells were present. Primitive endoderm descendants (both visceral and parietal endoderm cells) were present.

Class II embryos had structures representing all 3 primary germ layers, but they had no axial organization. Embryonic ectoderm was often folded, and mesoderm was present between this layer and the surrounding endoderm cells. It was not possible to characterize the endoderm as either embryonic or extraembryonic. The trophoblast-derived tissue in class II parthenogenotes consisted mainly of giant cells, and the Reichert's membrane was thickened compared to normal embryos.

Class III parthenogenotes were severely disorganized, lacked mesodermal cells, and were surrounded by masses of extracellular matrix with embedded parietal endoderm cells. Only a few trophoblast giant cells were found attached to these embryos.

Class IV parthenogenotes consisted solely of parietal endoderm cells embedded in extracellular matrix, accompanied by a few trophoblast giant cells; embryonic ectoderm, mesoderm, and other endoderm cell types were absent. The identity of the parietal endoderm cells and the associated matrix in classes III and IV was verified by immunohistological staining with antibodies against laminin and collagen type IV, both characteristic products of parietal endoderm, which is the apparent source of Reichert's membrane (28).

In summary, the postimplantation parthenogenetic phenotype revealed extensive perturbations of both embryonic and extraembryonic cell lineages. In particular, the mesoderm was absent in the more severely affected forms (classes III and IV), and ectoderm was missing from the most severely affected form (class IV). Trophectoderm differentiation was abnormal: Polyploid cells replaced diploid trophoblast cells in the ectoplacental cone. The primitive endoderm lineage was also severely disturbed in its differentiation: Visceral extraembryonic endoderm was disorganized in moderately affected parthenogenotes (class II), and parietal endoderm cells replaced visceral endoderm in the more severely affected forms (most of class III and all of class IV). The most frequent phenotypes were the severely disturbed forms (classes III and IV), rather than the nearly normal or moderately affected forms (classes I and II), thus dispelling the stereotype that parthenogenetic embryos are generally capable of forming a normal, but small fetus (4).

Gene Expression in Parthenogenotes

These disturbances implied that in parthenogenotes, the regulation of growth and differentiation was abnormal. Since peptide growth factors and their receptors are known to play roles in these processes, we examined several growth-related genes that are known to be expressed in preimplantation mouse embryos. We used the sensitive *reverse transcription-polymerase chain reaction* (RT-PCR) technique that can detect as few as 10–100 mRNA transcripts in a single cell (29, 30). We used this approach to examine the transcription of members of the insulin family of growth factors and their receptors, including IGF-II, *IGF-II receptor* (IGF-II-R), and IGF-I-R. Because the IGF-II gene is transcribed only when paternally inherited (20, 21, 32, and Chapter 11, this volume), if imprinting is manifested during preimplantation stages, this gene would not be expressed in parthenogenotes.

We found that this is the case: While fertilized embryos transcribed IGF-II from early cleavage stages throughout the peri-implantation period, there was no detectable transcription of this gene in parthenogenotes. Embryos in the same experimental series were examined for the transcription of β-actin, which was expressed in the same stages in parthenogenotes as in fertilized embryos. Embryos of this series were then examined for the transcription of the IGF-I-R, which we considered to be the signaling receptor for IGF-II in the preimplantation mouse embryo, as it is in other systems. Although we could not rule out the possibility that IGF-II also acts through the insulin receptor, its lower affinity for IGF-II made it a less likely candidate than the IGF-I-R (31, reviewed in 33). We found that preimplantation parthenogenotes did not transcribe detectable amounts of the IGF-I-R gene, in contrast to fertilized embryos that expressed this gene from the 2-cell to early outgrowth stages. Finally, we examined parthenogenotes for expression of the IGF-II/mannose-6-phosphate receptor gene that strongly binds IGF-II and activates G-protein-mediated signaling in response to ligand binding (34, 35). In accord with the prediction from previous work on late fetal stages and adult mice, when this gene is transcribed only from the maternal allele (22), we found that parthenogenotes transcribed the IGF-II/mannose-6-phosphate receptor gene from the 2-cell stage to the blastocyst stage, as did normally fertilized embryos.

Discussion

Morphological Observations

The morphological work undertaken in these studies was based on the premise that detailed knowledge of the phenotype of developing parthenogenetic embryos will be useful in predicting the molecular basis of the effects of genomic imprinting during embryogenesis.

Our analysis of preimplantation development of parthenogenetically activated mouse oocytes showed that a substantial fraction (>50%) of diploid parthenogenotes are capable of normal preimplantation development to the blastocyst stage. This result confirms observations (9–11, 36–38) that diploid parthenogenotes differentiate normally into both of the early cell lineages that constitute the mouse blastocyst, the trophectoderm, and the ICM. Assuming that the abnormal phenotype of parthenogenotes is the consequence of differential expression of specific imprinted genes, then the maternal alleles that are inactivated by imprinting do not encode functions that are essential for preimplantation development. Alternatively, the imprinted alleles that have 2 active copies in parthenogenotes are not lethal at preimplantation stages. These observations of the gross morphology of preimplantation parthenogenotes do not rule out, however, subtle perturbations of cell proliferation or embryo physiology that may have profound cumulative effects as development proceeds. In view of the imprinting of specific growth factor and receptor genes, such effects would be predicted (see below).

The phenotype of postimplantation parthenogenetic development ranges from the nearly normal (class I) to the vestigial (class IV). In attempting to explain the origin of this array of phenotypes, we first considered the possibility that the parent inbred strains (C57BL/6J and CBA/J) might represent extremes of parthenogenetic development, such that the F_2 parthenogenotes were segregating for a set of imprinted genes that gave highly variable development. Because we found representatives of all classes in both inbred strains, we concluded that genetic segregation was not a sufficient explanation for the diversity we observed (25). Therefore, we have attempted to account for the range of parthenogenetic phenotypes by invoking a series of epigenetic thresholds that individual parthenogenotes pass with varying probabilities of success.

Accordingly, we envision class IV parthenogenotes as arising from embryos in which all ICM cells are allocated to the primitive endoderm, leading to their differentiation as parietal endoderm, consequent cessation of trophoblast proliferation, and a deficiency in all other ICM-derived cell lineages; however, we cannot rule out cell death as a contributing factor. Class III parthenogenotes may arise from embryos that retain a population of primitive ectoderm cells, but in which parietal endoderm predominates in the primitive endoderm lineage, and the absence of normal visceral endoderm cells precludes the elaboration of normal levels of mesoderm-inducing signals, thus leading to the absence of mesoderm. Class II parthenogenotes may arise from embryos that undergo mesoderm induction, but in which the quality or quantity of inducing substance(s) needed for axial differentiation is lacking. Class I parthenogenotes may arise from embryos that undergo mesoderm induction and axial differentiation, but lack nutritional support due to the lack of sufficient cell proliferation in the trophoblast lineage as the embryo proper grows. These scenarios could account for the range of developmental abnormalities seen in the parthenogenotes studied. More detailed scrutiny of the abnormalities in specific lineages, in particular the extraembryonic lineages, provides further insight into the molecular basis of the phenotypes observed.

The trophectoderm lineage of severely affected parthenogenotes (classes II, III, and IV) consisted exclusively of trophoblast giant cells; this indicates that there would be little, if any, further trophoblast cell proliferation in these parthenogenotes. The stem cell population for the trophectoderm lineage probably consists of the extraembryonic ectoderm layer plus the diploid cells of the ectoplacental cone, as proposed by Rossant and Tamura-Lis (39). The absence of a diploid trophectoderm cell population can be taken as a clear indication that the differentiation of this lineage is affected by parthenogenesis.

A similar phenotype emerges in the primitive endoderm lineage of the severely affected parthenogenotes. Parthenogenotes have either a morphologically abnormal visceral extraembryonic endoderm layer (class II) or they actually lack this layer (classes III and IV), whereas the parietal endoderm cell population is abundantly represented. Parietal endoderm cells can be considered the terminally differentiated cell type of the primitive endoderm lineage, for which the visceral extraembryonic endoderm is the stem cell at postimplantation stages (40, 41). In the more extreme case of class IV parthenogenotes, parietal endoderm cells are the only remnant of the ICM-derived lineages. Thus, as in the trophoblast lineage, the terminal differentiation of primitive endoderm cells may occur at the expense of maintaining a pluripotent stem cell population capable of further cell proliferation.

The phenotypes of the epiblast-derived cell lineages are also perturbed by parthenogenesis, but their abnormalities are less remarkable than those of the trophectoderm and primitive endoderm lineages. The embryonic ectoderm and epiblast remain proliferatively active even in moderately and severely affected (classes II and III) parthenogenotes, although they are disorganized. While extraembryonic mesoderm is absent from severely affected parthenogenotes (class III), it is abundant in the allantois and yolk sac of class I parthenogenotes, and mesodermal cells are present in class II parthenogenotes. Likewise, embryonic mesoderm is abundant in the heart structures of class I parthenogenotes and may also be represented in the mesoderm of class II parthenogenotes. Embryonic endoderm is clearly present in class I parthenogenotes, but cannot be distinguished from visceral embryonic endoderm in the endodermal layer of class II parthenogenotes. While these observations document the effects of parthenogenesis on embryonic lineages, they do not rule out the possibility that such effects arise indirectly from the effects on the extraembryonic lineages. Such a determination requires analysis of the development of parthenogenetic embryonic lineages in chimeras possessing normally fertilized extraembryonic lineages (42, and our unpublished observations).

In sum, the phenotype of the extraembryonic lineages in parthenogenetic embryos reveals a syndrome in which the balance between stem cell proliferation and terminal differentiation is perturbed. As a result, both trophectoderm and primitive endoderm lineages are characterized by preferential accumulation of the terminal cell type in the lineage (trophoblast giant cells and parietal endoderm cells, respectively) with depletion of the putative stem cell population. While this syndrome may have a complex etiology involving many genes, one possible mechanism involves diffusible growth and differentiation-regulating factors and their receptors, as well as their cell autonomously acting receptors and other elements in the signal transduction pathway(s). Such factors are known to alter the proliferative and differentiative behavior of cells (reviewed in 43). Moreover, several of the genes already identified as being imprinted are included in this category.

Molecular Observations

Our observations on the transcriptional state of the genes for IGF-II and its two receptors in parthenogenotes are the first indication that imprinting is already manifested at preimplantation stages. Our results indicate that the IGF-II gene is imprinted earlier than shown in the previous studies (20, 21), which were limited to 9.5 days of gestation and later stages. Because the maternal allele of IGF-II failed to be transcribed from the earliest stage at which transcription is detected in normally fertilized embryos (2-cell stage), there is no time during preimplantation development when the maternal allele is expressed. Our observations that the IGF-I-R gene is transcribed weakly or not at all in parthenogenotes at any preimplantation stage suggest that it is also imprinted and expressed only from the paternal allele. Verification of this observation will require analysis of androgenetic embryos, as well as parthenogenotes at later stages of gestation.

This pattern of imprinting differs from the expression of the maternal and paternal homologues of the X-chromosome. Both X-chromosomes are expressed during preimplantation development of mouse embryos, before preferential inactivation of the paternal X-chromosome occurs in the extraembryonic lineages (reviewed in 44). In this respect, X-chromosome inactivation cannot be considered an appropriate model for the expression of imprinted autosomal genes. Observations on the transcription of other autosomal genes provide further evidence that autosomal imprinting is distinct from X-chromosome imprinting. The imprint status of autosomal genes can differ from closely linked, or even adjacent genes. The H19 gene is expressed when inherited maternally, although it is located on mouse chromosome 7 near the paternally expressed IGF-II gene (45). Similarly, the maternally expressed IGF-II/mannose-6-phosphate receptor gene located on mouse chromosome 17 is flanked by genes that are not imprinted (22). Therefore, closely adjacent autosomal genes can be oppositely imprinted or not imprinted, unlike most X-linked genes that are coordinately affected by X-inactivation (44). When X-chromosome inactivation occurs, the vast majority (but not all) of the genes are inactivated. This phenomenon appears to be under the regulation of the X-chromosome inactivation center (Xce) that has effects that spread throughout most of the chromosome (44).

The identification of specific growth factor and receptor genes as being imprinted has provided a compelling insight into the evolutionary origin of genomic imprinting. These observations led Haig and Graham (46) to the proposal that imprinting arose in mammals because of the selective advantage conferred by paternal mutations leading to rapid embryonic growth or by maternal mutations attenuating such growth. Their assumptions are (i) that mammals evolved with mating strategies that emphasized polyandry, and (ii) that larger size of offspring conferred a selective advantage to fathers, while larger litter size conferred a selective advantage to mothers.

Given these assumptions, the expression of the paternal allele of the growth stimulating factor IGF-II in the mouse yolk sac and placenta (20, 21) supports the proposal (17) that differential expression of growth-related genes arose during the evolution of species in which there is asymmetric provision of resources by the two parents. Similarly, expression of the maternal allele of the IGF-II/mannose-6-phosphate receptor gene also supports their proposal, assuming that this protein acts as a sink for IGF-II, rather than a signaling receptor (46). Haig and Graham's prediction of paternal expression of the IGF-I-R is further supported by our observations of gene expression in preimplantation parthenogenetic mouse embryos (31). Moreover, additional predictions based on the Haig and Graham hypothesis are testable. Because marsupials are viviparous mammals, they should also exhibit effects of imprinting; amphibia, birds, and reptiles, by contrast, would not be expected to show effects of imprinting because their eggs are provisioned before fertilization, and paternal gene expression cannot alter prenatal resource allocation. The examples of viable parthenogenesis in nonmammalian vertebrate classes (8, 48) and the phenomenon of preferential paternal X-chromosome inactivation in marsupials (49) support these predictions. It would be interesting to determine whether expression of genes that have been identified as imprinted in mammals is also dependent on parental origin in nonmammalian species.

What are the identities of other endogenous imprinted genes, and how many genes are imprinted? An extensive genetic analysis of parent-specific chromosome duplication/deficiency phenomena has been carried out using reciprocal and Robertsonian translocations and has revealed regions of the mouse genome that harbor developmentally important imprinted genes (50–52, and reviewed in 54). Five autosomes were identified that showed developmental perturbations (ranging from early gestation to perinatal death) as a result of maternal duplication/paternal deficiency (chromosomes 2, 6, 7, 8, and 11); and 4 autosomes were similarly identified in paternal duplication/maternal deficiency syndromes (chromosomes 2, 7, 11, and 17) (Fig. 16.1). That is, individuals inheriting both homologues of these autosomes from the specified parent and none from the other parent developed abnormally. Because the available translocation stocks encompassed large autosomal regions, there is presently no detailed mapping of the boundaries between imprinted and nonimprinted autosomal regions. At minimum, 8–12 imprinted genes could account for the observed mapping data; at the other extreme, greater than 10% of the genome could be affected by imprinting, based on the extent of the translocations used to identify noncomplementing phenomena. Numerous genes involved in growth and morphogenesis have been mapped to the noncomplementing regions of the autosomes (53) and can be the objects of further studies on imprinting of endogenous genes.

Conclusions

Our morphological observations implicate growth- and differentiation-regulating factors in the developmental perturbations of parthenogenesis. In addition to the IGF-II and IGF-II/mannose-6-phosphate receptor

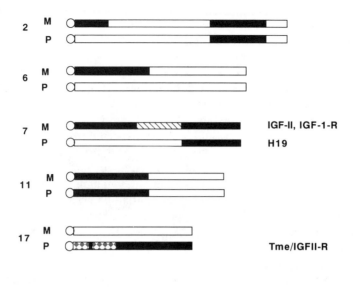

FIGURE 16.1. Regions of mouse autosomes showing defective complementation when both segments are derived from the same parent (M = maternal; P = paternal). Defective phenotypes include prenatal and neonatal deaths. Other autosomes are designated as being normal, having unknown regions, or having regions leading to differential recovery of offspring as a result of maternal or paternal duplication with corresponding paternal/maternal deficiency. Genes with differential expression of maternal and paternal alleles are indicated to the right of their chromosomal location. Adapted with permission from Cattanach and Beechey (54).

genes, we have identified the IGF-I-R gene as a candidate for imprinted endogenous genes. While the abnormal expression of these genes does not explain the abnormal phenotypes of parthenogenetic and androgenetic embryos, the hypothesis of Haig and Graham (46), plus chromosomal mapping data for growth-related genes, may lead to identification of other imprinted genes (53). The analysis of preimplantation parthenogenotes and androgenotes should be valuable in any effort to determine the imprinting status of any such genes.

Acknowledgments. The work performed in the authors' laboratories was supported by NIH Grant P01-HD-26732 and by USDOE/OHER Contract No. DE-AC03-76-SF-01012.

References

1. Surani MA, Barton SC, Norris ML. Development of reconstituted mouse eggs suggests imprinting of the genome during gametogenesis. Nature (London) 1984;308:548–50.
2. McGrath J, Solter D. Nuclear transplantation in the mouse embryo by microsurgery and cell fusion. Science 1983;220:1300–3.
3. Surani MA, Barton SC, Norris ML. Nuclear transplantation in the mouse: heritable differences between parental genomes after activation of the embryonic genome. Cell 1986;45:127–36.
4. Surani MA. Evidences and consequences of differences between maternal and paternal genomes during embryogenesis in the mouse. In: Rossant J, Pedersen RA, eds. Experimental approaches to mammalian embryonic development. Cambridge, UK: Cambridge University Press, 1986:401–35.
5. Surani MA. Mechanism and consequences of genomic imprinting and genetic disorders. In: Edwards RG, ed. Establishing a successful human pregnancy. New York: Raven Press, 1990:171–84.
6. Solter D. Differential imprinting and expression of maternal and paternal genomes. Annu Rev Genet 1988;22:127–46.
7. McGrath J, Solter D. Completion of mouse embryogenesis requires both the maternal and paternal genomes. Cell 1984;37:179–83.
8. Beatty RA. Parthenogenesis in vertebrates. In: Metz CB, Monroy A, eds. Fertilization; vol. I. New York: Academic Press, 1967:413–41.
9. Clarke HJ, Varmuza S, Prideaux VR, Rossant J. The developmental potential of parthenogenetically derived cells in chimeric mouse embryos: implications for action of imprinted genes. Development 1988;104:175–82.
10. Thomson JA, Solter D. Chimeras between parthenogenetic or androgenetic blastomeres and normal embryos: allocation to the inner cell mass and trophectoderm. Dev Biol 1989;131:580–3.
11. Thomson JA, Solter D. The developmental fate of androgenetic, parthenogenetic, and gynogenetic cells in chimeric gastrulating mouse embryos. Genes Dev 1988;2:1344–51.
12. Fundele R, Norris ML, Barton SC, Reik W, Surani MA. Systematic elimination of parthenogenetic cells in mouse chimeras. Development 1989;106:29–35.
13. Nagy A, Sass M, Markkula M. Systematic non-uniform distribution of parthenogenetic cells in adult mouse chimeras. Development 1989;106:321–4.
14. Paldi A, Nagy A, Markkula M, Barna I, Dezso L. Postnatal development of parthenogenetic ⟨−⟩ fertilized mouse aggregation chimeras. Development 1989:115–8.

15. Anderegg C, Markert CL. Successful rescue of microsurgically produced homozygous diploid uniparental mouse embryos via production of aggregation chimeras. Proc Natl Acad Sci USA;83:6509–13.
16. Mann JR, Gadi I, Harbison ML, Abbondanzo SJ, Stewart CL. Androgenetic mouse embryonic stem cells are pluripotent and cause skeletal defects in chimeras: implications for genetic imprinting. Cell 1990;62:251–60.
17. Mann JR, Stewart CL. Development to term of mouse androgenetic aggregation chimeras. Development 1991;113:1325–33.
18. Barton SC, Ferguson-Smith AC, Fundele R, Surani MA. Influence of paternally imprinted genes on development. Development 1991;113:679–88.
19. Surani MA, Barton SC, Howlett SK, Norris ML. Influence of chromosomal determinants on development of androgenetic and parthenogenetic cells. Development 1988;103:171–8.
20. DeChiara TM, Efstratiadis A, Robertson EJ. A growth-deficiency phenotype in heterozygous mice carrying an insulin-like growth factor II gene disrupted by targeting. Nature (London) 1990;345:78–80.
21. DeChiara TM, Robertson EJ, Efstratiadis A. Parental imprinting of the mouse insulin-like growth factor II gene. Cell 1991;64:849–59.
22. Barlow DP, Stöger R, Herrmann BG, Saito K, Schweifer N. The mouse insulin-like growth factor type-2 receptor is imprinted and closely linked to the Tme locus. Nature (London) 1991;349:84–7.
23. Cruz YP, Pedersen RA. Origin of embryonic and extraembryonic cell lineages in mammalian embryos. In: Pedersen RA, McLaren A, First NL, eds. Animal applications of research in mammalian development. Cold Spring Harbor, NY: Cold Spring Harbor Laboratory Press, 1991:147–204.
24. Hall JG. Genomic imprinting: review and relevance to human diseases. J Hum Genet 1990;46:857–73.
25. Sturm KS, Flannery ML, Pedersen RA. Abnormal development of embryonic and extraembryonic cell lineages in parthenogenesis mouse embryos. Development (submitted).
26. Cuthbertson KSR. Rapid communication: parthenogenetic activation of mouse oocytes in vitro with ethanol and benzyl alcohol. J Exp Zool 1983;226:311–4.
27. Surani MA, Barton SC, Norris ML. Experimental reconstruction of mouse eggs and embryos: an analysis of mammalian development. Biol Reprod 1987;36:1–16.
28. Hogan BLM, Cooper AR, Kurkinen M. Incorporation into Reichert's membrane of laminin-like extracellular proteins synthesized by parietal endoderm cells of the mouse embryo. Dev Biol 1980;80:289–300.
29. Rappolee DA, Brenner CA, Schultz R, Mark D, Werb Z. Developmental expression of PDGF, TGF-α, and TGF-β genes in preimplantation mouse embryos. Science 1988;241:1823–5.
30. Rappolee DA, Wang A, Mark D, Werb Z. Novel method for studying mRNA phenotypes in single or small numbers of cells. J Cell Biochem 1989;39:1–11.
31. Rappolee DA, Sturm KS, Behrendtsen O, Schultz GA, Pedersen RA, Werb Z. Insulin-like growth factor II, acting through the IGF-I receptor, forms an endogenous growth circuit regulated by imprinting in early mouse embryos. Genes Dev 1992;6:939–52.

32. Ferguson-Smith AC, Cattanach BM, Barton SC, Beechey CV, Surani MA. Embryological and molecular investigations of parental imprinting on mouse chromosome. Nature (London) 1991;351:667–70.
33. Czech MP. Signal transmission by the insulin-like growth factors. Cell 1989; 59:235–8.
34. Okamoto T, Katada T, Murayama Y, Ui M, Ogata E, Nishimoto I. A simple structure encodes G protein-activating function of the IGF-II/mannose 6-phosphate receptor. Cell 1990;62:709–17.
35. Okamoto T, Nishimoto I, Murayama Y, Ohkuni Y, Ogata E. Insulin-like growth factor-II/mannose 6-phosphate receptor is incapable of activating GTP-binding proteins in response to mannose 6-phosphate, but capable in response to insulin-like growth factor-II. Biochem Biophys Res Commun 1990;168:1201–10.
36. Kaufman MH. Early mammalian development: parthenogenetic studies. In: Barlow PW, Green PB, Wylie CC, eds. Developmental and cell biology series. Cambridge, UK: Cambridge University Press, 1983:84–110.
37. Kaufmann MH, Barton SC, Surani MA. Normal postimplantation development of mouse parthenogenetic embryos to the forelimb bud stage. Nature (London) 1977;265:53–5.
38. McGrath J, Solter D. Nucleocytoplasmic interactions in the mouse embryo. J Embryol Exp Morphol 1986;97:277–89.
39. Rossant J, Tamura-Lis W. Effect of culture conditions on diploid to giant-cell transformation in postimplantation mouse trophoblast. J Embryol Exp Morphol 1981;62:217–27.
40. Gardner RL. Investigation of cell lineage and differentiation in the extra-embryonic endoderm of the mouse embryo. J Embryol Exp Morphol 1982; 68:175–98.
41. Hogan BLM, Tilly R. Cell interactions and endoderm differentiation in cultured mouse embryos. J Embryol Exp Morphol 1981;62:379–94.
42. Gardner RL, Barton SC, Surani MAH. Use of triple tissue blastocyst reconstitution to study the development of diploid parthenogenetic primitive ectoderm in combination with fertilization-derived trophectoderm and primitive endoderm. Genet Res 1990;56:209–22.
43. Nilsen-Hamilton M, ed. Growth factors and development: current topics in developmental biology; vol 24. San Diego: Academic Press, 1990.
44. Grant SG, Chapman VM. Mechanisms of X-chromosome regulation. Annu Rev Genet 1988;22:199–233.
45. Bartolomei MS, Zemel S, Tilghman SM. Parental imprinting of the mouse H19 gene. Nature (London) 1991;351:153–5.
46. Haig D, Graham C. Genomic imprinting and the strange case of the insulin-like growth factor II receptor. Cell 1991;64:1045–6.
47. Haig D, Westoby M. Parent-specific gene expression and the triploid endosperm. Am Nat 1988;134:147–55.
48. Bell G. The masterpiece of nature: the evolution and genetics of sexuality. Berkeley: University of California Press, 1982:324–31.
49. Kaslow DC, Migeon BR. DNA methylation stabilizes X chromosome inactivation in eutherians but not in marsupials: evidence for multistep maintenance of mammalian X dosage compensation. Proc Natl Acad Sci USA 1987;84:6210–4.

50. Searle AG, Beechey CV. Complementation studies with mouse translocations. Cytogenet Cell Genet 1978;20:282–303.
51. Searle AG, Beechey CV. Noncomplementation phenomena and their bearing on nondisjunctional effects. In: Dellarco VL, Vojtek PE, Hollaender A, eds. Aneuploidy, aetology and mechanisms. New York: Plenum Press, 1985: 363–76.
52. Cattanach BM. Parental origin effects in mice. J Embryol Exp Morphol 1986;97:137–50.
53. Searle AG, Peters J, Lyon MF, et al. Chromosome maps of man and mouse, IV. Ann Hum Genet 1989;53:89–140.
54. Cattanach BM, Beechey CV. Autosomal and X-chromosome imprinting. Development 1990 (Suppl.);63–72.

Part V

Embryo-Maternal Interactions

17

Uterine Secretory Activity and Embryo Development

R. MICHAEL ROBERTS, WILLIAM E. TROUT,
NAGAPPAN MATHIALAGAN, MELODY STALLINGS-MANN,
AND PING LING

The uterus is an exocrine organ active in secretion and responsive to steroid hormones. In mature nonpregnant females, the uterine endometrium undergoes cyclical changes of secretory activity and cellular regeneration and loss in response to steroid and possibly other hormones, while during pregnancy it usually comes under the dominant long-term control of progesterone. It is clear that in all species studied, steroid hormones strongly influence the pattern of endometrial protein synthesis and secretion. Dozens of descriptive studies have been published, and there has been considerable speculation (but little proof) that these proteins are somehow involved in events critical to the reproductive process; for example, in promoting conceptus growth, providing nourishment, controlling implantation, and dampening maternal immune responses.

Uterine Secretory Proteins

The failure to understand function is best illustrated with *uteroglobin* (originally called blastokinin), a progesterone-induced secretory protein of the rabbit uterus first described by Beier (1) and Krishnan and Daniel (2) over 20 years ago. Since that time, uteroglobin has been well characterized biochemically (3), crystallized, its cDNA and its gene cloned, the promoter regions of the gene analyzed in great detail (4), and transgenic mice expressing uteroglobin created (5). It has become clear that the protein is synthesized in the male (6) as well as in the female reproductive tract and produced in the lung of both sexes (7). Its synthesis can be promoted by estrogen, testosterone, and glucocorticoids in addition to progesterone, depending upon the cellular environment in which the gene is transcribed (4). Despite all this information, the function of uteroglobin, which is the dominant secretory product of the rabbit uterus for a few days in early pregnancy, remains elusive. Originally touted as a factor necessary for blastocyst development (2), uteroglobin has been assigned roles as a progesterone carrier protein (7), as a factor that masks embryo antigenicity by acting as a substrate for uterine transglutaminase (8), and, most recently, as an anti-inflammatory agent by inhibiting phospholipase A_2 (4). Similar uncertainty exists about several other well-studied uterine proteins that have been recognized as major steroid-induced components synthesized and secreted by endometrium, but whose precise functions are not yet delineated. Among these are lactoferrin of the mouse (9) and complement component C-3 of the rat (10), both of which are induced by estrogen. Other well-studied progesterone-modulated components are mouse colony stimulating factor 1 (11), a binding protein for insulin-like growth factor from the human deciduum (12), a β-lactoglobulin-like protein also from the human (13), cathepsin L from the cat (14), and a member of the serine protease inhibitors (serpins) from the sheep (15). What has been particularly disquieting about these studies is the general lack of uniformity in patterns of secretory activity across species, so that no general concepts have emerged. The proteins may be good models for studying steroid control of gene expression, but their study has not, in general, been insightful to biologists interested in embryonic development. Perhaps when some of the minor components have been defined and more is known about the concentrations of major ions and micronutrients bathing the conceptus during pregnancy, the role of uterine secretions in maintaining early embryo will become better understood.

Is Uterine Secretory Activity Necessary for Embryonic Development?

The facts that ectopic pregnancies can occur in the human (but not in most other species) and that embryos from many species, once they have passed an early developmental block, can thrive successfully in culture and develop to the blastocyst stage and beyond in media identical to or only minimally different from those used to culture fibroblasts suggest that uterine secretions might not constitute a unique embryotrophic milieu. The presence of development blocks (16) or retarded in vitro development can often, it appears, be overcome by simple changes in medium composition (17), provision of growth factors (18), or coculture with epithelial cells (19). Therefore, it could be argued that the reproductive tract provides little more than a generally permissive environment in which the potentially invasive fetal allograph can be confined, nurtured optimally, and physically protected. There may be multiple ways in which this state can be achieved, and species have evolved different strategies to achieve it. In addition, the time at which the blastocyst implants and the type of placentation ultimately achieved must strongly influence the extent to which the embryos of different species rely upon uterine secretory activity for their continuing needs. In those species in which implantation occurs early and where the absorptive face of the trophoblast soon encounters maternal blood, a reliance on uterine secretory activity from the surface and glandular epithelium may be short lived. Here, the long-term progesterone dominance of the uterus may be most important in providing a period of constancy in which the cyclic remodeling of the endometrium and muscular contractions of the uterus are temporarily suspended.

However, in those species in which the trophoblast does not invade the endometrium or even erode the uterine epithelium and where there is extensive development of the embryo and its associated membranes before placentation begins, uterine secretions seem much more likely to play a more extended role in the maintenance of pregnancy (20). Thus, the pig, with its diffuse type of epitheliochorial placentation (Fig. 17.1),

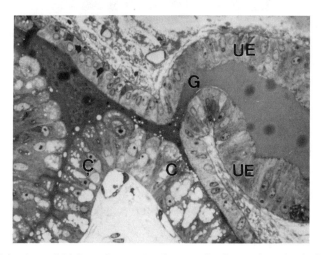

FIGURE 17.1. A semithick section across the mouth of a uterine gland of the pig at day 110 of pregnancy. The cells of the chorion (bottom, C) that develop opposite such gland (G) openings form specialized domelike structures called areolae. The domed surfaces of each areola consist of complex folds of cells. The chorionic cells themselves are full of large endocytotic vacuoles and appear to transport (transcytose) uterine secretory material to the capillaries beneath their basal surfaces (38). The uterine epithelial (UE) gland cells, particularly those that are more basally located and not shown on this section, are very active in secretion and release of their product into the lumen of the gland. In the interareolar regions, the cells of the uterine epithelium and the chorion (outer trophoblast layer) make close contact via interdigitation of microvilli. However, the uterine epithelium is never eroded throughout pregnancy (1200×). Micrograph provided by Professor Armin Friess, Universitat Bern.

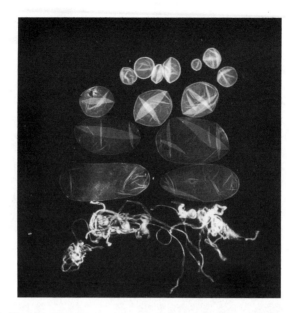

FIGURE 17.2. Pig blastocysts at representative stages of development flushed from gilts between days 10 and 13 of pregnancy. The smallest spherical blastocysts (top) average 3–5 mm. Before elongation, the blastocysts enlarge to 8–10 mm in diameter (row 2), then become tubular (~2 cm in length, rows 3 and 4). Finally, they become filamentous (row 5). The transition from tubular to threadlike form occurs within 6 h (23), and the whole development sequence occurs within about 48 h. Indeed, spherical, tubular, and filamentous forms can often be recovered from the same animal at day 12 of pregnancy. Reprinted with permission from Roberts and Bazer (31).

provides an excellent model for studying the role of uterine secretions in supporting the conceptus for a prolonged period, though it would be ingenuous to propose that the pig model could be extended to all other mammalian species.

Placentation, Early Conceptus Development, and the Requirement for Uterine Histotroph in the Pig

Pig conceptuses undergo remarkable changes in morphology (Fig. 17.2) before they become firmly attached to the uterine wall (21–23). By day 10 the hatched blastocysts are still spherical, ranging from 3 to 8 mm or so in diameter. Between days 10 and 12, the majority of these blastocysts undergo a transition first to tubular (10–50 mm long) and then to filamentous forms that within 48 h can attain lengths of up to 1 m. Not all conceptuses develop at the same rate, however, and a range of forms can often be found in a single uterus at day 12 (Fig. 17.2). The filamentous blastocysts become arranged end to end within each uterine horn, occupying no more than 10–20 cm of uterine length by following a serpentine course along the villous folds of the endometrium. Firm adhesion to the uterine epithelium begins around day 13 (24). Since embryonic losses are high during this period (around 25%), it is suspected that less-developed blastocysts cannot establish themselves in this pattern and are lost (25).

Pig conceptuses first begin to synthesize appreciable amounts of estrogen during elongation (23, 26). Synthesis peaks about the time the blastocysts become filamentous and subsequently declines by day 14. The onset of blastocyst estrogen synthesis is also accompanied by a synchronous release of secretory material by exocytosis from the uterine glandular epithelial cells into the lumen of the gland (23). This dumping of secretions, which can be mimicked in the nonpregnant animal by a single injection of estradiol (27), results in a marked change in both the amount and qualitative makeup of the uterine fluids (23). In particular, a group of proteins can be detected whose synthesis is regarded as being primarily under the control of progesterone (see next section). Before day 11 these proteins are not detectable, and the amount of protein in uterine flushings is low. It is assumed that this sudden change in the uterine milieu of proteins, ions, and bioactive factors (such as prostaglandins) is important to the nurturing of the conceptuses as they go through this remarkable period of growth and cellular reorganization.

Secretion of proteins is not confined to the early period around day 12 since the main components of the secretions remain predominantly under the control of progesterone. Therefore, they continue to be synthesized throughout gestation, although other hormones, and particularly estrogens whose serum concentrations change as pregnancy progresses, probably continue to play a modulatory role (28). It should also be added that the rates at which these proteins are synthesized are extraordinarily high. Uteroferrin production, for example, probably exceeds 2 g/day at midpregnancy (29) and most likely reflects the function of this protein in supplying an essential nutrient (iron) to the fetuses (30). Because of the diffuse epitheliochorial placentation of the pig, essential macromolecules or nutrients (such as iron) that are not readily diffusible across several layers of cells are probably supplied to conceptuses in the form of such uterine secretions. These macromolecules are taken up by endocytosis through the trophectoderm in the early stages of development and by structures called areolae after about day 30 (Fig. 17.1). The areolae consist of specialized absorptive epithelial cells that form opposite the mouths of each uterine gland and constitute the main gateway for uptake of iron and, most likely, other macromolecular-borne nutrients to the fetus (31).

Progesterone-Responsive Proteins of Porcine Histotroph

Uterine secretions from the pig can be readily obtained by flushing each uterine horn with saline (32). The amount of recoverable protein is low (<10 mg) during the first 8 days after estrus, and this phase of pregnancy has been poorly studied. Amounts increase as progesterone begins to dominate the endometrium, and by day 15 of both the cycle and pregnancy, up to 50 mg of protein can be recovered per horn (33). The flushings are colored purple because of the presence of uteroferrin. The amount of protein in uterine flushings does not differ significantly between pregnant and nonpregnant animals until about day 11 when the estrogen-triggered exocytosis of the contents of secretory vesicles, as discussed earlier, begins in pregnant animals. Release of secretions into the nonpregnant uterus occurs more gradually, but the total recovery over the next 4 days is not much different than in pregnancy (23). However, the secretory activity of the pregnant endometrium is maintained beyond day 15, whereas output of protein from the endometrium of nonpregnant gilts falls to low levels as the end of the cycle approaches and the corpora lutea regress (34).

Because of the gram quantities of protein that can be collected from pseudopregnant or unilaterally pregnant gilts in midgestation, uterine flushings have provided the most usual source for purification of the individual progesterone-induced components of porcine histotroph (35). However, the composition of these flushings does not appear to differ qualitatively from those obtained in early pregnancy at the time the conceptuses elongate (23, 36), and four major groups of progesterone-responsive proteins have been studied in detail (Fig. 17.3). Each of the proteins in these groups is synthesized locally by the endometrium, and the majority have been purified, partially sequenced, and, in most instances, their representative cDNA cloned. Interestingly, these same proteins are represented in allantoic fluid of the fetus, where they are believed to accumulate temporarily after transplacental transport and clearance from fetal serum (37, 38).

Group 1

This basic glycoprotein (M_r 35,000) has become known as uteroferrin. It is purple due to the presence of a bi-iron center and has potent acid phosphatase activity (30). Probably because it has such unusual properties, uteroferrin has been studied intensively as a prototype for what now appears to be an extensive class of iron-containing phosphohydrolases (30, 39). Its amino acid sequence is now known (40) Though secreted in very large quantities in the uterus, it is identical to a lysosomal acid phosphatase that we have cloned from porcine spleen (Ling, Roberts, unpublished results). Even though uteroferrin has an undoubted role in

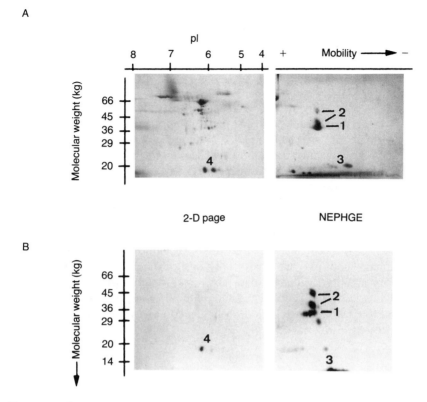

FIGURE 17.3. *A:* Two-dimensional electrophoresis of uterine flushings from a pseudopregnant gilt analyzed by two-dimensional electrophoresis and stained with Coomassie blue. The separation on the left is by standard two-dimensional electrophoresis. That on the right was performed by nonequilibrium pH gradient electrophoresis (NEPHGE) that allowed analysis of more basic components. The numbers refer to groups of polypeptides discussed in more detail in the text. *B:* Two-dimensional electrophoresis of the radioactive polypeptides secreted into the medium during culture of endometrial explants from a day 60 pseudopregnant gilt for 24 h in the presence of L-[^{35}S]methionine, as described in reference 36. Separations were performed as in *A*, and radioactive proteins synthesized by the endometrium were detected by fluorography. Groups of polypeptides 1–4 are again marked.

iron transport to the pig fetus (reviewed in 30, 31), it has been difficult to assess whether it constitutes the only transport mechanism at play. Its iron is rapidly utilized for fetal hemoglobin synthesis, and its rate of synthesis by the uterus until at least midpregnancy could meet the iron needs of the developing young. One other interesting role for uteroferrin—that of a hematopoietic growth factor—has also been proposed (41). Purple phosphatases resembling uteroferrin are also present in uterine secretions from at least two other species: the horse (42) and cattle (Ketcham, Clark, Bazer, Roberts, unpublished results).

Group 2

This trio of glycoproteins arises from differential glycosylation and proteolytic processing of a 45,000-M_r precursor polypeptide (43). They form a relatively weak heterodimeric association with some of the uteroferrin in uterine secretions, the significance of which is presently unclear (44). Their inferred amino acid sequences (45) and gene organization (Mathialagan, Roberts, unpublished results) have shown that they belong to the serpin superfamily of proteins whose members include a large number of proteinase inhibitors, at least 2 hormone carrier proteins, and proteins of unknown function, such as egg white ovalbumin (45 and references therein). It is also of interest that the uterine serpin, like uteroferrin, carries the lysosomal recognition marker—mannose 6-phosphate—on its carbohydrate chains (43). As yet, no proteinase inhibitory activity, acid hydrolase activities, or any other biological role for this interesting molecule has been defined. A serpin similar to the one from the pig has been identified in the sheep (15), while another is likely to be present in the cow (46).

Group 3

These low-molecular weight basic proteins include lysozyme and a family of serine protease inhibitors of the Kunitz class. The total activity of the former increases markedly in response to progesterone administration to gilts, but its specific activity—that is, activity relative to the total amount of protein secreted—does not change markedly as progesterone treatment is extended significantly (47, 48). The role of lysozyme is assumed to be antibacterial. Recently, we have purified and obtained an NH_2 terminal amino acid sequence for lysozyme from uterine secretions (Echetebu, Roberts, unpublished results) and cloned and sequenced a cDNA (Nevils, Trout, Roberts, also unpublished). It does not appear to constitute a uterine-specific form.

The Kunitz inhibitors have properties quite similar to bovine pancreatic trypsin inhibitor (aprotinin) (49) to which they have considerable sequence similarity at their NH_2 termini (50). However, they are considerably larger than aprotinin ($M_r \sim$ 14,000 vs. 6000) and consist of a group of at least 4 distinct isoforms. Because they are potent inhibitors of plasmin, it has been hypothesized that these serine proteinase inhibitors control the potential invasiveness of the pig trophoblast. The trophoblast actively releases plasminogen activator as it begins to elongate, and plasminogen, the zymogen form of plasmin, is present in uterine flushings at that time (51). Thus, a potentially damaging cascade of proteolytic activity could result. Even though the pig conceptus is noninvasive within the uterus, trophoblast tissue transplanted to ectopic sites outside the uterus exhibits invasive properties (51). Therefore, the superficial type of implantation seen in the pig may in part be the result of inhibitor production by the endometrium. Application of protease inhibitors, for example, prevents implantation in mice (52) and rabbits (53). Protease inhibitors may also be important in protecting the conceptus from the potentially damaging effects of an inflammatory response by mast cells and other components of the maternal immune system. Again, however, these speculations remain to be proved.

Expression of the Kunitz inhibitors has not been examined in detail, although their synthesis is induced by progesterone in ovariectomized gilts (49). Further studies will demand a cDNA probe, but molecular cloning and sequencing of such cDNA, though underway, is not yet completed.

Evidence for the uterine expression of a low-molecular weight neutral serine proteinase inhibitor, closely related or identical to antileukoproteinase and, therefore, potentially capable of inhibiting cathepsin G and elastase, has recently been presented (54). It appears to be responsive to steroid hormones, particularly estrogen, but may be concentrated in the myometrium rather than the endometrium.

Group 4

The components of group 4 are *retinol-binding proteins* (RBP). The presence of polypeptides that bound retinol in progesterone-treated gilts was first reported by Adams et al. (55); they were subsequently purified from uterine secretions of pseudopregnant gilts (56). At least 4 isoelectric forms of RBP could be identified by high-performance ion-exchange chromatography. This procedure also provided a means of separating the vitamin-loaded polypeptide from the retinol-free form of each RBP. Amino terminal sequencing of different isoforms after they had been purified by polyacrylamide gel electrophoresis and transferred to PVDF membranes has indicated that each is closely related to the retinol-binding protein of serum (Stallings-Mann, Trout, Roberts, unpublished results). The basis of the diversity in isoelectric points remains unclear, but may be due to charge modifications that occur after the protein has been secreted. Several cDNA for the uterine RBP have recently been cloned and sequenced and the primary structure of the protein inferred (57). These data confirm that the uterine RBP is probably identical to pig serum RBP.

This cDNA probe has been used to detect RBP mRNA on Northern blots (57). Messenger RNA was first detected around day 12 of pregnancy and expressed throughout the rest of gestation. Its expression in endometrium was completely dependent upon progesterone, thus confirming the results of Adams et al. (55). It has also become evident that the presence of conceptuses on or about day 12 markedly increases RBP expression in endometrium, possibly due to production of estrogen (57). Injections of estrogen on days 10 and 11 of the estrous cycle, for example, raised RBP mRNA levels on day 12 approximately 12-fold compared to controls, so that values became comparable to those noted in pregnant animals. Similarly, the content of RBP and retinol in uterine flushings from pregnant gilts carrying filamentous embryos was at least 7-fold higher than from gilts whose conceptuses had not yet begun to elongate. Thus, as conceptuses elongate, they directly control the amount of retinol available to them through their own production of estrogen. Presumably, very high amounts of retinol are required at about this stage of development.

Brief and Chew (58) demonstrated that injections of supplemental vitamin A in the form of retinol and β-carotene increased litter size and decreased embryonic mortality in pigs. Coffey and Britt (59) also reported that injections of β-carotene increased litter size when given to multiparous sows approximately 5 days prior to breeding. Because the majority of embryonic losses are known to occur within the first 2–3 weeks of pregnancy (60) and since supplemental injections of estradiol on days 12–13 of pregnancy also increase embryo survival (61), we suggest that the two effects are linked. Either injection of vitamin A or estradiol might be expected to increase the amount of retinol available to the conceptuses. Systemic estradiol, in particular, might be most effective in this regard by intervening for embryos that are lagging in development and possibly in danger of being lost. It should be stressed, however, that proper timing of such estradiol injections is crucial since their administration on days 9 and 10 leads to embryonic death within 3–4 days (62, 63).

Conclusions

In this review concerning the role of uterine secretory proteins in embryonic development, attention has been focused on the major components of porcine histotroph that are synthesized in response to progesterone. The majority of these proteins have been well characterized, and in most instances a biological activity has been recognized. Whether such activity— for example, the acid phosphatase activity of uteroferrin—is in any way relevant to the role of that component in pregnancy is not always clear. Nevertheless, it seems evident that at least some of the secretory proteins are involved in supplying such necessary nutrients as iron and fat-soluble vitamins to the conceptus. Others, such as the protease inhibitors, may play a part in controlling the growth and development of the conceptus and the uterus itself. Uterine secretions appear to contain a range of growth factors, such as the insulin-like growth factors and acidic and basic fibroblast growth factors, even though these are not among the most prominent components present (64). Other functions of uterine proteins might include antibacterial, antiviral, or immunoprotective activities. Lysozyme clearly falls into the first of these categories, and there have been several reports that components of uterine secretions are "immunosuppressive" (65, for example) in the sense that they inhibit incorporation of [^{3}H]thymidine into mitogen-stimulated lymphocytes.

Acknowledgments. The authors acknowledge Gail Foristal for editorial assistance. This is Article Number 11,473 of the Missouri Agricultural Experiment Station and was supported by Grants HD-21980 from NIH and 89-37240-4586 from the U.S. Department of Agriculture.

References

1. Beier HM. Uteroglobin: a hormone-sensitive endometrial protein involved in blastocyst development. Biochim Biophys Acta 1968;160:289-91.
2. Krishnan RA, Daniel JC. Blastokinin: inducer and regulator of blastocyst development in the rabbit uterus. Science 1967;158:490.
3. Beato M. Hormonal control of uteroglobin synthesis. In: Johnson MH, ed. Development in mammals; vol 2. Amsterdam: Elsevier/North Holland Biomedical Press, 1977;173-98.
4. Miele L, Cordella-Miele E, Mukherjee AB. Uteroglobin: structure, molecular biology, and new perspectives on its function as a phospholipase A2 inhibitor. Endocr Rev 1987;8:474-90.
5. DeMayo FJ, Damak S, Hansen TN, Bullock DW. Expression and regulation of the rabbit uteroglobin gene in transgenic mice. Mol Endocrinol 1991;5:311-8.
6. López de Haro MS, Alvarez L, Nieto A. Testosterone induces the expression of the uteroglobin gene in rabbit epididymis. Biochem J 1988;250:647-51.
7. Beato M, Beier HM. Characteristics of the purified uteroglobin-like protein from rabbit lung. J Reprod Fertil 1978;53:305-14.
8. Mukherjee AB, Ulane RE, Agrawal AK. Role of uteroglobin and transglutaminase in masking the antigenicity of implanting rabbit embryos. Am J Reprod Immunol 1982;2:135-41.
9. Pentecost BT, Teng CT. Lactotransferrin is the major estrogen-inducible protein of mouse uterine secretions. J Biol Chem 1987;262:10134-9.
10. Sundstrom SA, Komm BS, Ponce-de-Leon H, Yi Z, Teuscha C, Lyttle CR. Estrogen regulation of tissue-specific expression of complement C3. J Biol Chem 1989;254:16941-7.
11. Pollard JW, Bartocci A, Arceci R, Orlofsky A, Ladner MB, Stanley ER. Apparent role of the macrophage growth factor, CSF-1, in placental development. Nature (London) 1987;330:484-6.
12. Koistinen R, Kalkkinen N, Huhtala ML, Seppala M, Bohn H, Rutanen EM. Placental protein 12 is a decidual protein that binds somatomedin and has an identical amino acid sequence with somatomedin-binding protein from human amniotic fluid. Endocrinology 1986;118:1375-8.

13. Garde J, Bell SC, Eperon IC. Multiple forms of mRNA encoding human pregnancy-associated endometrial α_2-globulin, a β-lactoglobulin homologue. Proc Natl Acad Sci USA 1991;88:2456–60.
14. Jaffé RC, Donnelly KM, Mavrogianis PA, Verhage HG. Molecular cloning and characterization of a progesterone-dependent cat endometrial secretory protein complementary deoxyribonucleic acid. Mol Endocrinol 1989;3:1807–14.
15. Ing NH, Roberts RM. The major progesterone-modulated proteins secreted into the sheep uterus are members of the serpin superfamily of serine protease inhibitors. J Biol Chem 1989;264:3372–9.
16. Telford NA, Watson AJ, Schultz GA. Transition from maternal to embryonic control in early mammalian development: a comparison of several species. Mol Reprod Dev 1990;26:90–100.
17. Schini SA, Bavister BD. Two-cell block to development of cultured hamster embryos is caused by phosphate and glucose. Biol Reprod 1988;39:1183–92.
18. Paria BC, Dey SK. Preimplantation embryo development in vitro: cooperative interactions among embryos and role of growth factors. Proc Natl Acad Sci USA 1990;87:4756–60.
19. Gandolfi F, Moor RM. Stimulation of early embryonic development in the sheep by co-culture with oviduct epithelial cells. J Reprod Fertil 1987;81:23–8.
20. Amoroso EC. Placentation. In: Parkes AS, ed. Marshall's physiology of reproduction; vol 2. London: Longmans, 1951:127–311.
21. Perry JS, Rowlands IW. Early pregnancy in the pig. J Reprod Fertil 1962;4:175–88.
22. Anderson LL. Growth, protein content and distribution of early embryos. Anat Rec 1978;190:143–54.
23. Geisert RD, Renegar RH, Thatcher WW, Roberts RM, Bazer FW. Establishment of pregnancy in the pig, I. Interrelationships between preimplantation development of the pig blastocyst and uterine endometrial secretions. Biol Reprod 1982;27:925–39.
24. Keys JL, King GJ. Morphological evidence for increased uterine vascular permeability at the time of embryonic attachment in the pig. Biol Reprod 1988;39:473–87.
25. Pope WF. Uterine asynchrony: a cause of embryonic loss. Biol Reprod 1988;39:999–1003.
26. Perry JS, Heap RB, Amoroso EC. Steroid hormone production by pig blastocysts. Nature (London) 1973;245:45–7.
27. Geisert RD, Thatcher WW, Roberts RM, Bazer FW. Establishment of pregnancy in the pig, III. Endometrial secretory response to estradiol valerate administered on day 11 of the estrous cycle. Biol Reprod 1982;27:957–65.
28. Knight JW, Bazer FW, Wallace MD, Wilcox CJ. Dose response relationships between exogenous progesterone and estradiol and porcine uterine secretions. J Anim Sci 1974;39:747–51.
29. Basha SM, Bazer FW, Roberts RM. The secretion of a uterine specific, purple phosphatase by cultured explants of porcine endometrium: dependency upon the state of pregnancy of the donor animal. Biol Reprod 1979;20:431–41.

30. Roberts RM, Raub TJ, Bazer FW. The role of uteroferrin in transplacental iron transport in the pig. Fedn Proc 1986;45:2513–8.
31. Roberts RM, Bazer FW. The functions of uterine secretions. J Reprod Fert 1988;82:875–92.
32. Bazer FW, Roberts MJ, Sharp DC. Collection and analysis of female genital tract secretions. In: Daniel JC, ed. Methods in mammalian reproduction. New York: Academic Press, 1978:503–28.
33. Murray F, Bazer FW, Wallace HD, Warnick AC. Quantitative and qualitative variation in the secretion of protein by the porcine uterus during the estrous cycle. Biol Reprod 1972;7:314–20.
34. Zavy MT, Roberts RM, Bazer FW. Acid phosphatase and leucine aminopeptidase activity in the uterine flushings of nonpregnant and pregnant gilts. J Reprod Fertil 1984;72:503–7.
35. Basha SMM, Bazer FW, Geisert RD, Roberts RM. Progesterone-induced uterine secretions in pigs: recovery from pseudopregnant and unilaterally pregnant gilts. J Anim Sci 1980;50:113–23.
36. Basha SMM, Bazer FW, Roberts RM. Effect of the conceptus on quantitative and qualitative aspects of uterine secretion in pigs. J Reprod Fertil 1980;60:41–8.
37. Buhi WC, Ducsay CA, Bartol FF, Bazer FW, Roberts RM. A function of the allantoic sac in the metabolism of uteroferrin and maternal iron by the fetal pig. Placenta 1983;4:455–70.
38. Renegar RH, Bazer FW, Roberts RM. Placental transport and distribution of uteroferrin in the fetal pig. Biol Reprod 1982;27:1247–60.
39. Vincent JB, Averill BA. An enzyme with double identity: purple acid phosphatase and tartrate-resistant acid phosphatase. FASEB J 1990;4:3009–14.
40. Simmen RCM, Srinivas V, Roberts RM. cDNA sequence, gene organization and progesterone induction of messenger RNA for uteroferrin, a porcine uterine iron transport protein. DNA 1989;8:543–54.
41. Bazer FW, Worthington-White D, Fliss MFV, Gross S. Uteroferrin: a progesterone-induced hematopoietic growth factor of uterine origin. Exp Hematol 1991;19:910–15.
42. McDowell KJ, Sharp DC, Fazleabas A, Roberts RM, Bazer FW. Partial characterization of the equine uteroferrin-like protein. J Reprod Fertil 1982;(suppl)32:329–34.
43. Murray MK, Malathy PV, Bazer FW, Roberts RM. Structural relationship, biosynthesis and immunocytochemical localization of uteroferrin-associated basic glycoproteins. J Biol Chem 1989;264:4143–50.
44. Baumbach GA, Ketcham CM, Richardson DE, Bazer FW, Roberts RM. Isolation and characterization of a high molecular weight stable pink form of uteroferrin from uterine secretions and allantoic fluid of pigs. J Biol Chem 1986;261:12869–78.
45. Malathy P-V, Imakawa K, Simmen RCM, Roberts RM. Molecular cloning of the uteroferrin-associated protein, a major progesterone-induced serpin secreted by the porcine uterus, and the expression of its mRNA during pregnancy. Mol Endocrinol 1990;4:428–40.
46. Ing NH, Francis H, McDonnell JJ, Amann JF, Roberts RM. Progesterone induction of the uterine milk proteins: major secretory products of sheep endometrium. Biol Reprod 1989;41:643–54.

47. Roberts RM, Bazer FW, Baldwin N, Pollard WE. Progesterone induction of lysozyme and peptidase activities in the porcine uterus. Arch Biochem Biophys 1976;177:499-507.
48. Hansen PJ, Bazer FW, Roberts RM. Appearance of β-hexosaminidase and other lysosomal-like enzymes in the uterine lumen of gilts, ewes and mares in response to progesterone and oestrogens. J Reprod Fertil 1985;73:411-24.
49. Fazleabas AT, Bazer FW, Roberts RM. Purification and properties of progesterone-induced plasmin/trypsin inhibitor from uterine secretions of pigs and its immunocytochemical localization in the pregnant uterus. J Biol Chem 1982;257:6886-97.
50. Roberts RM, Murray MK, Burke MG, Ketcham CM, Bazer FW. Hormonal control and function of secretory proteins. In: Leavitt WW, ed. Cell and molecular biology of the uterus. New York: Plenum Press, 1987:137-50.
51. Fazleabas AT, Geisert RD, Bazer FW, Roberts RM. The relationship between release of plasminogen activator and estrogen by blastocysts and secretion of plasmin inhibitor by uterine endometrium in the pregnant pig. Biol Reprod 1983;29:225-38.
52. Dabitch D, Andary TJ. Prevention of blastocyst implantation in mice with proteinase inhibitors. Fertil Steril 1974;25:954-7.
53. Denker HW. Role of proteinases in implantation. Prog Reprod Biol 1980; 7:28-42.
54. Farmer SJ, Fliss AE, Simmen RCM. Complementary DNA cloning and regulation of expression of the mRNA encoding a pregnancy-associated porcine uterine protein related to human antileukoproteinase. Mol Endocrinol 1990;4:1095-104.
55. Adams KL, Bazer FW, Roberts RM. Progesterone-induced secretion of a retinol-binding protein in the pig uterus. J Reprod Fertil 1981;62:39-47.
56. Clawitter J, Trout WE, Burke MG, Araghi S, Roberts RM. A novel family of progesterone-induced, retinol-binding proteins from uterine secretions of the pig. J Biol Chem 1990;265:3248-55.
57. Trout WE, Stallings-Mann ML, Anthony RV, Hall JA, Galvin JM, Roberts RM. Steroid regulation of the synthesis and secretion of retinol-binding proteins by the uterus of the pig. Endocrinology 1992;130:2557-64.
58. Brief S, Chew BP. Effects of vitamin-A and β-carotene on reproductive performance in gilts. J Anim Sci 1985;60:998-1004.
59. Coffey MT, Britt JH. Effect of β carotene injection on reproductive performance of sows [Abstract 615]. J Anim Sci 1989;67(suppl 1).
60. Polge C. Embryo transplantation and preservation. In: Cole DJA, Foxcroft GR, eds. Control of pig reproduction. London: Butterworth Scientific, 1982: 277-91.
61. Pope WF, Lawyer MS, Butler WR, Foote RH, First NL. Dose-response shift in the ability of gilts to remain pregnant following exogenous estradiol-17β exposure. J Anim Sci 1986;63:1208-10.
62. Pope WF, First NL. Factors affecting the survival of pig embryos. Theriogenology 1985;23:91-105.
63. Morgan GL, Geisert RD, Zavy MT, Shawley RV, Fazleabas AT. Development of pig blastocysts in a uterine environment advanced by exogenous oestrogen. J Reprod Fertil 1987;80:125-31.

64. Simmen FA, Simmen RCM. Peptide growth factors and proto-oncogenes in mammalian conceptus development. Biol Reprod 1991;44:1–5.
65. Hansen PJ, Segerson EC, Bazer FW. Characterization of immunosuppressive substances in the basic protein fraction of uterine secretions from pregnant ewes. Biol Reprod 1987;36:393–404.

18

In Vitro Models for Implantation of the Mammalian Embryo

S.J. KIMBER, R. WATERHOUSE, AND S. LINDENBERG

In mammals, implantation of the embryo and development of the placenta are prerequisites for successful development since the maternal circulation provides the means of removal of waste products and the source of nutrition and gas exchange. Although the degree of interaction between blastocyst and uterus is highly species specific, in all cases three local cellular compartments need to be taken into account in any attempt to understand the mechanism and control of the implantation process and the significance of the changes taking place in uterus and embryo at this time. These are (i) the blastocyst, consisting of *inner cell mass* (ICM) and outer trophectoderm, within the uterine lumen; (ii) luminal and glandular epithelia of the uterus; and (iii) the mixed cell population of the endometrial stroma. None of these cellular compartments is a homogeneous collection of cells, but for convenience (and to some extent due to our ignorance) we will consider that they constitute three distinct interacting cell populations.

Implantation is initiated by apposition between trophoblast and endometrial epithelium, facilitated by the closing down of the uterine lumen, as progress of the embryo down the uterine tract is halted. Following this, firmer adhesion is established between the apical membranes of luminal epithelium and trophectoderm. Finally, where implantation is invasive, as in the mouse and the human, the trophoblast penetrates through the endometrial epithelium and into the stroma. The details of this latter process are characteristic of the species (1–3).

In the mouse embryo, implantation is initiated at 4.5 days of development when the abembryonic trophectoderm cells of the blastocyst make contact with the apical surfaces of epithelial cells lining the uterine lumen. Adhesion between these two cell types requires coordination between blastocyst and uterine epithelium. An *activated* blastocyst, hatched from the zona pellucida, must come into close contact with the endometrial epithelium, which is *receptive* for implantation. This receptive state is determined by the hormonal milieu, and implantation is under the control of the reproductive steroids estrogen and progesterone produced predominantly by cells of the ovary. Successful implantation, at least in rodents and probably the human, depends on a nidatory surge of estrogen superimposed on a period of progesterone priming (4).

The ability of embryos to attach at extrauterine sites in ectopic pregnancies in the human, or in an experimental situation in the mouse, irrespective of the hormonal milieu (5, 6), suggests that perhaps there is something fundamentally different about the nonreceptive luminal epithelium rather than something unique about the receptive epithelium. However, adhesion between the apical surfaces of two epithelial cells is somewhat unusual, and a temporary loss of apical characteristics by the luminal epithelium at this time has been suggested to contribute to the receptive "window" of between 24 and 30 h in the mouse (1). The reality of such a window in the human has still to be proved. Signals pass between all three cell compartments at this time, but whether communication between the embryo and the stroma is direct by way of molecular signals passing unmodified through the epithelium, via transduction by the epithelial layer, or perhaps by both means remains for future research to clarify (1).

By an as yet undefined mechanism, the embryo triggers a transformation in the underlying stromal cells known as the *decidual reaction*. The first change observed is an increase in vascular permeability in the stromal tissue with consequent edema. Second, a change in the composition of the *extracellular matrix* (ECM) of the stroma can be demonstrated, at least in rodents and the human. These cells differentiate into decidual cells, switching from synthesis predominantly of interstitial matrix molecules—such as collagen types 1 and 3 (7)—to production of molecules characteristic of basement membranes (8) such as fibronectin, laminin, entactin, collagen type 4, and heparan sulphate proteoglycan, as well as chondroitin sulphate proteoglycans (9–12). These are localized in a pericellular basement membrane-like layer. Degradation of type 6 collagen, proposed to crosslink other major collagen species (13), has been suggested to allow a decrease in matrix rigidity in the human decidua and promote trophoblast invasion (14). Finally, capillary ingrowth takes place in preparation for development of the placenta.

The site of implantation in vivo is extremely inaccessible to observation and experimental manipulation. Consequently, most earlier studies and those in the human have concentrated on morphological examination of fixed material (2, 15–19). However, by developing models for implantation in vitro, it has been possible to examine experimentally various aspects of the implantation process in isolation from the maternal environment.

In this chapter, we will start by describing different approaches to achieving a physiological in vitro model for implantation and our recent progress in this area. We will then review our evidence for the role of a carbohydrate-receptor interaction in blastocyst attachment. Finally, we will describe recent observations on the mode of attachment of human embryos in vitro.

Implantation In Vitro

Model Systems to Study Implantation In Vitro: A Brief Review

There have been many attempts to produce models for implantation that reflect the interactions of cells in vivo. These models fall into four main categories. First, in the earliest studies, the approach was to maintain all tissues as close as possible to their state in vivo. Blastocysts were cultured with strips of uterus (20–22) or whole uterine horns (23–25), but these studies were subject to problems of interpretation, partly a result of artifacts due to tissue necrosis and attachment of embryos to degenerating cells or serum clots. A major aim of demonstrating hormonal control of attachment and invasion into the uterine tissues by the blastocyst, as in vivo, was not achieved. Giant cell formation and invasion were promoted by progesterone in the study of Grant and colleagues, but no decidualization of the stroma developed, and the morphology of the embryo quickly became abnormal (summarized in 26, 27). The complexity of these systems means that dissecting the various reciprocal molecular interactions involved in the implantation process is unlikely to be achieved using these models.

More recently, vesicles of endometrium have been cocultured with embryos in hanging drops (27–29). Here, only the epithelium is present, with its luminal surface facing outward towards the blastocyst. The problem with this system is that it is impossible to mimic the close apposition phase found in vivo, and attachment occurs with rather a low success rate, while again there is the problem of attachment to dead cells. Related three-dimensional model systems have been developed consisting of cocultures of endometrial fragments with either trophoblast in the human (30) or blastocysts in the rabbit (26) using gyratory shaking. In the human, trophoblast attached only when endometrium from day 19 of the cycle was used. Day 19 falls within the expected period for embryo attachment in utero, and this supports the idea that there is a window during which the endometrium is receptive for implantation in the human as in other species. This restricted period appears to be maintained in vitro. In the rabbit model, it was found necessary to devise a system to keep the endometrium and embryo in close contact, as would occur in vivo.

Second, a more reductionist approach has been taken allowing limited aspects of blastocyst behavior to be examined, such as the attachment of the embryo to solid substrates (e.g., tissue culture plastic). In many cases, these surfaces have been coated with macromolecules. This approach has borne much fruit where the behavior of the trophectoderm has been examined on surfaces coated with ECM components known to be present in the epithelial basement membrane or the decidua (31–39).

From this type of system, we know, for instance, that mouse trophoblast can degrade ECM substrates by a mechanism independent of plasminogen (37), a protease previously implicated in this process. Fibronectin, laminin, and various collagens have been shown to support attachment and/or migration of trophoblast, and the time course of attachment varies with the molecule employed (31–33, 35, 40). Inhibition of trophoblast outgrowth on collagen 4 and fibronectin with peptides containing the Arg-Gly-Asp sequence or antibody against β_1 integrin suggests that the integrin family of cell adhesion receptors for ECM function in the interactions of trophoblast with these matrix components (31, 36, 40). Immunolocalization and immunoprecipitation of β_1 integrin support these data in both mouse and human (36). Thus, the evidence is good that both mouse and human embryos use an integrin-collagen/fibronectin mechanism of interaction where and when these ECM molecules are available.

However, these mechanisms probably come into force following the initial adhesion between trophectoderm and endometrial epithelium, during invasion through the stroma and development of the placenta. In addition, migration through the stroma may involve a mechanism employing the addition of galactose to laminin oligosaccharides by galactosyltransferase. Such galactosylation is mediated by secondary trophoblast cells in vitro (39). Trophoblast cell migration, but not attachment, on laminin is inhibited by α-lactalbumin and UDP-gal that interfere with the activity of this enzyme in other systems. Hyaluronate, synthesis of which increases dramatically at implantation in the mouse, may also facilitate penetration of the stroma (41).

Heparin has been shown to slow the rate of attachment of murine trophoblast to matrix molecules in vitro (35). The related molecule *heparan sulphate proteoglycan* (HSPG) is present on the apical surface of the uterine endometrial epithelial cells and secreted by these cells (36, 42). There is also some evidence that heparin/heparan sulphate-like molecules may be present on the trophectoderm surface (35, 42), as are laminin and nidogen (15). Blastocysts attach in vitro to a fraction of proteins derived from mouse uterine epithelial cells that bind HSPG, but they do not spread and outgrow on them, suggesting that HSPG is involved in the initial attachment process rather than in infiltration through the epithelium (34).

Third, the attachment, spreading, and outgrowth of trophectoderm on monolayers of cells grown in culture allows the investigation of trophoblast-cell interactions. Nonendometrial cells have been used (44), but a more physiological model uses monolayers of uterine endometrial epithelial cells (45–50). By studying the behavior of mouse blastocysts with the cells that they will contact in vivo, information about the reciprocal relationship between these cell types can be obtained. Uterine endometrial epithelial cells can conveniently be grown as monolayers on tissue culture plastic, as demonstrated in the above-mentioned studies. From this model, we have gained significant information about trophoblast behavior and trophoblast-epithelial adhesion (47, 51). It is encouraging that the mouse embryo generally attaches to monolayers of murine endometrial epithelial cells by the abembryonic pole (opposite the ICM), while the human embryo invariably attaches to murine monolayers by the embryonic (polar) trophoblast that covers the ICM (47, 51, 52). These are exactly the orientations that have been observed from morphological studies of the implanting embryo in utero (17, 18), suggesting that this does indeed reflect normal attachment even if it does not allow study of later events.

However, the model is limited in application because the cells have a flattened morphology and do not show the polarized arrangement of cell surface features or cytoplasmic organelles found in the epithelium in vivo. It is difficult to achieve a complete monolayer, and spaces are frequently present. The attachment of embryos in regions where cells are absent complicates the assessment of experiments designed to examine the interaction of embryos with the cells. Moreover, the spreading murine trophoblast appears to clear adjacent cells from the dish and grow out on the underlying surface. In any case, only the initial interactions between trophoblast and endometrial cells can be examined because although the epithelial cells carry associated extracellular matrix components such as laminin (unpublished observation), they do not form a morphologically distinct basement membrane, and there is, of course, no underlying stroma.

All these models are useful for studying restricted aspects of endometrial biology or the implantation process. However, they are inadequate for studying the sequential cellular interactions occurring during implantation or for analyzing the influence of stromal cells on blastocyst-epithelial interaction or epithelial cells on the blastocyst-stromal interaction. Furthermore, the cells tend to lose hormone responsiveness, a crucial difference from the situation in vivo.

A fourth method has been to approach the three-dimensional architecture observed in vivo by growing uterine endometrial epithelial cells in (53–55) or on (54, 56) collagen gels. In the study of Sengupta et al. (54), the stromal cells were cocultured in the gels. The period of normal organization was restricted when epithelial cells were grown on the gel surface, and the epithelial cells were not found to survive well inside the gel. However, development of this system has great potential. For instance, blastocysts can be cultured on top of the gel so that trophoblast invasion into the gel can be monitored.

One of the most reproducible physiological methods available for growing endometrial epithelial cells is to use a suspended semipermeable membrane incorporating ECM material. In the rat, Glasser and colleagues (57) have developed a culture system using Millicell culture plate inserts (Millipore) coated with Matrigel, the ECM material produced by EHS sarcoma cells. This mimics the basement membrane that normally delimits the basal surface of the cells in vivo. On such a substrate the cells exhibit a polarized morphology with apical microvilli and a basal lamina. They develop a transcellular resistance, suggesting the presence of a fairly complete tight junctional network. Their morphology is similar to that of uterine endometrial epithelium in vivo; they exhibit polarized secretory activity and maintain hormone responsiveness (57, 58). Rabbit (59) and human (60) endometrial epithelial cells have also been grown in this way and similarly exhibit a polarized morphology and, at least for the rabbit, hormone responsiveness. However, in our hands (and Glasser, personal communication) murine epithelial cells do not show the well-developed polarity of the rat and rabbit epithelial cells using this system. It is interesting that by TEM, the mouse cells do not appear quite so elongated as those of the rat in utero either (Waterhouse, Kimber, unpublished; 61).

One disadvantage of this culture system is that it is difficult to control the thickness of the Matrigel layer coating the membrane, a feature that is critical for the successful attachment of the cells. Furthermore, murine endometrial epithelial cells frequently "pull" the matrix layer off the membranes and round up into vesicles. This can be remedied to some extent by using only very thin layers of Matrigel, but it can make the system inconvenient for studying the interaction of the epithelial cells with the embryo.

Characterization of an In Vitro Model

We have cultured murine endometrial epithelial cells on Cellagen membranes, semipermeable suspended membranes made of crosslinked rat tail collagen (ICN). On these membranes the cells divide and grow as monolayers, or monolayers with regions containing overlapping cells, in appropriate medium containing fetal calf serum plus the serum extender NuSerum (Flow). The membranes are of a good standard of uniformity, as examined by transmission electron microscopy (TEM). This allows us to manipulate the system experimentally without the problem of variability in the substrate. We isolated endometrial epithelial cells by the method of McCormack and Glasser (62). Although we obtained large numbers of cells from the uteri of mature female mice on day 2 of pregnancy, these did not grow well in culture. In contrast, the uteri from immature mice, around day 19–21 of postnatal development, that had not yet entered the estrous cycle yielded fewer cells and reached confluence only on the 4th

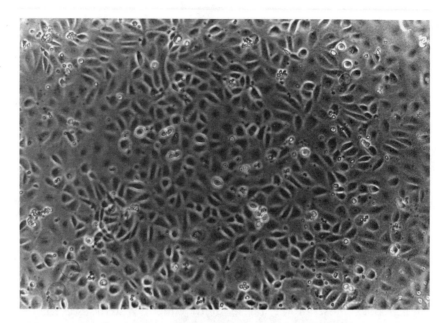

FIGURE 18.1. Phase contrast micrograph showing confluent monolayer of murine endometrial epithelial cells 3 days after explanting on Cellagen membrane.

day of culture when grown on the Cellagen membranes. The cells had little RER, few microvilli, were rather flattened, and did not have a strikingly epithelial appearance.

The most successful cultures were obtained from immature female mice that were injected with pregnant mare serum gonadotropin and 46 h later with human chorionic gonadotropin to stimulate follicle growth and ovulation. The uteri from these mice gave rise to large numbers of cells that grew well in culture and reached confluence on the 3rd day (Fig. 18.1). After 1 day of culture, cell number fell to 1/4 of the initial seeding density. It then reached a plateau of 8 times this value between 4 and 6 days in culture. Thereafter, numbers dropped slightly, but the cultures remained viable for at least 10 days. Although this is not as long as the 22 days reported for immature rat endometrial epithelial cells by Glasser et al. (57), it allows a reasonable time for experimental manipulation and analysis of the embryo-endometrial epithelial interaction. By SEM the majority of cells showed relatively large numbers of microvilli on their apical surfaces, although a few relatively "bare" cells were always found. By TEM the cells were polarized and contained RER and Golgi vesicles, indicating that the cells were synthetically active. Gap junctions present between the cells suggested intercellular communication. However, the polarity was still not as well developed as that achieved in the rat by Glasser and colleagues (57, 58). We have examined the interaction of the mouse and human embryo with endometrial epithelial cells grown in different ways to gather information about the initial events of implantation.

Adhesion of the Mouse Embryo In Vitro

Recently, there has been renewed interest in carbohydrates as ligands for lectin-like cell surface receptors or in modulating the interaction of receptors with cell and matrix ligands. For instance, there have been a larger number of papers reporting adhesive interactions of cells of the immune system. Among these are mechanisms responsible for the attachment of neutrophils, monocytes, and lymphocytes to endothelial cells lining blood vessels that have some regional specificity for venule location (63). The receptors fall into three main families: (i) the immunoglobulin superfamily that binds to members of a (ii) second family of integrin-like molecules, or homophilically, and (iii) the LECAM or Selectin family of lectin-like cell adhesion molecules that bind carbohydrate ligands (64–66). Fucosylated carbohydrates have been implicated as the ligands for all three known members of this family (64, 66, 67). There is evidence that the trisaccharide Galβ_1-4 [Fucα_1-3] GlcNAc (Le^x), an epitope related to the Lewis blood group antigens, may be part of the ligand for LECAM-3 (68). Its sialylated derivative has similarly been suggested to be the ligand for LECAM-2 from experiments using glycosylation mutants and transfectants (69, 70). The Le^x epitope was previously implicated in the stabilization of cell-cell adhesion at compaction in the 8-cell embryo (71, 72). These are probably minimum structures necessary for cell interaction, and the actual ligands are likely to be more complex. Opinion is still not unanimous that these structures form the receptor binding sites, but these data certainly indicate that fucosylated oligosaccharides based on a Gal-GlcNAc backbone may have fundamental importance in cell interactions.

Using immunofluorescence with specific *monoclonal antibodies* (MAbs), we identified a number of blood group-related carbohydrate antigens that are present on the murine endometrial epithelium and not on stromal cells in the uterus (73). Among these is an H-type 1 oligosaccharide determinant, lacto-N-fucopentaose 1 (LNF-1: Fucα_1-2Galβ_1-3GlcNAcβ_1-3Galβ_1-Glc). This sugar is recognized by MAb 667/9E9 and is present on almost all luminal and glandular epithelial cells up until day 4 of pregnancy. In ovariectomized mice, expression of some of the endometrial epithelial carbohydrate epitopes (including LNF-1) is maximum in the presence of both estrogen and progesterone (74), indicating hormonal control.

Between days 4 and 5 of pregnancy, LNF-1 became restricted to patches of 10–50 epithelial cells interspersed among unstained cells and disappeared from the luminal epithelium by the day following implantation. This expression pattern indicated that it might have a role in the implantation process. We therefore tested whether the oligosaccharide

TABLE 18.1. Effect of oligosaccharides on attachment of murine blastocysts to endometrial epithelial cells in vitro.

Sugar[a]	Total no. in test/ total no. in control	Implantation (% attached in test/ % attached in control)			Sig. 72 h
		24 h	48 h	72 h	
2FL	130/115	38/37	74/61	91/97	NS*
LDFT	91/73	38/37	77/62	76/74	NS
LNT	129/146	35/30	66/66	71/71	NS
LNF-I	228/175	21/17	38/51	59/95	$P < 0.001$
LNF-II	134/178	37/26	78/72	88/86	NS
LND-I	194/162	38/27	74/72	75/81	NS
LNF-III	153/160	21/26	60/73	72/86	NS

* NS = not significant.
[a] Oligosaccharide used at 0.1 mM.
Source: Modified with permission from Lindenberg, Sundberg, Kimber, and Lundblad (47).

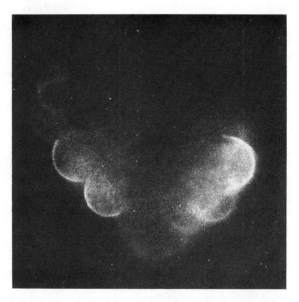

FIGURE 18.2. Blastocyst on day 5 of development stained with an LNF-1-human serum albumin neoglycoprotein conjugate (15–20 sugar units/protein molecule) labeled with FITC. Note that only the abembryonic trophectoderm binds the conjugate.

LNF-1 purified from human milk had any effect on attachment of mouse blastocysts to endometrial epithelial cells cultured as monolayer on tissue culture plastic. Only initial adhesion could be analyzed in this system, but carbohydrates seem frequently to function in cell interactions by contributing a transient component (of variable duration) to adhesion in a process involving a sequence of adhesive steps (63, 65, 71, 75, 76). LNF-1 was found to inhibit attachment of blastocysts to the epithelial monolayers in a dose-dependent manner (47). Seven other related oligosaccharides had no effect in the same concentration range, 0.1 to 5 mM (Table 18.1).

The simplest view of the inhibitory effect of free LNF-1 is that it competes with epithelial cell-surface LNF-1 for a receptor on the trophoblast. To test if the trophoblast expressed such a receptor, neo-glycoproteins comprising 10–20 sugar moieties conjugated via a *spacer* group to human or bovine serum albumin were constructed. These were linked to fluorescein isothiocyanate to provide a signal that allowed us to monitor binding. We found that LNF-1-BSA/HSA bound to the abembryonic pole of the blastocyst (Fig. 18.2), the region that first contacts the epithelium in vivo, suggesting that a receptor for this sugar was present on the trophectoderm (Fig. 18.3). The receptor was first expressed shortly before implantation, which is in keeping with a role in attachment (67).

The blood group-related epithelial antigens continue to be expressed by a proportion of epithelial cells grown on tissue culture plastic and a larger proportion of cells grown on Cellagen membranes in serum-containing

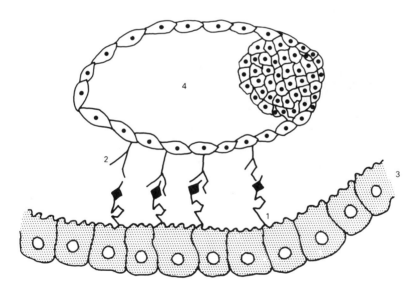

FIGURE 18.3. Simplified model for initial attachment of the blastocyst (labeled 4) to the uterine endometrial epithelium (labeled 3). Receptors (labeled 2) on the trophoblast surface interact with lacto-N-fucopentaose 1 (LNF-1) determinants (labeled 1) present on the luminal epithelial surface, thus anchoring the embryo by its abembryonic surface. Work from other laboratories (11, 20) suggests that 1 could stand for HSPG binding proteins and 2 for HSPG (or possibly vice versa) as well. Reprinted with permission from Lindenberg, Kimber, and Kallin (67).

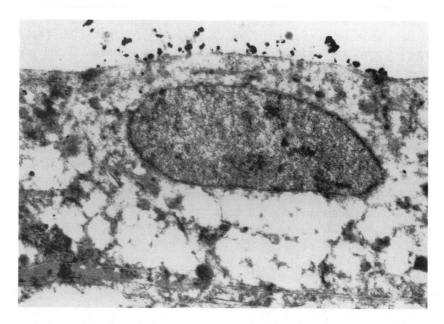

FIGURE 18.4. Transmission electron micrograph showing binding of MAb 667/9E9 to apical microvilli and glycocalyx of an endometrial epithelial cell grown on a Cellagen membrane. The cells were cultured on Cellagen membranes for 5 days and fixed in 1% paraformaldehyde. They were incubated with MAb 667/9E9 for 30 min at room temperature followed by a gold-labeled goat antimouse IgM. After silver enhancement, the monolayers were fixed in 2.5% glutaraldehyde and processed and embedded for TEM.

medium for at least 7 days. Therefore, these carbohydrates can be used as markers of the epithelium, although hormone dependence of expression in vitro has yet to be shown. The trophoblast makes initial contact with the glycocalyx as it adheres to the luminal epithelium, and the LNF-1 determinant can be demonstrated on the apical microvilla membrane and glycocalyx of cultured endometrial epithelial cells using EM immunocytochemistry (Fig. 18.4). The glycocalyx is known to be carbohydrate rich and to undergo thinning and qualitative changes in sugar residues prior to implantation (77).

Originally, we thought that the LNF-1 oligosaccharide might be present on a glycolipid because staining with 667/9E9 disappeared after extraction of frozen sections with chloroform:methanol (2:1) under conditions when other epithelial carbohydrate antigens were not affected (73). However, 667/9E9 did not recognize any components of a lipid extract from endometrial epithelial cells separated on TLC plates. Other glycolipid antigens could be recognized in the same extracts, while 667/9E9 recognized glycolipids in extracts of tumor cells using the same technique (Paulson,

Kimber, Lindenberg, unpublished). We have recently identified a number of endometrial epithelial glycoproteins carrying the 667/9E9 determinant. Immunoblotting has revealed that while two of these glycoproteins can be separated from endometrial epithelial cells using a buffered salt solution, the major crossreactive components, represented by a cluster of glycoproteins of between 120 and 180 kd, require detergent extraction (McCubbin, Kimber, unpublished). Efforts are now under way to purify these molecules by affinity chromatography.

The observation that the binding to the H-type 1 structure by a trophectoderm receptor is involved in attachment of the blastocyst to endometrial cells in vitro indicates that similar binding may be involved in halting progress of the blastocyst down the uterine tract. This gives rise to an image of the embryo "mooring up" at an LNF-1–rich patch of endometrial epithelium. However, the distribution of the 667/9E9 reactive patches appears random and cannot account either for embryonic spacing down the length of the uterine horn or the antimesometrial polarity of implantation in the mouse.

The sugar-receptor concept (Fig. 18.3) is analogous to the adhesion mechanism of recirculating lymphocytes attaching to the surface of high endothelial venules (above). Here, cells moving through the lumen of the blood vessel attach via a lectin-carbohydrate interaction to the endothelial lining. As with implantation, this is only the start of a sequence of interactions involving a number of other molecules, in particular integrins, that allow the penetration of the lymphocytes through the vessel wall and into the underlying connective tissue. The LECAM-carbohydrate interaction seems to be expressed transiently as a first step in arresting neutrophils to cytokine-stimulated endothelial cells (78, 79). A similar mechanism may obtain for arrest of the blastocyst.

Other cell interactions come into force shortly after this LNF-1-mediated initial adhesion of the blastocyst. It is likely that these include the interactions with ECM molecules, particularly HSPG while the trophoblast is interacting with epithelial cells and laminin, collagen, fibronectin, and perhaps hyaluronate once the infiltrating trophectoderm has penetrated the lining epithelium and reached the basement membrane and stroma. We need further information on the precise time course of these changing interactions in utero.

Loss of the type 2 H-antigen (Fucα_1-2Galβ_1-4 GlcNAc) has been associated with invasive behavior of oral squamous cell carcinoma, a migratory rather than adhesive phenotype being expressed (80). In contrast, a monoclonal antibody with very catholic specificity for terminal Fucα_1-2Gal in a variety of contexts had powerful antimotility properties against various tumor cell lines (81), suggesting that this structure may be necessary for tumor cell motility and metastasis formation. A particular carbohydrate antigen may function differently depending on the nature of the molecule carrying it (steric influences, distance from the bilayer, adjacent charged groups, etc.) and the other membrane molecules in its immediate environment. It will also be influenced by associated ECM and cytoskeletal interactions. However, these observed alterations in $\alpha_1-\alpha_2$ linked fucose associated with changes in cell interactions in tumorigenesis lend support to our data suggesting a role in adhesion between embryo and uterine lumen.

Human Embryo-Epithelial Interactions In Vitro

The human embryo has been a potential subject of in vitro study since the advent of the IVF technique made available morphologically suboptimal or "spare" embryos. Thus, it has now become possible to investigate the mode of attachment of the human blastocyst to human endometrial epithelial cells in vitro. As in the mouse, hatching of the human blastocyst, a prerequisite for trophoblast attachment, is initiated from a slit in the zona pellucida in vitro (52, 82, unpublished observations). Attachment to monolayers of human uterine epithelial cells grown on tissue culture plastic occurs via the polar (embryonic) trophectoderm (47, 51, 52). Thus, the polarity of attachment between trophectoderm and epithelial cells is maintained. The polar trophoblast cells in the human embryo intrude between the endometrial cells, and the latter appear to be relocated to the periphery of the attachment site, in contrast to the behavior shown by bovine trophoblast cells that remain on top of the monolayer, hardly disturbing it at all (51). Even when attached embryos were fixed 24h following attachment, the trophoblast had replaced

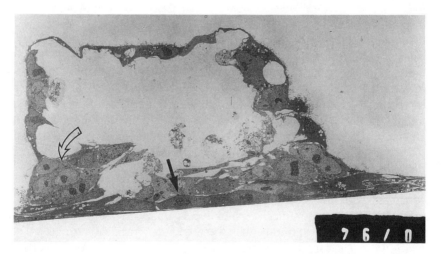

FIGURE 18.5. Low-power TEM of a human embryo interacting with endometrial epithelial cells grown on tissue culture plastic. The trophoblast cells (hollow arrow) have displaced the epithelial cells from the plastic, and the latter are piled up at the periphery of the embryo. (Solid arrow = ICM.)

18. In Vitro Models for Implantation of the Mammalian Embryo 257

FIGURE 18.6. TEM showing interaction of human trophoblast with collagen of a Cellagen disc following intrusion between endometrial epithelial cells. Note the membrane infoldings adjacent to the substrate.

endometrial cells at the center of the attachment site and was in direct contact with the dish (Fig. 18.5). The endometrial cells appear to respond to the embryo by piling up on one another, forming a multilayer at the periphery of the trophoblast-endometrial interaction site, but there was no sign of cell death. TEM reveals that this behavior is seen both on plastic- and Cellagen-grown cultures (Lindenberg, Kimber, unpublished). In contrast, in the mouse, there seems to be a direct displacement of the endometrial epithelial cells and, ultimately, the sloughing of dying cells from the dish by the trophoblast.

Interestingly, the human trophoblast was observed to show an active interaction with the collagen where the endometrial cells had vacated it, producing multiple membranous infoldings adjacent to the substrate that were never observed when the cells came into contact with plastic (Fig. 18.6). In both plastic and Cellagen culture systems, the interaction between the human trophoblast and endometrial epithelium is clearly an active one. TEM of the interface between the human polar trophoblast and the endometrial cells (51; Lindenberg, Kimber, unpublished) revealed that fewer microvilli were present on cells adjacent to the trophoblast than distant from it. Intense endocytotic activity could be demonstrated

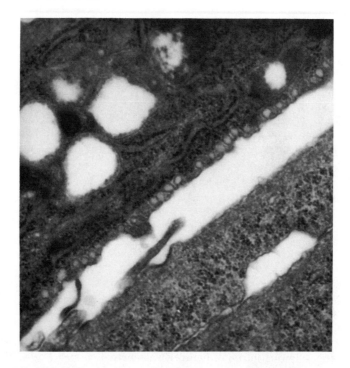

FIGURE 18.7. TEM of human trophoblast-endometrial epithelial interface in vitro. There is intense endocytotic activity at the cell surface of the endometrial cell adjacent to the trophoblast.

along the endometrial plasma membrane facing the infiltrating trophoblast (Fig. 18.7), and this same activity was seen at the bovine trophoblast-endometrial interface in vitro. The endometrial cell cytoplasm contained abundant distended RER in all regions, while the trophoblast lacked RER, but contained abundant smooth endoplasmic reticulum.

Future Prospects

A combination of in vitro models will allow different types of information to be gleaned about the trophoblast-endometrial interactions. As long as the natural condition of implantation within the complex environment of the maternal uterus is kept in mind and data obtained from less than perfect models are realistically interpreted, the future is full of promise for this approach. Many exciting possibilities present themselves. For instance, specific oligosaccharides can be incorporated into liposomes, and the interaction of these vesicles with embryonic and uterine cells can be examined. This approach could be applied to the human embryo-epithelial interaction where competitive inhibition with free oligosaccharides cannot be used to analyze large populations of embryos. We are currently perfecting techniques to coculture stromal and epithelial cells on opposite sides of suspended membranes with hormone supplementation to either or both sides and examining blastocyst interaction in this system.

References

1. Denker H-W. Trophoblast-endometrial interactions at embryo implantation: a cell biological paradox. Trophoblast Res 1990;4:3–29.
2. Schlafke S, Enders AC. Cellular basis of interaction between trophoblast and uterus at implantation. Biol Reprod 1975;12:41–65.
3. Wimsatt WA. Some comparative aspects of implantation. Biol Reprod 1975; 12:1–40.
4. Psychoyos A. Uterine receptivity for nidation. Ann N Y Acad Sci 1986;476: 36–42.
5. Kirby DRS. Ectopic autografts of blastocysts in mice maintained in delayed implantation. J Reprod Fertil 1967;14:512–7.
6. Kirby DRS. The extra-uterine mouse egg as an experimental model. In: Raspe G, ed. Schering symposium on mechanisms involved in conception. Oxford: Pergamon Press/Vieweg, 1970:255–73.
7. Kisalus LL, Herr JC, Little CD. Immunolocalisation of extracellular matrix proteins and collagen synthesis in first trimester human decidua. Anat Rec 1987;218:402–15.
8. Timpl R. Structure and biological activity of basement membrane proteins. Eur J Biochem 1989;180:487–502.
9. Charpin C, Kopp F, Pourreau-Schneider N, et al. Laminin distribution in human decidua and immature placenta. An immunoelectron microscopic study (avidin-biotin-peroxidase complex method). Am J Obstet Gynecol 1985;151:822–7.
10. Glasser SR. Biochemical and structural changes in uterine endometrial cell types following natural or artificial deciduogenic stimuli. Trophoblast Res 1990;4:377–416.
11. Glasser SR, Lampclo S, Munir MI, Julian J. Expression of desmin, laminin and fibronectin during in situ differentiation (decidualization) of rat uterine stroma cells. Differentiation 1987;35:132–42.
12. Wewer UM, Damjanov A, Weiss J, Liotta LA, Damjanov I. Mouse endometrial stromal cells produce basement membrane components. Differentiation 1986;32:49–58.
13. Bruns RR, Press W, Engvall E, Timpl R, Gross J. Type VI collagen in extracellular, 100 nm periodic fibrils and filaments: identification by immuno-electron microscopy. J Cell Biol 1986;103:393–404.
14. Aplin JD, Charlton AK, Ayad S. An immunohistochemical study of human endometrial extracellular matrix during the menstrual cycle and first trimester of pregnancy. Cell Tissue Res 1988;253:231–40.

15. Dziadek M, Timpl R. Expression of nidogen and laminin in basement membranes during mouse embryogenesis and in teratocarcinoma cells. Dev Biol 1985;111:372–82.
16. Enders AC, Schlafke S. Surface coats of the mouse blastocyst and uterus during the preimplantation period. Anat Rec 1974;180:31–46.
17. Hertig AT, Rock J, Adams EC. A description of 34 human ova within the first 17 days of development. Am J Anat 1956;98:435–91.
18. Nilsson O. The morphology of blastocyst implantation. J Reprod Fertil 1974; 39:187–94.
19. O'Rahilly R, Muller F. Developmental stages in human embryos. Carnegie Institute of Washington pub. no. 637, 1987.
20. Glenister TW. Organ culture as a new method for studying the implantation of mammalian blastocysts. Proc R Soc Lond (Biol) 1961;154:428–31.
21. Glenister TW. Organ culture and its combination with electron microscopy of the nidation process. In: Westsin B, Wiqvist N, eds. Fertility and sterility; Ser 133. Amsterdam: Exerpta Medica Foundation, 1967:385–94.
22. Glenister TW. Methods for ovoimplantation and early embryo-placental development in vitro. In: Daniel JC Jr, ed. Methods in mammalian embryology. San Francisco: W.H. Freeman, 1971:320–33.
23. Grant PS. The effect of progesterone and oestradiol on blastocysts cultured within the lumina of immature mouse uteri. J Embryol Exp Morphol 1973; 29:617–38.
24. Grant PS. The effect of progesterone and oestradiol on immature mouse uteri maintained as organ cultures. J Endocrinol 1973;57:171–4.
25. Grant PS, Ljungkvist I, Nilsson O. The hormonal control and morphology of blastocyst invasion into the mouse uterus in vitro. J Embryol Exp Morphol 1975;34:299–310.
26. Hohn H-P, Denker H-W. A three-dimensional organ culture model for the study of implantation of rabbit blastocysts in vitro. Trophoblast Res 1990; 4:71–95.
27. Morris JE, Potter SW. An in vitro model for studying interactions between mouse trophoblast and uterine epithelial cells. In: Denker HW, Aplin JD, eds. Trophoblast invasion and endometrial receptivity. New York: Plenum Press, 1990:51–69.
28. Morris JE, Potter SW, Buckley PW. Mouse embryos and uterine epithelia show adhesive interactions in culture. J Exp Zool 1982;222:195–8.
29. Morris JE, Potter SW, Rynd LS, Buckley PM. Adhesion of mouse blastocysts to uterine epithelium in culture: a requirement for mutual surface interactions. J Exp Zool 1983;225:467–79.
30. Kliman HJ, Feinberg RF, Haimowitz JE. Human trophoblast-endometrial interactions in an in vitro suspension culture system. Placenta 1990;11: 349–67.
31. Armant DR, Kaplan HA, Lennarz WJ. Fibronectin and laminin promote in vitro attachment and outgrowth of mouse blastocysts. Dev Biol 1986;116: 519–23.
32. Armant DR, Kaplan HA, Mover H, Lennarz WJ. The effect of hexapeptides on attachment and outgrowth of mouse blastocysts cultured in vitro: evidence for the involvement of the cell recognition tripeptide Arg-Gly-Asp. Proc Natl Acad Sci USA 1986;83:6751–5.

33. Carson DD, Tang J-Y, Gay S. Collagen support embryo attachment and outgrowth in vitro: effects of the Arg-Gly-Asp sequence. Dev Biol 1988; 127:368–75.
34. Carson DD, Wilson OF, Dutt A. Glycoconjugate expression and interactions at the surface of mouse uterine epithelial cells and periimplantation-stage embryos. Trophoblast Res 1990;4:211–41.
35. Farach MC, Tang JP, Decker GL, Carson DD. Heparin/heparan sulphate is involved in attachment and spreading of mouse embryos in vitro. Dev Biol 1987;123:401–10.
36. Fisher SJ, Sutherland A, Moss L, et al. Adhesive interactions of murine and human trophoblast cells. Trophoblast Res 1990;4:115–38.
37. Glass RH, Aggeler J, Spindle A, Pedersen RA, Werb Z. Degradation of extracellular matrix by mouse trophoblast outgrowths: a model for implantation. J Cell Biol 1983;96:1108–16.
38. Negami AI, Tominaga T. Gland and epithelium formation in vitro from epithelial cells of the human endometrium. Hum Reprod 1989;4:620–4.
39. Romagno L, Babiarz B. The role of murine cell surface galactosyl transferase in trophoblast: laminin interactions in vitro. Dev Biol 1990;141:254–61.
40. Sutherland AE, Calarco PG, Damsky CH. Expression and function of cell surface extracellular matrix receptors in mouse blastocyst attachment and outgrowth. J Cell Biol 1988;106:1331–48.
41. Carson DD, Dutt A, Tang J-Y. Glycoconjugate synthesis during early pregnancy: hyaluronate synthesis and function. Dev Biol 1987;120:228–35.
42. Tang J-P, Julian J, Glasser SR, Carson DD. Heparan sulfate proteoglycan synthesis and metabolism by mouse uterine epithelial cells in vitro. J Biol Chem 1987;262:12832–42.
43. Dziadek M, Fujiwara S, Paulsson M, Timpl R. Immunological characterization of basement membrane type of heparan sulphate proteoglycan. EMBO J 1985;4:905–12.
44. Glass H, Spindle AI, Pedersen RA. Mouse embryo attachment to substratum and interaction of trophoblast with cultured cells. J Exp Zool 1979;208: 327–36.
45. Chavez DJ, Van Blerkom J. In vitro attachment and outgrowth of mouse trophectoderm. In: Glasser SR, Bullock DW, eds. Cellular and molecular aspects of implantation. New York: Plenum Press, 1981:457–60.
46. Lindenberg S, Nielsen MH, Lenz S. In vitro studies of human blastocyst implantation. Ann N Y Acad Sci 1985;442:368–74.
47. Lindenberg S, Sundberg K, Kimber SJ, Lundblad A. The milk oligosaccharide, lacto-N-fucopentaose 1, inhibits attachment of mouse blastocysts on endometrial monolayers. J Reprod Fertil 1988;83:149–58.
48. Sherman MI. Implantation of mouse blastocysts in vitro. In: Daniel JC Jr, ed. Methods in mammalian reproduction. New York: Academic Press, 1978: 247–57.
49. Sherman MI, Wudl LR. The implanting mouse blastocyst. In: Poste G, Nicholson GI, eds. The cell surface in animal embryogenesis and development. Amsterdam: North Holland, 1976:81–125.
50. Van Blerkom J, Chavez DJ. Morphodynamics of outgrowth of mouse trophoblast in the presence and absence of a monolayer of uterine epithelium. Am J Anat 1981;162:143–55.

51. Lindenberg S, Hyttel P, Sjogren A, Greve T. A comparative study of attachment of human, bovine and mouse blastocysts to uterine epithelial monolayers. Hum Reprod 1989;4:446–56.
52. Lindenberg S, Hyttel P, Lenz S, Holmes PV. Ultrastructure of the early human implantation in vitro. Hum Reprod 1986;1:533–8.
53. Iguchi T, Uchima F-DA, Ostrander PL, Hamamoto ST, Bern HA. Proliferation of normal mouse uterine luminal epithelial cells in serum-free collagen gel culture. Proc Jpn Acad 1985;61:292–5.
54. Sengupta J, Given RL, Carey JB, Weitlauf HM. Primary culture of mouse endometrium in floating collagen gels: a potential in vitro model for implantation. Ann N Y Acad Sci 1986;476:75–94.
55. Uchima F-DA, Edery M, Iguchi T, Bern HA. Growth of mouse endometrial luminal epithelial cells in vitro: functional integrity of the oestrogen receptor system and failure of oestrogen to induce proliferation. J Endocrinol 1991; 128:115–20.
56. Tomooka Y, DiAugustine RP, McLachlan JA. Proliferation of mouse uterine epithelial cells in vitro. Endocrinology 1986;118:1011–8.
57. Glasser SR, Julian J, Decker GL, Tang J-P, Carson DD. Development of morphological and functional polarity in primary cultures of immature rat uterine epithelial cells. J Cell Biol 1988;107:2409–23.
58. Carson DD, Tang J-Y, Julian J, Glasser SR. Vectorial secretion of proteoglycans by polarised rat uterine epithelial cells. J Cell Biol 1988; 107:2425–35.
59. Mani SK, Decker GL, Glasser SR. Hormonal responsiveness of immature rabbit uterine epithelial cells polarized in vitro. Endocrinology 1991;128: 1563–73.
60. Schatz F, Gordon RE, Laufer N, Gurpide E. Culture of human endometrial cells under polarizing conditions. Differentiation 1990;42:184–90.
61. Enders AC. Anatomical aspects of implantation. J Reprod Fertil 1976; 25:1–15.
62. McCormack SA, Glasser SR. Differential response of individual uterine cell types from immature rats treated with estradiol. Endocrinology 1980;106: 1634–49.
63. Springer TA. The sensation and regulation of interactions with the extracellular environment: the cell biology of lymphocyte adhesion receptors. Ann Rev Cell Biol 1990;6:359–402.
64. Brandley BK, Swiedler SJ, Robbins PW. Carbohydrate ligands of the LEC cell adhesion molecules. Cell 1990;63:861–3.
65. Springer TA. Adhesion receptors in the immune system. Nature (London) 1990;346:425–34.
66. Springer TA, Lasky LA. Sticky sugars for selectins. Nature (London) 1991; 349:196–7.
67. Lindenberg S, Kimber SJ, Kallin E. Carbohydrate binding properties of mouse embryos. J Reprod Fertil 1990;89:431–9.
68. Larsen E, Palabrica T, Sajer S, et al. PADGEM-dependent adhesion of platelets to monocytes and neutrophils is mediated by a lineage-specific carbohydrate LNF III (CD15). Cell 1990;63:467–74.
69. Lowe JB, Stoolman LM, Nair RP, Larsen RD, Berhend TL, Marks RM. ELAM-1-dependent cell adhesion to vascular endothelium determined by transfected human fucosyltransferase. Cell 1990;63:475–84.

70. Phillips ML, Nudelman E, Gaeta FCA, et al. ELAM-1 mediates cell adhesion by recognition of a carbohydrate ligand sialyl Le^x. Science 1990;250:1130-2.
71. Bird JM, Kimber SJ. Oligosaccharides containing fucose linked α(1-3) and α(1-4) to N-acetyl glucosamine cause decompaction of mouse morulae. Dev Biol 1984;104:449-60.
72. Fenderson BA, Zehavi U, Hakomori S-I. A multivalent lacto-N-fucopentaose III conjugate decompacts preimplantation mouse embryos while the free oligosaccharide is ineffective. J Exp Med 1984;160:1591-6.
73. Kimber SJ, Lindenberg S, Lundblad A. Distribution of some Galβ1-3(4)GlcNAc related carbohydrate antigens on the mouse uterine epithelium in relation to the peri-implantation period. J Reprod Immunol 1988;12:297-313.
74. Kimber SJ, Lindenberg S. Hormonal control of a carbohydrate epitope involved in implantation in mice. J Reprod Fertil 1990;89:13-21.
75. Kimber SJ. Glycoconjugates and cell-surface interactions in pre- and peri-implantation mammalian development. Int Rev Cytol 1990;120:53-167.
76. Lasky LA. Lectin cell adhesion molecules (LEC-CAMS): a new family of cell adhesion proteins involved with inflammation. J Cell Biochem 1991;45:139-46.
77. Chavez DJ, Anderson TL. The glycocalyx of the mouse uterine luminal epithelium during estrus, early pregnancy, the peri-implantation period and delayed implantation, 1. Acquisition of *Ricinus communis* 1 binding sites during pregnancy. Biol Reprod 1985;32:1135-42.
78. Smith CW, Kishimoto TK, Abbass O, et al. Chemotactic factors regulate lectin adhesion molecule 1 (LECAM-1)-dependent neutrophil adhesion to cytokine stimulated endothelial cells in vitro. J Clin Invest 1991;87:609-18.
79. Lawrence MB, Springer TA. Leukocytes roll on a selectin at physiologic flow rates: distinction from and prerequisite for adhesion through integrins. Cell 1991;65:859-73.
80. Bryne M, Thrane PS, Dabelsteen E. Loss of expression of blood group antigen H is associated with cellular invasion and spread of oral squamous cell carcinomas. Cancer 1990;67:613-8.
81. Miyake M, Hakomori S-I. A specific cell surface glycoconjugate controlling cell motility: evidence by functional monoclonal antibodies that inhibit cell motility and tumour cell metastasis. Biochem J 1991;30:3328-34.
82. Perona RM, Wassarman PM. Mouse blastocysts hatch in vitro by using a trypsin-like proteinage associated with cells of mural trophectoderm. Dev Biol 1986;114:42-52.

19
Chemical Signals in Embryo-Maternal Dialogue: Role of Growth Factors

S.K. Dey and B.C. Paria

In 1947, Corner described that the uterine chamber is actually a less favorable place for early embryos than, say, the anterior chamber of the eye, except when the hormones of the ovary act upon it and change it to a place of superior efficiency for its new functions (1). More recent is the realization that initiation and establishment of pregnancy result from an intimate interaction between the developing embryo and the differentiating uterus. However, although conceptually realized, the nature and timing of such a two-way dialogue between the blastocyst and the uterus are still challenging questions. The establishment of early pregnancy is a conglomeration of several synchronized and precisely controlled embryonic and maternal components: (i) the migration, spacing, and orientation of the embryo within the uterus, (ii) development of the embryo to the blastocyst stage and differentiation of the uterus to the receptive state, (iii) apposition and adhesion of the trophoblast with the uterine epithelium followed by attachment to the uterus, and (iv) increased capillary permeability and blood flow in the uterine vascular bed (2). These events are primarily dependent on temporal and cell type-specific interactions between progesterone and estrogen. However, the molecular and cellular mechanisms involved in these steroid hormone-regulated processes are not clearly understood. An emerging concept is that progesterone/estrogen effects in the uterus and embryo are mediated by growth factors in an autocrine/paracrine manner (3–7). This is consistent with the findings of synthesis of growth factors and expression of their receptors in the embryo and uterus (8–23). This chapter focuses on growth factors, especially on *epidermal growth factor* (EGF) and *transforming growth factor* α (TGFα) in embryo-uterine interactions during implantation.

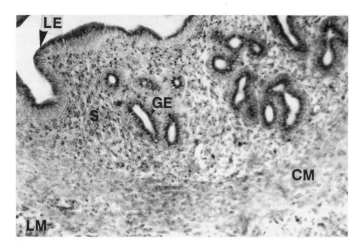

FIGURE 19.1. Immunohistological localization of TGFα in the mouse uterus on day 4 of pregnancy. Bouin's-fixed paraffin-embedded uterine sections (7 μm) were mounted onto poly-l-lysine-coated slides. After deparaffinization and hydration, sections were incubated in primary antibody at a concentration of 4 μg/mL for 24 h at 4°C. Affinity purified polyclonal antibody to TGFα was purchased from BIOTX, Biosciences Corporation of Texas, Houston, Texas. This antibody was raised in sheep against a synthetic peptide corresponding to amino acids 30–50 at the −COOH terminus of rat TGFα. Immunostaining was performed by employing the avidin-biotin-peroxidase complex technique. Red deposits indicate sites of immunoreactive TGFα. A photomicrograph of uterine sections on day 4 of pregnancy is shown at 200×. (LE = luminal epithelium; GE = glandular epithelium; S = stroma; CM = circular muscle; LM = longitudinal muscle.)

Expression of TGFα and EGF in the Preimplantation Mouse Uterus

Results of immunohistochemistry as well as Northern blots and in situ hybridization provide evidence for the synthesis of TGFα in the luminal and glandular epithelia on days 1–4 of pregnancy (day 1 = vaginal plug) as well as in stromal cells on days 3 and 4 of pregnancy (7). The immunoreactive TGFα on day 4 of pregnancy was primarily located at the apical region of the luminal epithelium (Fig. 19.1). As of now, it is not known whether this immunoreactive uterine TGFα on day 4 is the mature TGFα or the unprocessed transmembrane form of TGFα. Similarly, the immunoreactive EGF at or near the apical border of the luminal epithelium on day 4 of pregnancy has not been characterized (3). Since either the mature or the transmembrane form of TGFα can interact with *EGF receptor* (EGF-R) and can initiate cell-cell signaling processes (24),

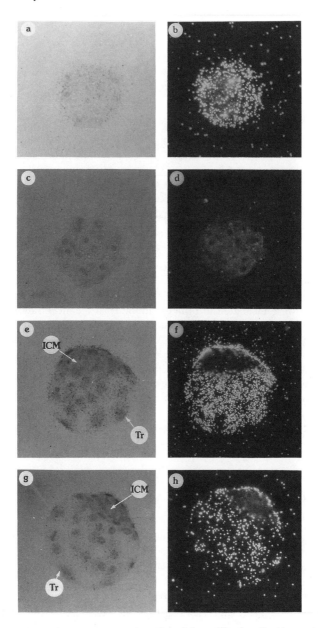

FIGURE 19.2. Microphotographs of autoradiographic localization of ^{125}I-EGF binding in the morula and blastocyst (200×). While binding in a morula is shown in (a) bright and (b) dark fields, nonspecific binding (in the presence of 500-fold molar excess of unlabeled EGF) is shown in (c) bright and (d) dark fields. Localization of binding in a blastocyst is shown in (e) bright and (f) dark fields; that of nonspecific binding is shown in (g) bright and (h) dark fields. (ICM = inner cell mass; Tr = trophectoderm.) Reprinted with permission from Paria and Dey (22).

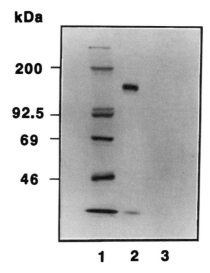

FIGURE 19.3. In situ crosslinking of EGF binding in the mouse blastocyst. Blastocysts were incubated with ^{125}I-EGF in the absence (lane 2) or presence (lane 3) of 500-fold molar excess of unlabeled EGF. Crosslinking was performed by the use of disuccinimidyl suberate. Extracts were analyzed by SDS-PAGE followed by autoradiography. (Lane 1 = radioactive molecular weight markers.) Reprinted with permission from Paria, Tsukamura, and Dey (23).

it will be necessary to determine the predominant form of TGFα in the uterus on day 4 of pregnancy in the mouse.

Expression of EGF-R in the Preimplantation Embryo

Autoradiographic Localization of EGF Binding in the Blastocyst

Specific autoradiographic signals for ^{125}I-EGF binding to the embryonic cell surface were first detected at the 8-cell stage (22). The binding increased at the morula and blastocyst stages (Fig. 19.2). The employment of in situ crosslinking showed that EGF binding in the day 4 blastocyst represents approximately a 170-kd protein (Fig. 19.3). The functionality of this EGF-R was tested by determining autophosphorylation of EGF-R and subcellular *protein kinase* (PTK) activity (23). A 170-kd protein in the blastocyst homogenate was phosphorylated when exposed to EGF in the presence of *adenosine triphosphate* (ATP) (Fig. 19.4). Furthermore, EGF induced about a 2-fold increase in PTK activity when blastocyst homogenates were incubated in the presence of a peptide

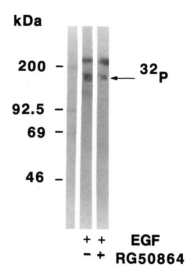

FIGURE 19.4. Autophosphorylation of EGF-R in the blastocyst. Eight-cell embryos were cultured for 24 h to blastocysts in the presence or absence of RG 50864 (40 μM). Blastocyst homogenates were preincubated with EGF (1 μg/mL) for 10 min on ice. The reaction was run on ice for 2 min by adding 5 μCi of (γ-^{32}P) ATP (1 μM). Extracts were subjected to SDS-PAGE followed by autoradiography. The gel was treated with 1N NaOH to determine that the phosphorylation of the 170-kd protein was at the tyrosine moiety. Reprinted with permission from Paria, Tsukamura, and Dey (23).

substrate with tyrosine moiety and ATP (Fig. 19.5). The specificity of EGF-dependent protein phosphorylation in the blastocyst was determined by the use of a tyrphostin compound, RG 50864, a specific inhibitor of EGF-specific PTK activity (25–27). This inhibitor decreased EGF-R phosphorylation and completely blocked PTK activity in the blastocyst (Figs. 19.4 and 19.5) without altering EGF binding to the embryonic cell surface (Fig. 19.6). In contrast, the inactive compound, RG 50862, did not inhibit PTK activity in the blastocyst (Fig. 19.5).

Regulation of EGF Binding in the Blastocyst

In the mouse and rat, the removal of the source of estrogen/progesterone by either ovariectomy or hypophysectomy during the preimplantation period induces delayed implantation (2, 28). Under these conditions, the blastocyst undergoes dormancy and remains viable in the uterus if maintained on progesterone. However, implantation can be induced in these progesterone-treated delayed-implanting animals by a single injection of estrogen (2, 28). In order to delineate whether EGF/TGFα has any role in the process of implantation, regulation of EGF binding was studied in the delayed and activated mouse blastocyst. Within 8 h of

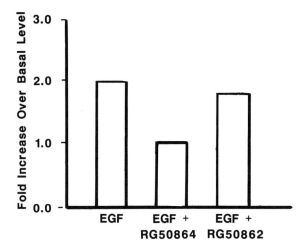

FIGURE 19.5. EGF-dependent PTK activity in the blastocyst. Tyrosine kinase activity was assayed using a synthetic peptide, Arg-Arg-Leu-Ile-Glu-Asp-Ala-Glu-Tyr-Ala-Ala-Arg-Gly, as a substrate. Eight-cell embryos were cultured for 24 h to blastocysts in the presence or absence of the active or inactive tyrphostin (40 μM). Blastocyst homogenates were preincubated with or without EGF (16.5 nM). The enzyme reaction was initiated by the addition of 20 μM ATP containing 10 μCi of (γ-^{32}P) ATP. Basal activity (without the substrate) was subtracted from the experimentals (with the substrate) to calculate the enzyme activity (fold increase over basal level/embryo). A value of 1.0 is equal to the basal level. Experiments with EGF alone were repeated 5 times and with EGF + RG 50864 or with EGF + RG 50862 2 times. Reprinted with permission from Paria, Tsukamura, and Dey (23).

ovariectomy on day 4 of pregnancy, EGF binding completely disappeared from the blastocyst cell surface, and the levels of binding remained almost undetectable even when the animals were maintained on progesterone. However, EGF binding rapidly reappeared on the blastocyst cell surface within 8 h of a single injection of estrogen. EGF binding, however, was not induced in delayed blastocysts cultured in vitro for 24 h in the presence of estrogen. The data suggest that the effects of estrogen in the regulation of EGF binding to the blastocyst cell surface are mediated via the uterus.

Effects of EGF/TGFα on Preimplantation Embryo Development and Zona Hatching In Vitro

A cooperative interaction among preimplantation mouse embryos and the role of growth factors on their development and growth has recently been described (22, 23). It was noted that 2-cell embryos cultured singly

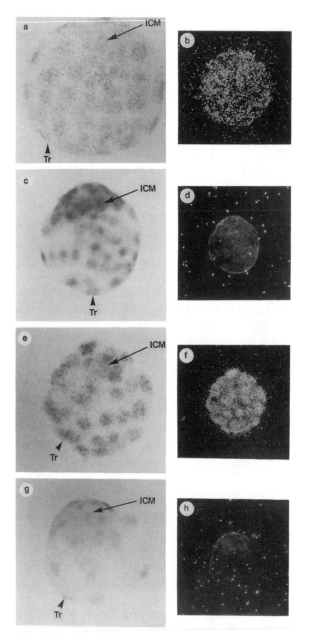

FIGURE 19.6. Autoradiographic localization of EGF binding in the mouse blastocyst. ^{125}I-EGF binding was performed on blastocysts developed from 8-cell embryos in the presence or absence of RG 50864 (40 µM). Binding in control or RG 50864-treated blastocysts are shown in a and e (bright field 400×) and b and f (dark field 200×), respectively, and that of nonspecific binding in c and g (bright field) and d and h (dark field), respectively. Blastocysts were stained with hematoxylin following autoradiographic development. (ICM = inner cell mass; Tr = trophectoderm.) Reprinted with permission from Paria, Tsukamura, and Dey (23).

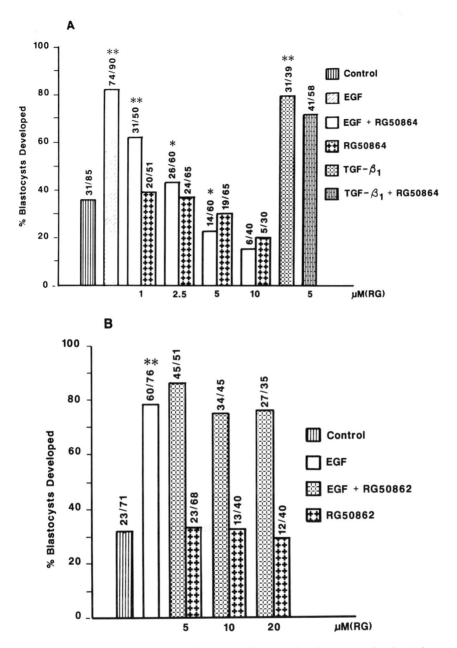

FIGURE 19.7. Effects of growth factors and/or tyrphostins on preimplantation embryo development. Two-cell embryos were cultured singly in 25 µL of medium for 72 h. The numbers on each bar indicate number of blastocysts developed/number of embryos cultured. Each experiment was repeated 4–7 times with the exception of numerous replicates of controls included in each experimental repetition. EGF was used at 4 ng/mL and TGFβ$_1$ at 2 ng/mL. Statistical comparisons were made between the treatment groups as control and EGF; EGF and EGF + RGs; control and RGs; control and TGFβ$_1$; and TGFβ$_1$ and TGFβ$_1$ + RG. (*P < 0.01; **P < 0.001 [chi square test].) *A:* Effects of RG 50864; *B:* effects of RG 50862. Reprinted with permission from Paria, Tsukamura, and Dey (23).

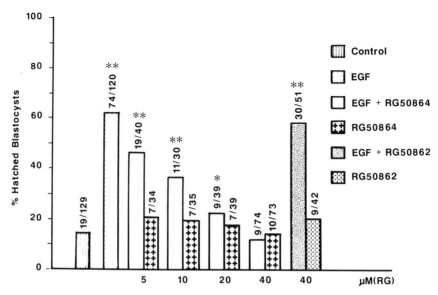

FIGURE 19.8. Effects of EGF and/or tyrphostins on blastocyst hatching in vitro. Eight-cell embryos were cultured singly in 25 µL of medium for 72 h. EGF was used at 10 ng/ml. Each experiment was repeated at least 4 times with the exception of numerous controls. Statistical comparisons followed the same fashion as described in Figure 19.7. (*$P < 0.01$; **$P < 0.001$ [chi square test].) Percentages of blastocysts that hatched were used for analysis. Reprinted with permission from Paria, Tsukamura, and Dey (23).

showed inferior development to blastocysts compared to a remarkable improvement in development when embryos were cultured in groups or singly with EGF or TGFα. However, this EGF-induced improvement of singly cultured embryos to blastocysts was impaired by RG 50864, an inhibitor of EGF-specific protein tyrosine phosphorylation, in a concentration-dependent manner (Fig. 19.7A). The developmental arrest was mostly at the 8-cell stage, which is consistent with the appearance of EGF binding at this stage (22, 23). On the other hand, a structurally similar inactive tyrphostin compound, RG 50862, did not interfere with EGF-induced embryonic development (Fig. 19.7B). Moreover, RG 50864 failed to impair TGFβ$_1$-induced embryonic development (Fig. 19.7A). EGF also influenced zona hatching of blastocysts in vitro, and the beneficial effect of EGF in this response was obliterated by RG 50864 (Fig. 19.8). Taken together, the results show that protein tyrosine phosphorylation by EGF or TGFα ligand-receptor interaction is important for preimplantation embryo development and zona hatching in vitro.

Effects of EGF on Implantation

Initiation of implantation in the mouse occurs on the evening of day 4. Ovariectomy in the morning of day 4 (0600 h) interferes with the implantation process if the animal is not treated with progesterone and estrogen. However, in the absence of estrogen, the progesterone-treated mice undergo delayed implantation. Under this condition, as described above, EGF binding from the blastocyst cell surface disappeared within 8 h. It was postulated that EGF could induce implantation in the absence of estrogen as long as EGF binding on the blastocyst cell surface persisted. Indeed, a single intraperitoneal injection of EGF (500 ng/mouse) induced implantation in most of the mice within 4 h of ovariectomy in the absence of estrogen; EGF was not effective when given 8 h after ovariectomy. The response to EGF appears to be specific since IGF-I was not effective in inducing implantation under these conditions (unpublished results).

Conclusions

The work described above presents evidence that growth factors, especially EGF or TGFα, can participate in preimplantation embryo development and blastocyst zona hatching in vitro and initiation of implantation. The source of these growth factors could be the embryo or the reproductive tract. While growth factors produced and secreted by an embryo can affect its functions in an autocrine manner, the possibility also exists that these factors influence the functions of neighboring embryos and/or the reproductive tract in a paracrine manner. Similar autocrine/paracrine functions could be envisioned for the growth factors produced and secreted by specific cell types of the reproductive tract (oviduct and uterus) that can influence embryonic or cell type-specific functions of the reproductive tract. Although evidence for synthesis of growth factors in the embryo and the uterus is continually forthcoming, very little is known about the synthesis of growth factors in the oviduct. Our recent preliminary studies, however, indicate the presence of several growth factors in the mouse oviduct (unpublished results). Since successful initiation and establishment of pregnancy results from a complex interaction between the developing embryo and the uterus, the relative contribution of growth factors of embryonic or reproductive tract origin in this interaction remains obscure. However, one interesting hypothesis that requires thorough investigation is the interaction between the precursor forms of EGF or TGFα and the EGF-R in embryo-uterine interaction during implantation. If it is found that the luminal epithelial TGFα in the mouse uterus on day 4 of pregnancy accumulates primarily in the precursor form, then it could be envisioned that blastocyst-uterus signaling could be mediated by the TGFα precursor interacting with EGF-R on blastocyst cell surface. Work is in progress in our laboratory to test this hypothesis.

References

1. Corner GW. In: Corner GW, ed. The hormones in human reproduction. Princeton, N.J.: Princeton University Press, 1947:259.
2. Psychoyos A. Endocrine control of egg implantation. In: Greep RO, Astwood EG, Geiger SR, eds. Handbook of physiology. Washington, D.C.: American Physiological Society, 1973:187-215.
3. Huet-Hudson YM, Andrews GK, Dey SK. Epidermal growth factor and pregnancy in the mouse. In: Heyner S, Wiley LM, eds. Early embryo development and paracrine relationships. New York: Wiley-Liss, 1990:125-36.
4. Huet-Hudson YM, Chakraborty C, De SK, Suzuki Y, Andrews GK, Dey SK. Estrogen regulates synthesis of EGF in mouse uterine epithelial cells. Mol Endocrinol 1990;4:510-23.
5. Tamada H, McMaster MT, Flanders KC, Andrews GK, Dey SK. Cell type-specific expression of transforming growth factor-β1 in the mouse uterus during the periimplantation period. Mol Endocrinol 1990;4:965-72.
6. DiAugustine RP, Petruz P, Bell GI, et al. Influence of estrogens on mouse uterine epidermal growth factor receptor mRNA. Mol Endocrinol 1988; 2:230-5.
7. Tamada H, Das SK, Andrews GK, Dey SK. Cell type-specific expression of transforming growth factor-α in the mouse uterus during the periimplantation period. Biol Reprod 1991;45:365-72.
8. Murphy LJ, Murphy LC, Friesen HG. Estrogen induces insulin-like growth factor-I expression in the rat uterus. Mol Endocrinol 1987;1:445-50.
9. Norstedt G, Levinoritz A, Eriksson H. Regulation of uterine insulin-like growth factor I mRNA and insulin-like growth factor II mRNA by estrogen in the rat. Acta Endocrinol 1989;8120:466-72.
10. Rappolee DA, Brenner CA, Schultz R, Mark D, Werb Z. Developmental expression of PDGF, TGFα and TGFβ genes in preimplantation mouse embryos. Science 1988;241:1823-5.
11. Simmen FA, Simmen RCM. Peptide growth factors and proto-oncogenes in mammalian conceptus development. Biol Reprod 1991;44:1-5.
12. Chakraborty C, Tawfik OW, Dey SK. Epidermal growth factor binding in rat uterus during the periimplantation period. Biochem Biophys Res Commun 1988;154:564-9.
13. Lingham RB, Stancell GM, Loose-Mitchell DS. Estrogen induction of the epidermal growth factor receptor mRNA. Mol Endocrinol 1988;2:230-5.
14. Ghahary A, Murphy LJ. Uterine insulin-like growth factor-I receptors: regulation by estrogen and variation throughout the estrous cycle. Endocrinology 1989;125:597-604.
15. Corps AN, Brigstock DR, Littlewood CJ, Brown KD. Receptors for epidermal growth factor and insulin-like growth factor-I on preimplantation trophoderm of the pig. Development 1990;110:221-7.
16. Lin T-H, Mukku VR, Verner G, Kirkland JL, Stancel GM. Autoradiographic localization of epidermal growth factor receptors to all major uterine cell types. Biol Reprod 1988;38:403-11.

17. Mukku VR, Stancel GL. Regulation of epidermal growth factor receptors by estrogen. J Biol Chem 1985;260:9820–4.
18. Adamson ED, Deller MJ, Warshaw JB. Functional EGF receptors are present on mouse embryo tissues. Nature 1981;291:656–9.
19. Adamson ED, Meek J. The ontogeny of epidermal growth factor receptors during mouse development. Dev Biol 1984;103:62–70.
20. Adamson ED. EGF-receptor activity in mammalian development. Mol Reprod Dev 1990;27:16–22.
21. Brown MJ, Zogg JL, Schultz GS, Hilton FK. Increased binding of epidermal growth factor at the implantation sites in mouse uteri. Endocrinology 1989;124:2882–8.
22. Paria BC, Dey SK. Preimplantation embryo development in vitro: cooperative interactions among embryos and role of growth factors. Proc Natl Acad Sci USA 1990;87:4756–60.
23. Paria BC, Tsukamura H, Dey SK. Epidermal growth factor specific protein tyrosine phosphorylation in preimplantation embryo development. Biol Reprod 1991;45:711–8.
24. Derynck R. Transforming growth factor-α. Mol Reprod Dev 1990;27:3–9.
25. Yaish P, Gazit A, Gilon C, Levitzk A. Blocking of EGF-dependent cell proliferation by EGF receptor kinase inhibitors. Science 1988;242:933–5.
26. Lyall RM, Zilberstein A, Gazit A, Gilon C, Levitzki A, Schlessinger J. Tyrphostins inhibit epidermal growth factor (EGF)-receptor tyrosine kinase activity in living cells and EGF-stimulated cell proliferation. J Biol Chem 1989;264:14503–9.
27. Margolis B, Rhee SG, Felder S, et al. EGF induces tyrosine phosphorylation of phospholipase C-II: a potential mechanism for EGF receptor signalling. Cell 1989;57:1101–7.
28. Yoshinaga K, Adams CE. Delayed implantation in the spayed, progesterone-treated adult mouse. J Reprod Fertil 1966;12:593–5.

20

Regulation of Chorionic Gonadotropin Secretion by Cultured Human Blastocysts

ALEXANDER LOPATA AND KAREN OLIVA

In human *in vitro fertilization* (IVF) and *embryo transfer* (ET), the successful implantation rate, leading to term pregnancy, is currently believed to be approximately 10% per embryo placed in the uterine cavity (1, 2). Thus, one of the main problems that has to be solved is why about 90% of replaced IVF embryos fail to implant. In treatment cycles with ovarian hyperstimulation, the major causes of implantation failure have been attributed to endocrine-induced anomalies of endometrial development and follicular growth. It has been proposed that ovarian overstimulation, the resulting multiple follicular aspiration, and associated high plasma progesterone and estradiol levels, may induce one or more of the following endometrial conditions that are unfavorable for implantation (3-7): (i) morphological and secretory abnormalities (3, 8) associated with surface changes unfavorable for receptivity (7, 9), (ii) asynchronous development of the uterine environment with respect to embryonic development (3, 10), and (iii) an altered duration or setting of the implantation window (11).

Alternatively, implantation may fail as a result of embryonic abnormalities. These may arise as a result of anomalies of meiotic maturation (12), penetration by abnormal sperm (13), or other fertilization defects.

Embryonic Signals

It is not widely recognized, however, that every embryo placed in the uterine cavity must ultimately be capable of releasing a range of survival signals to ensure that its presence is recognized and, at the same time, to avoid being rejected as a foreign graft. Within the uterine cavity, the mother's cellular immune response is apparently inhibited by the release of a range of immunosuppressive signals from the blastocyst. To date, the following embryonic signals that have the potential to modulate the maternal immune system have been identified: *human chorionic gonadotropin* (hCG) (14, 15); early pregnancy factor (16); low-molecular weight immunosuppressive substances (17, 18); prostaglandin E_2 (19, 20); interleukin-1 and interleukin-6 (21–23); interferon α (24); and platelet activating factor (25, 26).

In this chapter, we will primarily examine hCG secretion by the blastocyst and some of the factors that influence it. At this point, however, it is worth pointing out that it is only at the blastocyst stage that a full spectrum of immunosuppressive signals is likely to be released. Linked with this notion is the concept that embryos placed in the uterine cavity in an IVF and ET program may be better adapted to surviving immunological rejection if they are transferred at the blastocyst stage rather than at early cleavage stages. This proposal may be particularly applicable to women who have failed to conceive following embyro replacements in at least three previous treatment cycles (27).

Role of hCG Secreted by the Blastocyst

Studies on the regulation of hCG secretion by blastocysts are pertinent to understanding the factors involved in establishing and maintaining early pregnancy for several reasons. Apart from the gonadotropin's known role in rescuing the corpus luteum (28), it has already been mentioned that hCG may play a part as an immunosuppressive agent that protects the embryo as it becomes grafted onto the endometrium at implantation. In addition, hCG has recently been shown to promote the growth of cytotrophoblasts (29), and thus it acts as an autocrine growth factor. The hormone may, therefore, be essential for establishing an adequate early placenta. As a glycoprotein associated with the cell membrane of trophoblasts, hCG may also play a part in the adhesion of polar trophectoderm (30) to receptive endometrial epithelial cells. Moreover, since hCG is known to have protease activity, it would be of interest to know whether its secretion in the region where polar trophectoderm faces the endometrium reaches a sufficiently high focal concentration to alter the integrity of the cytocalyx lining the surface of the uterine epithelium (31), hence promoting further apposition and, subsequently, the early stage of invasion by the blastocyst.

Methods

The protocols used for ovarian stimulation with a combination of clomiphene citrate and human menopausal gonadotropin and the procedures used for aspirating follicles and collecting oocytes under ultrasound control have been described previously (32). In preparation for IVF, the oocytes recovered from the follicular aspirate were rinsed in *human tubal fluid* (HTF) medium, buffered with HEPES, and then incubated in bicarbonate-buffered HTF containing 10% heat-inactivated *human cord serum* (HCS)—or, alternatively, the same level of patient's serum—in a humidified CO_2 incubator at 37°C. After 4–6 h of maturation culture, the oocytes were inseminated with $5-10 \times 10^4$ mobile sperm/mL HTF. On the next day, oocytes were examined for the presence of pronuclei after mechanical removal of undispersed cumulus and corona cells. Fertilized oocytes were transferred into freshly equilibrated HTF containing 1% HCS or the patient's serum. After a further 24 h, cleaving embryos were assessed, and a maximum of 3 embryos per patient were transferred to the uterus at 2- to 6-cell stages of development. Any excess embryos that arose from bipronucleate oocytes and that were considered to be of suitable quality to survive freeze-thawing procedures were cryopreserved for replacement in subsequent natural cycles.

The studies described in this chapter were carried out with the remaining surplus embryos that were considered to be unsuitable for clinical use. These comprised embryos that arose from monopronucleate or tripronucleate oocytes and bipronucleate cleavage stages that were of poor quality due to pronounced cytoplasmic fragmentation (32, 33). All of these surplus embryos were removed from HTF + 1% serum on the second day after insemination and were subsequently cultured individually in 1-mL aliquots of serum-free *alpha minimum essential medium* (α-MEM) (32) in Nunc 4-well plates. The embryos were cultured in a humidified atmosphere of 5% CO_2 + 5% O_2 + 90% N_2 at 37°C until they either degenerated or attained the blastocyst stage.

Under these conditions, the embryos that continued to cleave produced expanded blastocysts between days 5 and 6 after insemination and hatched blastocysts by day 7. The blastocysts that failed to hatch produced a clump of cells that increased in size until, in most cases, the intrazonal space was filled completely. Studies on hCG secretion by hatched or intrazonal blastocysts commenced on day 7 and were terminated on day 14 after insemination. Each blastocyst tissue was transferred every 24 h to a freshly equilibrated 1 mL of α-MEM containing a supplement that was being studied or control medium. The spent culture medium was frozen at −80° each day and kept until a complete series was available for hCG assays. The levels of dimer hCG were measured in coded samples using an immunoradiometric assay kit, as described previously (32). Since the medium was replaced daily, the levels of hCG measured in every sample represent the 24-h output of gonadotropin for each blastocyst. The sensitivity, specificity, and precision of the assay have been reported previously (34).

Effect of Serum on hCG Secretion

Human cord serum was found to be one of the most effective stimulants of hCG production by the blastocyst tissues in culture. When used at a concentration of 1% (containing <2 mIU/mL hCG) in α-MEM, HCS induced an increase in hCG secretion between days 7 and 8 after fertilization (Fig. 20.1). This increase was observed in both hatched and intrazonal blastocysts, although there was considerable variation between tissues. Subsequently, there was a rapid rise in the daily output of hCG up to about day 11 postfertilization; then the 24-h gonadotropin production tended to plateau to day 14. It is of interest that the daily output of baboon chorionic gonadotropin, produced by blastocysts that were recovered from the uterine cavity and subsequently cultured in Ham's F10 + 20% fetal calf serum (35), was similar to the levels released by human blastocysts at equivalent stages of development. In contrast, marmoset blastocysts produced 100×–1000× less chorionic gonadotropin when cultured in Dulbecco's modification of Eagle's medium supple-

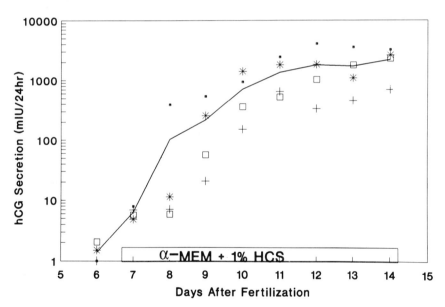

FIGURE 20.1. Effect of 1% HCS added to α-MEM on hCG secretion by cultured blastocyst tissues. The horizontal bar represents the time after fertilization at which the serum was added and the duration of culture in the presence of serum. Four blastocysts were cultured under these conditions, and the solid line shows the average daily production of hCG for this group of embryos. The symbols at each time interval represent daily hCG levels released by individual blastocysts.

mented with 10% human cord plasma (36) when compared with developmental stages similar to those of the human and baboon blastocysts.

Platelet Activating Factor (PAF) as an Embryonic Signal

Role of PAF in Embryo Development and Implantation

It has been reported that the presence of exogenous *platelet activating factor* (PAF) in the culture medium increased the growth rate of mouse embryos (37) and significantly improved the viability of both mouse and human embryos (37, 38). In addition, it has recently been shown that human embryos produced by IVF released an embryo-derived PAF into the culture medium (39, 40). It was also reported that the embryo-derived PAF was homologous to PAF-acether and that it acted as an embryonic paracrine growth factor (37). In view of the above findings, it was difficult to reconcile why PAF antagonists had no effect on embryo development in vitro nor in utero up to the blastocyst stage (40, 41). However, recent studies from the same group of investigators have suggested that the antagonists of embryo-derived PAF were inhibitors of development beyond the blastocyst stage, particularly of trophoblast cell outgrowth (37). On the basis of these results and the observation that PAF antagonists reduced pregnancy rates in the Quackenbush strain of mice (41, 42), it was postulated that embryo-derived PAF is essential mainly at the time of blastocyst implantation (40).

Role of PAF in hCG Secretion

On the basis of the investigations that implicated embryo-derived PAF in the autocrine control of trophoblast growth and blastocyst implantation, we have carried out experiments to determine whether this endogenous embryonic signal was also involved in hCG secretion. We were particularly interested to know whether the stimulatory effect of HCS on hCG secretion was promoted by embryo-derived PAF, possibly after it was released from the blastocyst tissues in response to serum. The positive influence of PAF on blastocyst hCG output when the PAF was added to the culture medium at $0.1-1.0\,\mu M$ concentrations was previously demonstrated in a separate series of experiments (see below). Thus, in the studies to evaluate the possible involvement of embryo-derived PAF in the serum-induced hCG secretion, we used a specific PAF antagonist to block the action of the putative embryonic PAF signal (37, 41).

In the first group of experiments, blastocyst tissues were grown in α-MEM + 1% HCS up to day 9 after fertilization, and the PAF antagonist, WEB 2086, was added at a concentration of $10\,\mu M$ on the 10th and 11th days of culture. Subsequently, the blastocyst tissues were again incubated in α-MEM + 1% HCS in the absence of WEB 2086, and the cultures were terminated on day 14 after fertilization. In these studies, the WEB 2086 was dissolved in *dimethyl sulphoxide* (DMSO), and the PAF antagonist was added to the culture medium from dilutions of the stock solution so that the final concentration of DMSO in the cultures was 0.1%. Control experiments were performed, therefore, in which blastocysts were cultured as described above, except that 0.1% DMSO was added to the medium on the 10th and 11th days of culture.

The hCG output by 4 different blastocysts cultured in the presence of 1% HCS and WEB 2086 on days 10 and 11 is shown in Figure 20.2. As may be seen, there was considerable variation in the levels of hCG released by the blastocysts during the first 3 days of culture. Although this probably reflects differences in embryonic quality, there was nevertheless a consistent rapid increase in the daily hCG output up to day 10 of culture from all 4 blastocysts. However, in the presence of 10-μM WEB 2086 and 0.1% DMSO over the next 2 days, there was an 80%–100% decline in hCG output from 3 of the 4 blastocysts studied. On removing the PAF antagonist and DMSO from the culture medium, the blastocysts that demonstrated a decline in hCG secretion appeared to regain the ability to increase their daily output of gonadotropin (Fig. 20.2).

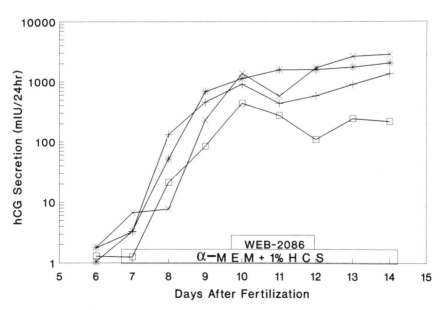

FIGURE 20.2. The hCG secretion by 4 blastocysts cultured in α-MEM + 1% HCS to which WEB 2086 (10 μM) and 0.1% DMSO were added on days 10 and 11. The graphs represent daily hCG levels released by individual blastocysts.

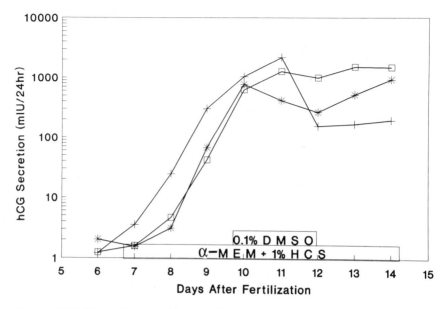

FIGURE 20.3. The hCG secretion by 3 blastocysts cultured in α-MEM + 1% HCS to which 0.1% DMSO was added on days 10 and 11. The graphs represent daily hCG levels released by individual blastocysts.

The next group of experiments was performed in the same way as those described above, except that the PAF antagonist was excluded and 0.1% DMSO was added on its own on days 10 and 11 of culture (Fig. 20.3). These control studies, using the vehicle only, showed that the low concentration of DMSO was able to induce a marked decline in hCG output in 2 of 3 blastocysts that were studied. This drop in hCG secretion was not due to degenerative changes in the embryonic tissues since removal of the DMSO was followed by an increase in gonadotropin output up to day 14 of culture (Fig. 20.3). It would appear, therefore, that the transient decline in hCG secretion observed in the first group of experiments could be attributed to the presence of 0.1% DMSO in the medium rather than an inhibitory effect exerted by the PAF antagonist. It is also noteworthy that in each experimental group, there was a blastocyst resistant to the effect of DMSO. In one of these (PAF antagonist group), the hCG secretion was also unaffected by the presence of 10-μM WEB 2086 on days 10 and 11 of culture.

To avoid the deleterious effects of DMSO on hCG secretion, the WEB 2086 was prepared in an aqueous solution (the WEB 2086 was dissolved in dilute HCl and the stock solution neutralized with NaOH) for the next series of studies. In these experiments, 4 blastocysts were incubated in α-MEM in the presence of 1% HCS and 10-μM WEB 2086 from days 7 to

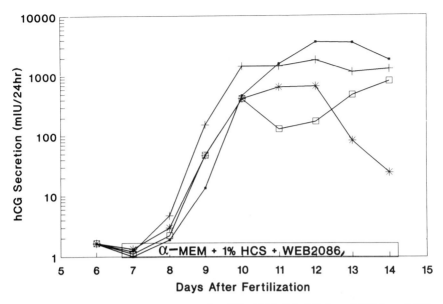

FIGURE 20.4. Effect of 1% HCS and WEB 2086 (10 μM) in α-MEM on hCG secretion by cultured blastocyst tissues. The horizontal bar shows the time after fertilization at which the serum and WEB 2086 were added and the duration of culture in the presence of these ingredients. The graphs represent daily hCG levels released by individual blastocysts.

14 of culture. As before, the blastocyst tissues were transferred to fresh medium each day, and the 24-h hCG outputs were measured in the spent supernatants. The pattern and levels of hCG secretion (Fig. 20.4) compared to cultures in the absence of PAF antagonist (Fig. 20.1) appeared to be unaffected by the presence of WEB 2086 up to day 12 of culture in 3 of 4 blastocysts studied. These 3 embryonic tissues subsequently showed a slight to severe decline in the daily hCG output, which may have been due to degenerative changes in the tissues. The pattern of hCG secretion in the fourth blastocyst was similar to that of the others up to day 10 of culture, but then a transient decrease followed by a rising hCG output was observed (Fig. 20.4). Overall, however, the weight of evidence suggests that the PAF antagonist did not influence the initiation and maintenance of hCG secretion during the early stages of blastocyst tissue culture.

The failure of PAF antagonist to influence hCG production by blastocysts could be explained by either a lack of responsiveness of embryonic tissues to PAF or a lack of production of effective levels of PAF by the cultured tissues. The former possibility was tested by determining whether exogenous PAF could stimulate hCG secretion by blastocysts cultured in the absence of serum. It was found that both 0.1-µM and 1.0-µM concentrations of PAF significantly increased hCG output from blastocyst tissues during at least the first 3 days of culture. However, experiments have not yet been done to determine whether blastocysts release endogenous PAF during this period of their growth. If they do not release PAF in vitro, then the serum-induced increase in hCG secretion is independent of stimulation by PAF. If, however, blastocyst-derived PAF is released by the cultured tissues, then it must play a minor role in promoting hCG secretion since the output of the gonadotropin was not affected by the specific PAF antagonist, WEB 2086.

Influence of Growth Factors on hCG Secretion

The published literature on the control of hCG secretion by the placenta is all based on cultured or explanted trophoblast cells or on choriocarcinoma cell lines used as models of placental tissue. The early data have been comprehensively reviewed by Hussa (43). Some of the recent reports have again emphasized that the response of trophoblast cells to growth factors and other stimulatory or inhibitory agents depends on whether the cytotrophoblasts were isolated from first- or third-trimester placentas (44), as well as on the conditions of culture (45). Moreover, there appear to be notable differences in the mechanisms that control hCG secretion in cultured cytotrophoblasts compared with choriocarcinoma cell lines (46). It is likely, therefore, that previously established studies will not provide clear guidelines on the factors that regulate the initiation and maintenance of hCG secretion by blastocyst tissues.

Effect of Insulin on hCG Secretion

The stimulatory action of insulin on hCG secretion by JEG-3 and JAR choriocarcinoma cells, but not by trophoblast cells derived from term placentas, is a clear example of the disparity in response between different cell types (46). It has been postulated that the effect of insulin may be mediated through the *insulin-like growth factor I receptors* (IGF-I-R) detected in the choriocarcinoma cells. At present, there is no information on whether trophoblast cells derived from early pregnancy possess either the insulin or IGF-I receptors. However, the IGF-II gene has been shown to be highly expressed in cytotrophoblasts of first-trimester placentas, but not in preimplantation human embryos (47). On the basis of the latter findings, it has been proposed that IGF-II is an endogenously produced growth factor that participates in the rapid proliferation of cytotrophoblasts after implantation has been established (47). It was of interest to determine, therefore, whether insulin was involved in the functional differentiation of the trophectoderm during the earlier periimplantation stages of development.

The role of insulin on blastocyst function was studied by determining the influence of the hormone on hCG output in serum-free α-MEM. Insulin was added to the medium at a concentration of 25 µg/mL at the end of the 6th day after fertilization. As before, 1-mL culture samples were collected daily for hCG assay and the blastocyst tissues were transferred to fresh α-MEM containing insulin at the end of each day. Five blastocysts were cultured in this way up to the end of day 12 after fertilization; subsequently, these tissues were incubated daily in α-MEM supplemented with 25-µg/mL insulin, 25-µg/mL transferrin, and 25-µg/mL selenium (ITS mixture) to the end of day 14.

Insulin on its own in serum-free medium induced a marked secretion of hCG from all blastocysts being studied by day 8 after fertilization (Fig. 20.5). As in our previous studies, there was a large range of responses between blastocysts, although the average daily output of gonadotropin was about 10-fold lower than that from blastocyst tissues that were stimulated with 1% HCS from the outset (compare Figs. 20.1 and 20.5).

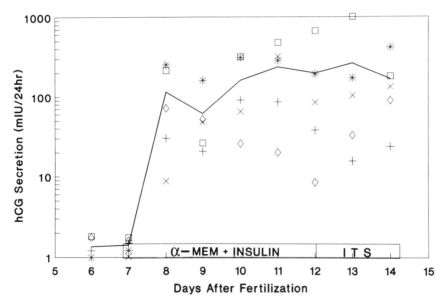

FIGURE 20.5. Effect of insulin (25 µg/mL) on hCG secretion by blastocyst tissues cultured in serum-free α-MEM. The insulin was present in the medium from day 7 to the end of day 11, and the tissues were subsequently cultured in insulin (25µg/mL) + transferrin (25 µg/mL) + selenium (25 ng/mL) to day 14. Five blastocysts were cultured under these conditions, and the solid line shows the average daily production of hCG. The symbols at each time interval represent daily hCG levels released by individual blastocysts.

It was of interest to note that 4 of the 5 embryonic tissues cultured in α-MEM + insulin alone to the end of day 12 responded by an increased output of gonadotropin when the medium was supplemented with the ITS mixture. These findings indicate that some of the hCG-producing tissues remained viable in the presence of insulin up to day 12 of culture under serum-free conditions. Moreover, the boost of hCG release during the last 2 days of culture in the presence of ITS indicates that the 3 components in this supplement acted synergistically to increase gonadotropin secretion.

Comparing Effects of Insulin with Serum on hCG Secretion

A comparison between the action of HCS and insulin on its own is shown in Figure 20.6. In these studies, individual blastocysts were cultured in 1 mL of serum-free α-MEM to the end of day 9 postfertilization before supplements were added. At the commencement of day 10, either 1% HCS or insulin (25 μg/mL) were included in the medium, and the embryos were cultured in each supplement to the end of day 13, whereupon, the tissues were returned to α-MEM without additives. The same culture

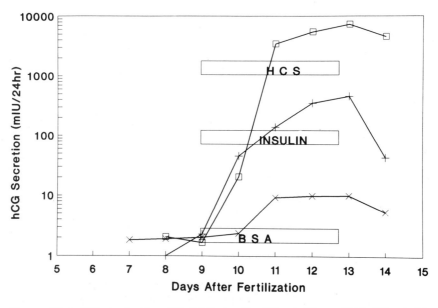

FIGURE 20.6. Effect of either 1% HCS, insulin (25 μg/mL), or BSA (25 μg/mL) on hCG secretion by blastocyst tissues cultured in α-MEM. The horizontal bars represent the time after fertilization at which each ingredient was added and the duration of exposure. Two blastocysts were studied in each group, and the values shown in the graphs represent the average daily output of hCG.

protocol was used to examine the influence of *bovine serum albumin* (BSA) on hCG secretion, the albumin being added at 25 μg/mL to serve as a protein-supplemented control for insulin. Two blastocysts were cultured to evaluate the effect of each supplement.

The results illustrated in Figure 20.6 represent the average daily output of hCG for the 2 blastocysts in each group. As may be seen, only basal levels of hCG were detected in the supernatants up to the end of day 9. Thus, in the absence of supplements in the medium, basal levels of hCG were maintained by blastocyst tissue stages that are capable of boosting gonadotropin secretion after day 7 of development. However, within 24 h of exposure to either HCS or insulin on day 10 of culture, the hCG output by these tissues increased by more than 10-fold. As observed previously, the average peak production of gonadotropin in the presence of serum was about 20 times higher than the peak output induced by insulin. A delayed and relatively low level of hCG secretion occurred in the presence of BSA. In all three groups, the transfer of blastocyst tissues to supplement-free α-MEM on day 14 of culture caused an immediate decline in hCG production (Fig. 20.6).

The failure of albumin when present in the medium at the same concentration as insulin to elicit a prompt and marked increase in hCG output indicates that the effect of insulin was specific. This specificity may be related to the presence of insulin and/or IGF-I receptors on the early trophectoderm. However, the factors that induce enhanced hCG secretion in blastocyst tissues exposed to HCS are yet to be defined. It is likely that multiple factors (22, 48), possibly acting synergistically via a number of signal-transducing pathways, are involved in producing the mRNA and other cytoplasmic mechanisms required for the biosynthesis, glycosylation, and release of hCG.

Effect of Transferrin on hCG Secretion

Transferrin receptors have been shown to appear on cultured cytotrophoblast cells derived from term placentas (49). It has been proposed that the process of cellular differentiation, particularly in response to inductive stimuli originating in ncighboring cells, involves the appearance of transferrin receptors on the surface of the differentiating cells (50). The expression of these receptors on the cell surface can occur within several minutes, probably as a result of the translocation of receptors from intracellular membrane compartments (50), or, alternatively, it could involve increased synthesis of the transferrin receptor over several days (49). The establishment of these surface receptors during the inductive process enables the cells to respond to transferrin both by the provision of iron needed for progression through the cell cycle and by transmembrane signaling unrelated to iron delivery (50). In the latter mechanism, the intracellular domain of the transferrin receptor may

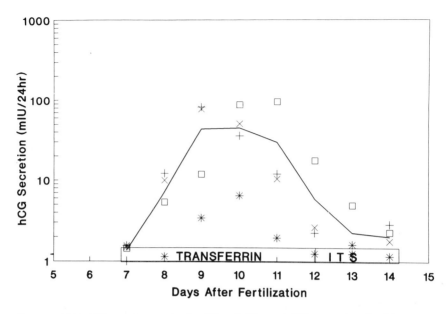

FIGURE 20.7. Effect of transferrin (25 μg/mL) on hCG secretion by blastocyst tissues cultured in serum-free α-MEM. The transferrin was present in the medium from day 7 to the end of day 11, and the tissues were subsequently cultured in insulin (25 μg/mL) + transferrin (25 μg/mL) + selenium (25 ng/mL) to day 14. Three blastocysts were cultured under these conditions, and the solid line shows the average daily hCG production. The symbols at each time interval represent daily hCG levels released by individual blastocysts.

function as a protein kinase (51) that is involved in mitogenic responses and other cell functions.

In the present studies, we wanted to determine whether transferrin was involved in hCG secretion during the early stages of trophectoderm differentiation in the blastocyst. As in the case of insulin, transferrin on its own was added to serum-free α-MEM at a concentration of 25 μg/mL. Blastocysts were cultured in this medium from day 7 to day 12 after fertilization. Subsequently, the blastocyst tissues were exposed to ITS supplement—at the same concentration as previously—on days 13 and 14 postfertilization. Again, the daily 1-mL aliquots of spent culture medium were kept frozen until hCG assays could be performed on the entire series.

The pattern of the average daily hCG output from 4 blastocysts responding to transferrin (Fig. 20.7) was quite different from that induced by insulin. As before, there was considerable variation in hCG secretion between blastocysts, but the overall pattern showed a slower initial rise, a lower average peak output, and a decline in gonadotropin production after day 10 of culture (Fig. 20.7). Moreover, the falling daily levels of hCG secretion were not affected by the presence of ITS on days 13 and 14. These results indicate that transferrin on its own induced a transient stimulation of hCG production by trophectoderm. This suggests that unlike insulin, transferrin in serum-free medium was not capable of maintaining the viability of the embryonic tissues that secreted hCG.

Effect of Epidermal Growth Factor (EGF) on hCG Secretion

High levels of EGF receptors were found predominantly on the syncytiotrophoblast of early (7–8 week) placentas (52) and relatively low levels in midterm and term tissues. In these studies, explants of villous placental tissues, derived from various stages of pregnancy, were cultured with 100 ng/mL of EGF. As would be expected, the growth factor augmented significantly the output of hCG from the receptor-rich early placental tissues, but not from explants of term pregnancies. It was of interest, however, that despite the presence of a high concentration of EGF receptors in the early placental tissues, their increased hCG response was observed after a lag period of 3 days. Other investigators, using cultures of human term cytotrophoblasts and syncytiotrophoblasts (53) exposed to 10-ng/mL EGF, observed a significant increase in hCG secretion, but again after a delay of 3 days.

In a more recent study, placental tissues derived from 7–10 week pregnancies were prepared in several different ways, employing either explants to evaluate short-term effects of EGF or dispersed cultured cells to investigate the long-term effects of the growth factor (54). In the latter studies, using two methods to purify the placental trophoblasts, EGF either prevented the rapid decline of hCG output observed in control cultures or induced a significant increase of hCG secretion after a lag of 3 days (54). It was only in short-term experiments lasting about 3 h, using perfused placental explants and pulsed delivery of EGF (50–100 ng/mL), that more immediate, but transient spikes of hCG release were observed (54).

In view of the studies that demonstrated the presence of high levels of EGF receptors in early placental trophoblasts and the reported role of EGF in inducing differentiation of trophoblast tissues in vitro (52), we examined the effects of 2 concentrations of EGF on hCG secretion by blastocysts. As in the previous studies, the polypeptide growth factor was added to serum-free culture medium containing expanded or hatching blastocysts at the end of the 7th day after fertilization. The blastocyst tissues were transferred to fresh α-MEM + EGF each day to the end of day 12; subsequently, they were cultured for 2 days in medium supplemented with 1% HCS. As before, the daily samples of spent medium were frozen until the entire series was available for hCG assays.

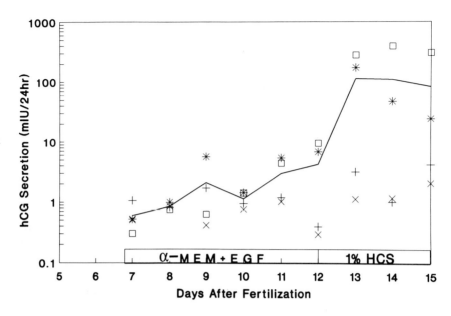

FIGURE 20.8. Effect of EGF (100 ng/mL) on hCG secretion by blastocyst tissues cultured in serum-free α-MEM. The EGF was present in the medium from day 7 to the end of day 11, and the tissues were subsequently cultured in α-MEM containing 1% HCS to day 14. Four blastocysts were cultured under these conditions, and the solid line shows the average daily hCG production. The symbols at each time interval represent daily hCG levels released by individual blastocysts.

Two blastocysts failed to respond to 10-ng/mL EGF even after 5 days of culture. The same tissues showed a small increase in hCG output in response to 1% HCS over the next 2 days (data not shown), indicating that the failure of response to EGF was not due to loss of viability. Similar experiments using 100-ng/mL EGF showed that 2 blastocysts responded by an 8- to 10-fold increase in hCG output after 2–3 days of culture (Fig. 20.8), and these embryonic tissues responded by a further marked increase in hCG production in response to 1% HCS. A further 2 blastocysts failed to respond to the higher concentration of EGF after 5 days of culture, although their viability was confirmed by an increase in hCG secretion in response to serum during the last 2–3 days of culture (Fig. 20.8). It would appear, therefore, that some human blastocyst tissues are able to respond to EGF, but like cultured trophoblast cells derived from early placentas, their response to the growth factor is expressed after a lag period of about 3 days.

Effect of Platelet-Derived Growth Factor (PDGF) on hCG Secretion

The human placenta has been shown to be a rich source of PDGF, and the corresponding PDGF receptors are expressed on the placental cytotrophoblast cells (55). It has been proposed, therefore, that PDGF may be an autocrine factor responsible for the rapid proliferation of invasive trophoblastic cells (55). In view of this function in the developing placenta, we are conducting studies to examine the role of PDGF in blastocyst growth and hCG secretion.

The experimental design used is similar to the protocol employed for studying the action of the other growth factors. Blastocysts are cultured in serum-free medium supplemented with PDGF from days 7 to 12 and, subsequently, in medium containing 1% HCS to day 14. To date, we have found that at a concentration of 2 ng/mL, PDGF induced a rapid increase in hCG secretion by 5 blastocyst tissues. This increasing output of gonadotropin was maintained to day 10 or 11, and it then declined until hCG secretion was again boosted by exposure of the embryonic tissues to HCS for the last 2 days of culture. The stimulation of hCG production by PDGF was similar to that induced by transferrin. Both of these factors appear to promote an increase in hCG secretion for 3 to 4 days followed by reduction of output in serum-free medium. It was only in the presence of insulin that blastocyst tissues maintained the initial increase of hCG secretion for the entire duration of serum-free culture.

Conclusions

The studies described in the present article suggest that multiple factors influence the synthesis and secretion of chorionic gonadotropin by the human blastocyst. It would appear that the trophectoderm of preimplantation blastocysts and of early postimplantation stages is capable of producing and releasing hCG in response to insulin, transferrin, and PDGF. It may be concluded that receptors for these growth factors are expressed by day 7 expanded blastocysts and by the more advanced periimplantation stages. In contrast, the failure of EGF to evoke hCG secretion until day 10 or 11 of development suggests that EGF receptors are not expressed until after implantation of the human embryo. Alternatively, factors that were required for inducing the expression of EGF receptors in the early trophectoderm were not present in the in vitro system. An interesting paradox reported in the literature, however, was the finding that EGF induced a significant increase in hCG secretion after a lag period of 3 days, even in early trophoblast tissues that contained high levels of EGF receptors (52). In considering these factors, it should also be noted that the embryos used in our studies were morphologically and genetically abnormal.

Of all the factors that influenced hCG production by the blastocyst, serum was found to be the most potent stimulant for long-term gonadotropin secretion. During the period of study, the levels of hCG output attained in the presence of 1% HCS were more than 10-fold higher than the increase induced by any other stimulant investigated. The ability of serum to maintain the viability of blastocyst tissues for up to 14 days after fertilization may partly account for the enhanced levels of hCG secretion. On the other hand, insulin on its own, which also appeared to be capable of maintaining trophectoderm viability, was not nearly as effective as serum in promoting hCG production.

The possibility that serum acted by releasing PAF from the blastocyst tissues was investigated. Our studies showed that the specific PAF antagonist, WEB 2086, did not dampen the initiation and maintenance of hCG production in response to serum. These studies suggested that it was not embryo-derived PAF that induced the enhanced hCG secretion in the presence of serum. We have also started to examine the likelihood that multiple stimulatory factors present in serum may act synergistically to boost hCG production. To date, we have shown that the presence of insulin and transferrin in combination did not enhance or maintain hCG output above the levels attained in the presence of insulin on its own (these studies will be reported elsewhere). The mechanism by which serum acts as a potent stimulant of blastocyst hCG secretion remains to be explained. In addition, the biological activity of the hCG released by blastocysts must also be determined.

References

1. National Perinatal Statistics Unit. IVF and GIFT pregnancies, Australia and New Zealand, for 1988: Sydney: National Perinatal Statistics Unit, 1990:4–7.
2. IVF-ET Registry US. In vitro fertilization-embryo transfer in the United States: 1988 results from the IVF-ET Registry. Fertil Steril 1990;53:13–20.
3. Sharma V, Whitehead M, Mason B, et al. Influence of superovulation on endometrial and embryonic development. Fertil Steril 1990;53:822–9.
4. Garcia JE, Acosta AA, Hsiu JG, Jones HW. Advanced endometrial maturation after ovulation induction with human menopausal gonadotropin/human chorionic gonadotropin for in vitro fertilization. Fertil Steril 1984;41:31–5.
5. Birkenfeld A, Navot D, Levij IS, et al. Advanced secretory changes in the proliferative human endometrial epithelium following clomiphene citrate treatment. Fertil Steril 1986;45:462–8.
6. Testart J, Forman R, Belaisch-Allart J, et al. Embryo quality and uterine receptivity in in-vitro fertilization cycles with or without agonists of gonadotrophin-releasing hormone. Hum Reprod 1989;4:198–201.
7. Paulson RJ, Sauer MV, Lobo RA. Embryo implantation after human in vitro fertilization: importance of endometrial receptivity. Fertil Steril 1990;53: 870–4.

8. Marchini M, Fedele L, Bianchi S, Losa GA, Ghisletta M, Candiani GB. Secretory changes in preovulatory endometrium during controlled ovarian hyperstimulation with buserelin acetate and human gonadotropins. Fertil Steril 1991;55:717–21.
9. Tur-Kaspa I, Confino E, Dudkiewicz AB, Myers SA, Friberg J, Gleicher N. Ovarian stimulation protocol for in vitro fertilization with gonadotropin-releasing hormone agonist widens the implantation window. Fertil Steril 1990;53:859–64.
10. Gidley-Baird AA, O'Neill C, Sinosich MJ, Porter RN, Pike IL, Saunders DM. Failure of implantation in human in vitro fertilization and embryo transfer patients: the effects of altered progesterone/estrogen ratios in humans and mice. Fertil Steril 1986;45:69–74.
11. Keenan D, Cohen J, Suzman M, Wright G, Kort H, Massey J. Stimulation cycles suppressed with gonadotropin-releasing hormone analog yield accelerated embryos. Fertil Steril 1991;55:792–6.
12. Papadopoulos G, Randall J, Templeton AA. The frequency of chromosomal anomalies in human unfertilized oocytes and uncleaved zygotes after insemination in vitro. Hum Reprod 1989;4:568–73.
13. Martin RH, Ko E, Rademaker A. Distribution of aneuploidy in human gametes: comparison between human sperm and oocytes. Am J Med Genet 1991;39:321–31.
14. Yagel S, Parhar RS, Lala PK. Trophic effects of first-trimester human trophoblasts and human chorionic gonadotropin on lymphocyte proliferation. Am J Obstet Gynecol 1989;160:946–53.
15. North RA, Whitehead R, Larkins RG. Stimulation by human chorionic gonadotropin of prostaglandin synthesis by early human placenta. J Clin Endocrinol Metab 1991;73:60–70.
16. Bose R, Cheng H, Sabbadini E, McCoshen J, Mahadevan MM, Fleetham J. Purified human early pregnancy factor from preimplantation embryo possesses immunosuppressive properties. Am J Obstet Gynecol 1989;160:954–60.
17. Daya S, Clark DA. Identification of two species of suppressive factor of differing molecular weight released by in vitro fertilized human oocytes. Fertil Steril 1988;49:360–3.
18. Clark DA, Lee S, Fishell S, et al. Immunosuppressive activity in human in vitro fertilization (IVF) culture supernatants and prediction of the outcome of embryo transfer: a multicenter trial. J In Vitro Fertil Embryo Transfer 1989;6:51–8.
19. Lala PK. Similarities between immunoregulation in pregnancy and in malignancy: the role of prostaglandin E_2. Am J Reprod Immunol Microbiol 1989; 9:105–10.
20. Holmes PV, Sjogren A, Hamberger L. Prostaglandin-E_2 released by pre-implantation human conceptuses. J Reprod Immunol 1990;17:79–86.
21. Yagel S, Lala PK, Powell WA, Casper RF. Interleukin-I stimulates human chorionic gonadotropin secretion by first trimester human trophoblast. J Clin Endocrinol Metab 1989;68:992–5.
22. Nishino E, Matsuzaki N, Masuhiro K, et al. Trophoblast-derived interleukin-6 (IL-6) regulates human chorionic gonadotropin release through IL-6 receptor on human trophoblasts. J Clin Endocrinol Metab 1990;71:436–41.
23. Masuhiro K, Matsuzaki N, Nishino E, et al. Trophoblast-derived interleukin-1 (IL-1) stimulates the release of human chorionic gonadotropin by activating

IL-6 and IL-6-receptor system in first trimester human trophoblasts. J Clin Endocrinol Metab 1991;72:594–601.
24. Chard T. Interferon in pregnancy. J Dev Physiol 1989;11:271–6.
25. Orozco C, Perkins T, Clarke FM. Platelet-activating factor induces the expression of early pregnancy factor activity in female mice. J Reprod Fertil 1986;78:549–55.
26. Clarke FM, Orozco C, Perkins AV, Cock I. Partial characterization of the PAF-induced soluble factors which mimic the activity of "early pregnancy factor." J Reprod Fertil 1990;88:459–66.
27. Boyer P, Guichard A, Risquez F, et al. Reducing the risks of multiple pregnancies by co-culture of human embryos. J In Vitro Fertil Embryo Transfer 1990;7:186.
28. Hearn JP. The embryo-maternal dialogue during early pregnancy in primates. J Reprod Fertil 1986;76:809–19.
29. Yagel S, Casper RF, Powell W, Parhar RS, Lala PK. Characterization of pure human first-trimester cytotrophoblast cells in long term culture: growth pattern, markers, and hormone production. Am J Obstet Gynecol 1989; 160:938–45.
30. Chavez DJ. Cellular aspects of implantation. In: Van Blerkom J, Motta PM, eds. Ultrastructure of reproduction. Boston: Martinus Nijhoff, 1984:247–59.
31. Anderson TL, Simon JA, Hodgen GD. Histochemical characteristics of the endometrial surface related temporally to implantation in the non-human primate (*Macaca fascicularis*). In: Denker H-W, Aplin JD, eds. Trophoblast research; vol 4. New York: Plenum Press, 1990:273–84.
32. Lopata A, Hay DL. The potential of early human embryos to form blastocysts, hatch from their zona and secrete HCG in culture. Hum Reprod 1989;4(suppl):87–94.
33. Bolton VN, Hawes SM, Taylor CT, Parsons JH. Development of spare human preimplantation embryos in vitro: an analysis of the correlations among gross morphology, cleavage rates, and development to blastocyst. J In Vitro Fertil Embryo Transfer 1989;6:30–5.
34. Hay DL. Discordant and variable production of human chorionic gonadotropin and its free alpha- and beta-subunits in early pregnancy. J Clin Endocrinol Metab 1985;61:1195–201.
35. Bambra CS, Tarara R. Immunohistochemical localization of chorionic gonadotrophin on baboon placenta, dispersed trophoblast cells and those derived from blastocysts grown in vitro. J Reprod Fertil 1990;88:9–16.
36. Hearn JP, Hodges JK, Gems S. Early secretion of chorionic gonadotropin by marmoset embryos in vivo and in vitro. J Endocrinol 1988;119:249–55.
37. Spinks NR, Ryan JP, O'Neill C. Antagonists of embryo-derived platelet-activating factor act by inhibiting the ability of the mouse embryo to implant. J Reprod Fertil 1990;88:241–8.
38. O'Neill C, Collier M, Ammit AJ, Ryan JP, Saunders DM, Pike IL. Supplementation of in-vitro fertilisation culture medium with platelet activating factor. Lancet 1989;769–72.
39. Collier M, O'Neill C, Ammit AJ, Saunders DM. Biochemical and pharmacological characterization of human embryo-derived activating factor. Hum Reprod 1988;3:993–8.

40. Collier M, O'Neill C, Ammit AJ, Saunders DM. Measurement of human embryo-derived platelet-activating factor (PAF) using a quantitative bioassay of platelet aggregation. Hum Reprod 1990;5:323–8.
41. Spinks NR, O'Neill C. Antagonists of embryo-derived platelet-activating factor prevent implantation of mouse embryos. J Reprod Fertil 1988;84:89–98.
42. Spinks NR, O'Neill C. Embryo-derived PAF activity is essential for the establishment of pregnancy in the mouse. Lancet 1987;1:106–7.
43. Hussa RO. Human chorionic gonadotropin, a clinical marker: review of its biosynthesis. Ligand Rev 1981;3(suppl 2):5–44.
44. Kato Y, Braunstein GD. Purified first and third trimester placental trophoblasts differ in in vitro hormone secretion. J Clin Endocrinol Metab 1990;70:1187–92.
45. Branchaud CL, Goodyer CG, Guyda HJ, Lefebvre Y. A serum-free system for culturing human placental trophoblasts. In Vitro Cell Dev Biol 1990;26:865–70.
46. Ren S-G, Braunstein GD. Insulin stimulates synthesis and release of human chorionic gonadotropin by choriocarcinoma cell lines. Endocrinology 1991;128:1623–9.
47. Ohlsson R, Larsson E, Nilsson O, Wahlstrom T. Blastocyst implantation precedes induction of insulin-like growth factor II gene expression in human trophoblasts. Development 1989;106:555–9.
48. Oike N, Iwashita M, Muraki TK, Nomoto T, Takeda Y, Sakamoto S. Effect of adrenergic agonists on human chorionic gonadotropin release by human trophoblast cells obtained from first trimester placenta. Horm Metab Res 1990;22:188–91.
49. Bierings MB, Adriaansen HJ, Van Dijk JP. The appearance of transferrin receptors on cultured human cytotrophoblast and in vitro-formed syncytiotrophoblast. Placenta 1988;9:387–96.
50. Mescher AL, Munaim SI. Transferrin and the growth-promoting effect of nerves. Int Rev Cytol 1988;110:1–26.
51. May WS Jr, Cuatrecasas P. Transferrin receptor: its biological significance. J Membr Biol 1985;88:205–15.
52. Maruo T, Matsuo H, Oishi T, Hayashi M, Nishino R, Mochizuki M. Induction of differentiated trophoblast function by epidermal growth factor: relation of immunohistochemically detected cellular epidermal growth factor receptor levels. J Clin Endocrinol Metab 1987;64:744–50.
53. Morrish DW, Bhardwaj D, Dabbagh LK, Marusyk H, Siy O. Epidermal growth factor induces differentiation and secretion of human chorionic gonadotropin and placental lactogen in normal human placenta. J Clin Endocrinol Metab 1987;65:1282–90.
54. Barnea ER, Feldman D, Kaplan M, Morrish DW. The dual effect of epidermal growth factor upon human chorionic gonadotropin secretion by the first trimester placenta in vitro. J Clin Endocrinol Metab 1990;71:923–8.
55. Goustin AS, Betsholtz C, Pfeifer-Ohlsson S, et al. Co-expression of the *sis* and *myc* proto-oncogenes in human placenta suggest autocrine control of trophoblast growth. Cell 1985;41:301–12.

Appendix: Poster Presentations

This appendix contains 80 abstracts given as poster presentations at the Serono Symposia, USA Symposium on Preimplantation Embryo Development, held August 15 to 18, 1991, in Newton, Massachusetts.

MOUSE PREIMPLANTATION EMBRYOS EXPRESS GLUCOSE TRANSPORTER MOLECULES

M. AGHAYAN[1], L.V. RAO[2], R.M. SMITH[3], L. JARETT[3], N. SHAH[3] AND S. HEYNER[1].
[1,3], Univ. Pennsylvania, Phila., PA. [2]Case Western Reserve Univ., Cleveland, OH.

Glucose becomes the preferred energy substrate at the late 8-cell to morula stage of the mouse preimplantation embryo (Biggers and Borland, Ann. Rev. Physiol. 39:95, 1976). At the same stage of development insulin receptors are detectable in significant numbers (Mattson et al. Diabetes 37:585, 1988), and insulin acts to stimulate embryonic metabolism (Heyner et al., Dev. Biol. 134:48, 1989). Two general mechanisms mediate glucose transport, one is a sodium-coupled glucose transporter found in the apical border of intestinal epithelia, while the other is a sodium-independent transport system. Of the latter, several facilitated transporters have been identified, including the erythrocyte/brain (GLUT 1), the liver (GLUT 2), and the adipose/muscle (GLUT 4) isoforms. In this study, we used Western blot analysis and high resolution immunoelectron microscopy to investigate the stage-related expression of GLUT 1, 2 and 4. The results demonstrate that GLUT 1 and GLUT 2 isoforms are expressed at different stages during mouse preimplantation development, while the GLUT 4 isoform could not be detected in the oocyte through blastocyst stages. They also confirm studies at the molecular level which demonstrate that mRNAs encoding the same GLUT isoforms are detectable at corresponding developmental stages (Hogan et al. Development, in press 1991). These observations suggest that glucose transport in preimplantation mouse embryos may be independent of insulin, although insulin has significant effects on metabolic activity in the early embryo. Supported by NIH grants DK 19525 and HD 23511.

ANTIBODIES TO SPERM SURFACE ANTIGENS AND c-myc PROTO-ONCOGENE PRODUCT INHIBIT EARLY EMBRYONIC DEVELOPMENT IN MICE

KHALIQ AHMAD AND RAJESH K. NAZ
Department of Ob/Gyn, Albert Einstein College of Medicine, Bronx, New York.

Antibodies against sperm antigens, lithium diiodosalicylate-solubilized murine sperm extract (LIS-sperm), fertilization antigen-1 (FA-1) and c-myc proto-oncogene product have been shown to bind to sperm and to inhibit *in vitro* fertilization in mice (Naz et al: Science 225:342, 1984; Proc Natl Acad Sci USA 83:5713, 1986; Biol Reprod 40:163, 1989; Biol Reprod 44:842, 1991). In this investigation, effects of these antibodies were investigated on early embryonic development in mice. Affinity-purified anti-LIS sperm and anti-FA-1 Fab's reduced (P<.01) blastulation rates of *in vitro* cultured 2-cell embryos mainly due to an arrest of development at morula stage. Similarly, the c-myc monoclonal antibody (mAb) affected embryonic development in a dose-dependent manner, and immunoabsorption of c-myc mAb with its respective peptide completely blocked the anti-embryonic effect. Anti-LIS sperm Fab' identified four protein bands of M_r 36, 29, 24.6 and 17.6 kD, respectively on Western blots of unfertilized and fertilized ova, a 68 kD band each on 4-8 cell embryo and morula, and a 53 kD band on blastocyst extract. Anti-FA-1 Fab' did not react with murine ova and pronuclear stage zygotes but specifically identified two protein bands of M, 53 and 25.7 kD on 2-cell embryo, a 25.7 kD band on morula, and a 53 kD band on blastocyst extracts. c-myc mAb did not recognize any band corresponding to c-myc protein on Western blots of unfertilized or fertilized ova, 2-cell embryo, 4-8 cell embryo, morula and blastocyst extracts indicating that the product of c-myc proto-oncogene is not abundantly expressed during early stages of embryogenesis. These results suggest that some of the cross-reacting sperm antigens that are expressed during early cleavages, and the product of c-myc proto-oncogene may have a role in preimplantation embryonic development.

BOVINE OOCYTE MATURATION IS REGULATED BY INTERACTIONS BETWEEN cAMP, CUMULUS AND GRANULOSA

H. Aktas, M.L. Leibfried-Rutledge and N.L. First
University of Wisconsin, Madison, WI 53706

Invasive adenylate cyclase (AC) reversibly maintains meiotic arrest in bovine oocytes by elevating cAMP ($[cAMP]_i$) in cumulus oocyte complexes (COC). Here we investigated contributions of follicle cell investments in maintenance of meiotic arrest by $[cAMP]_i$. Fluid was aspirated from 2-5 mm follicles. Follicular fluid of treatment groups was supplemented with 0.5 mM IBMX and 5 µg/ml AC. Oocytes with cumulus were divided into two populations, those with considerable granulosa (GCOC) in addition to cumulus or only cumulus (COC). Half of each population was cultured with or without 100 µg/ml AC. Presence of intact germinal vesicle (GV) after 20 h incubation was taken as successful maintenance of meiotic arrest. Percentage GV (r=4) was with AC: COC 11%, GCOC 80%; without AC: COC 2%, GCOC 10%. Only GCOC with AC showed meiotic inhibition. Next oocytes in each group were stripped of surrounding cells prior to culture and treated as in Exp. 1. Percentage GV (r=4) was with AC: COC 72%, GCOC 90%; without AC: COC 0%, GCOC 0%. Both populations of oocytes showed a maintenance of GV with AC over the controls. In a third experiment we evaluated developmental competence of COC and GCOC complexes as well as oocytes denuded of cumulus prior to maturation and cultured with (DO+C) or without (DO) stripped cumulus cells. The end point was blastocyst (BL) formation. Percentage BL/total ova fertilized was: COC 20%, GCOC 30%, DO+C 8%, DO 1%. Both intact groups had more blastocysts than denuded oocyte group. We measured oocyte $[cAMP]_i$ content of COC and GCOC cultured with AC for 2 h or denuded and cultured with AC for 2 h. No differences in $[cAMP]_i$ were found that explained differences in GV retention between groups. We measured $[cAMP]_i$ content of COCs and oocytes during spontaneous maturation at 0, 15, 30 and 60 min after the start of culture. A precipitous decline in $[cAMP]_i$ occurred in COC but not in oocytes at 15 and 30 min with partial recovery at 60 min. In summary, oocytes derived from GCOC and COC undergo maturation and embryonic development. Both groups are sensitive to AC when denuded and AC elevates $[cAMP]_i$ in both groups. We conclude that oocytes are maintained in meiotic arrest by high $[cAMP]_i$ and cumulus may produce a positive signal to initiate meiotic maturation. Granulosa may interfere with generation of this signal in the presence of high $[cAMP]_i$.

EFFECT OF AZASERINE AND HYPOXANTHINE ON MOUSE PREIMPLANTATION DEVELOPMENT

MARIA ALEXIOU & HENRY J. LEESE
Department of Biology, University of York, York YO1 5DD, U.K.

Azaserine (25µg/ml), a potent inhibitor of de novo purine synthesis, administered at the 1-cell stage, blocks mouse embryo development at the 2-cell stage. When administered to zygotes early in the cell cycle, the first cleavage division is also blocked. Exposure from the 2-cell stage prevents embryo development beyond the 4-cell stage. Compaction then occurs on Day 3, with the embryos still at the 4-cell stage. These data suggest that preimplantation embryo development in the mouse is dependent on de novo synthesis for its supply of purine precursors required for the nucleic acid formation.
Hypoxanthine added to cultures treated with azaserine, at concentrations high enough to stimulate the salvage pathway of purine nucleotide biosynthesis (0.1-0.05mM), failed to reverse the inhibitory effect of azaserine, indicating that the salvage pathway is unable on its own to fulfil the demand for nucleotides. Exogenous hypoxanthine is taken up by the embryos and partially inhibits development when it is administered from the 1-cell stage. No inhibition is observed when the embryos are treated from the 2-cell stage. HPLC (High performance liquid chromatography) analysis indicates that hypoxanthine uptake may be accounted for by the appearance in the culture medium of xanthine, an intermediate in the pathway of uric acid production. Xanthine excretion, as a result of hypoxanthine addition, increases slowly between the 1-cell and compacted morula stages, before reaching a maximum with blastocyst formation, coincident with a significant increase in nucleic acid synthesis and turnover.

M. Alexiou acknowledges support from the Hellenic State Scholarships Foundation.

A DEFICIENCY IN THE MECHANISM FOR p34^{CDC2} PROTEIN KINASE ACTIVATION IN MOUSE EMBRYOS ARRESTED AT THE 2-CELL STAGE

F. AOKI[1], T. CHOI[1], M. YAMASHITA[2], Y. NAGAHAMA[2] & K. KOHMOTO[1]

[1]Department of Animal Breeding, University of Tokyo, Tokyo, Japan. [2]Department of Developmental Biology, National Institute for Basic Biology, Okazaki, Japan.

In mice, most cultured embryos arrest their cell cycles at the G2 phase in the 2-cell stage, and this phenomenon is termed the "2-cell block." We examined the amounts of p34^{cdc2} and its protein kinase activity in mouse 2-cell embryos to investigate its relationship with the "2-cell block." Previously, we have detected three different migrating bands (upper, middle and lower bands) by immunoblotting following SDS-PAGE with anti-p34^{cdc2} antibody in 1-cell embryos. The upper and middle bands were phosphorylated forms of p34^{cdc2}. When the embryos entered into the M-phase of the first mitotic cell cycle, the upper and middle bands shifted to the lower one and p34^{cdc2} protein kinase was activated.

In most ddY mouse embryos inseminated in vitro, the first cleavage to the 2-cell stage occurred 16-20 h after insemination and thereafter second cleavage did not occur even 45 h after insemination. In these embryos arrested in the 2-cell stage, three migrating bands were also detected by anti-p34^{cdc2} antibody. Although the total amounts of the protein did not change all through the 1- and 2-cell stages, the phosphorylation states changed. After the embryos cleaved to 2-cells, the amounts of phosphorylated forms of p34^{cdc2} increased up to 33 h and thereafter did not show any change until 45 h after insemination. When the embryos arrested in the 2-cell stage were treated with okadaic acid, an inhibitor of protein phosphatase 1 and 2A, the upper and middle bands shifted to the lower one and histone H1 kinase activity increased. The DNA staining of these embryos with 4'-, 6'-diamino-2-phenylindol revealed chromosome condensation. These results suggest that 2-cell embryos contain enough p34^{cdc2} to induce mitotic events, and that the protein remains in a latent form. The mechanism for the activation of p34^{cdc2} is probably deficient.

PRONUCLEAR SYNTHESIS OF DNA IN HUMAN ZYGOTES

H. BALAKIER, N.J. MACLUSKY, R.F. CASPER
DIV. REPROD. SCIENCE, DEPT. OB/GYN, UNIV. OF TORONTO, CANADA

Recently, we showed that freezing mouse embryos in G2 rather than in S phase resulted in a higher rate osurvival and development. In an attempt to improve cryopreservation success of human 1-cell stage zygotes, the duration of S phase and length of the 1st cell cycle of human zygotes was studied. Forty four human zygotes were exposed to 3[H]-thymidine for 1 hr at different times after insemination. Incorporation of thymidine, indicating DNA synthesis, was detected in one or both pronuclei, 12-21 hr after insemination. Between 22-27 hr, DNA synthesis was not detected and by 21 hr, some zygotes entered the 1st mitotic division and exhibited condensing chromosomes. The length of the first mitotic division was 3-3.5 hr. Observations on living zygotes indicated that about 37% (83 out of 224) had started mitosis (pronuclei were not visible) and 8% already formed 2-cell stage embryos by 24-25 hr after insemination. The present findings suggest that DNA synthesis in human zygotes may start asynchronously and lasts about 5 hr. The G2 phase seems to be short, about 3-5 hr. Therefore, the optimal time for freezing human zygotes might be approximately 24 hr after insemination.

SPONTANEOUS AND INDUCED PARTHENOGENETIC ACTIVATION OF HUMAN OOCYTES

H. BALAKIER, J. SQUIRE, R.F. CASPER
DIV. REPROD. SCIENCE, DEPT. OB/GYN, UNIV. OF TORONTO, CANADA

About 3% of human oocytes (75/2617) underwent abnormal activation and exhibited a single pronucleus 18 hr after insemination. This type of activation occurred more often in immature or intermediate oocytes than in mature oocytes. The activation was independent of stimulation protocol or cause of infertility. The average age of patients was 34 years. Abnormally activated oocytes (38) that were cultured in vitro, were arrested at interphase or underwent fragmentation (45%). Others have reached 2-8 cell stage (21%), formed early morulae (12%) or developed up to blastocyst stage (13%). The embryos contained a haploid, intermediate or diploid number of chromosomes. Activated oocytes that were examined at the first mitotic division also exhibited varied numbers of chromosomes. In some oocytes (9 out of 19) in addition to a single nucleus other structures were found, such as decondensed sperm heads, small nucleus-like structures or some condensed chromosomes. Also in situ hybridization techniques demonstrated the presence of chromosome Y in two oocytes that entered the first mitosis (2 out of 13). This observation suggests that in some cases, abnormal activation is not parthenogenetic but is caused by sperm penetration with the possibility for gynogenetic and androgenetic development.

Our attempts to induce activation of human, unfertilized oocytes (17-48 hr after insemination) with ethanol, calcium ionophore and phorbol ester resulted in very low level of activation (0-20%). In contrast, almost 100% of oocytes treated for 8-24 hr with the protein synthesis inhibitor, puromycin, underwent parthenogenetic activation. Activated oocytes contained from one to several nuclei which were capable of DNA synthesis and were rarely capable of one division after treatment was withdrawn. This suggests that human oocytes are very resistant to parthenogenetic activation and their activation may depend mainly on inhibition of synthesis of chromosome condensation factor.

PARAMETERS OF RECEPTIVITY DURING THE LUTEAL PHASE TO PREDICT BLASTOCYST IMPLANTATION

H. M. BEIER, C. HEGELE-HARTUNG, U. MOOTZ, W. ELGER*
Dept. of Anatomy and Reproductive Biology, RWTH University of Aachen and
*Schering AG Berlin, Federal Republic of Germany

We describe a well-established approach to study the parameters and mechanisms of synchronization or desynchronization between the maternal and embryonic systems before implantation in the rabbit. It has become a useful experimental design to induce "delayed secretion" of the endometrium by different endocrine interventions, which dissociate the endometrial transformation from its control by the corpus luteum. The most fascinating tool has been achieved by means of direct progesterone antagonists (Mifepristonen, Lilopristone, Onapristone), which competitively bind to the progesterone receptor and, in turn, inhibit the physiological effects of progesterone. During the luteal phase, there are significant parameters which indicate the receptive stage of the endometrium: the secretory protein patterns. Substantial evidence is presented here, that these patterns, analyzed by electrophoretical means and densitometry, actually can serve to define the right point of time at which an embryo transfer is promising for implantation and establishment of pregnancy.

Progesterone antagonist treatment of rabbits during their early phase of pseudopregnancy inhibits the synthesis and release of uteroglobin for a short period. Synthesis resumes within 48 h after treatment. This behavior can be seen in all other preimplantation uterine secretion proteins. Since pseudopregnancy in the rabbit was used as a model for the luteal phase of humans, we conclude that a progesterone antagonist treatment at the beginning of the luteal phase may represent the first step in our attempt to extend or to delay the receptive stage of the luteal phase endometrium to meet all requirements for implantation. Such a treatment, particularly with compounds like Lilopristone, could be promising for improving the adaptation of the endometrium to the more slowly developing in-vitro-cultured and retransferred embryos. IVF/ET-programmes performing with the therapy of human sterility as well as programmes improving reproduction in domestic animals would benefit from such exogenous control of the luteal phase and possible significant increase of implantation rates.

THE SIGNIFICANCE OF PROTEIN PATTERNS IN HUMAN UTERINE SECRETIONS FOR IMPLANTATION

K. BEIER-HELLWIG, K. STERZIK*, B. BONN, J. O. BECKMAN, AND H. M. BEIER
Department of Anatomy and Reproductive Biology, Medical Faculty, University of Aachen; *Department of Gynecology and Obstetrics, University of Ulm/Donau, Federal Republic of Germany

Uterine secretion electrophoresis (USE) offers a direct parameter for the evaluation of endometrial response. By analyzing the endometrium's ability to secrete proteins, the normal menstrual cycle can be divided into three typical functional states (quiescent phase, proliferative- and secretory phase). The specific sequence and characteristic protein patterns of these phases correlate with the process of endometrial transformation and maturation. The aim of this study was to assess the typical secretory (luteal) phase pattern whose complete expression seems to be so vital for the endometrium's full receptivity and the embryo's ability to implant. A total of 180 patients from the University of Ulm's sterility program were examined after giving their written consent. Secretion samples were obtained from the cavum uteri during the luteal phase using a transfer catheter or a disposable Prevical® instrument. All patients were treated as outpatients. Their cycles were either untreated or stimulated following various protocols. All secretion samples were frozen and later subjected to a protein-biochemical analysis. Prior to the protein analysis the samples (50 to 70 µl) were thawed, treated with ultrasound for better solubility and centrifuged for 10 minutes at 3000 RPM. Following a microprotein analysis (Micro-Lowry) and lyophilization, the samples were fractionated by SDS-polyacrylamidgel-electrophoresis using thin-layer gradient gels with polyacrylamid gradients from 8.3 to 16.6%. The buffer system consisted of a tris glycin buffer set at pH 8.3. Samples were treated with 0.1% SDS prior to electrophoresis. After electrophores the polyacrylamidgels were stained with Coomassie-blue and photographed. USE diagnostics indicated adequate luteal phases (termed "USE-positive") in only 30% of the total collective of the University of Ulm's sterility program patients. Non-stimulated cycles presented adequate luteal phases in 47.3%, whereas the endometrium quality sank drastically in patients undergoing standard hormonal cycle stimulation (assisted reproduction therapy): HMG/HCG patients presented a USE-positive rate of 28.8%, and CC/HCG and CC/HMG/HCG patients of only 15.8%. Clomiphene citrate (CC) alone on the other hand did not seem to affect the endometriums' quality adversely compared to non-stimulated cycles (41.7% USE-positive cycles). We should therefore assume that both the dosage and timing of the HCG injection play an important role. We suggest that during assisted reproduction therapy one should avoid shortening the physiological length of the proliferation phase, thereby enabling the protein pattern to fully develop, before giving the signal to transform the proliferative into the luteal phase pattern which is essential for establishing endometrial receptivity. "Decoupling" (inadequate luteal phase in spite of normal follicle size and physiological progesterone levels) took place in 21.3% of the total patient collective (with and without stimulation). When decoupling takes place, USE protein patterns present aberrations to various degrees from the expected pattern. The pathogenesis of this phenomenon will be the subject of a follow-up study. An interesting aspect of this phenomenon, however, is that hormonal stimulation seems to offer a certain protection against decoupling (18.3% decoupled stimulated patients versus 25.3% decoupled in the non-stimulated group).

ULTRASTRUCTURAL STUDY OF NUCLEOLAR CHANGES IN RAT EMBRYOS BEFORE, DURING AND AFTER EMBRYONIC DIAPAUSE

H. BERGERON, W.A. KING & A.K. GOFF, C.R.R.A., Fac. Méd. Vét., Université de Montréal, C.P.5000, Saint-Hyacinthe (Québec), J2S 7C6

In order to determine whether structural changes of the nucleolus reflect the metabolic status of the cell, changes in nucleoli ultrastructure were studied in preimplantation rat embryos before diapause (day 5), during diapause (day 6, 7 and 8) and after reactivation. Reactivation was brought about either by in vitro culture for 24 hours (day 7 and 8) or by estradiol-17β treatment of the mothers in vivo (day 11). Before diapause, the nucleolus was fully developed and was composed of several low-density fibrillar centers each surrounded by a layer of dense fibrillar component with an abundant granular component. This corresponds to a compact type nucleolus and indicates a high activity of ribosomal RNA synthesis. Nucleolar morphology during diapause revealed a disorganization of the fibrillar and granular elements corresponding to a diminution of the transcriptional activity. No fibrillar center was seen. During reactivation, nucleoli progressively returned to a classical reticular configuration, characteristic of the beginning of intense transcriptional activity. These nucleoli were principally made up of a dense fibrillar component arranged in a three-dimensional network with some granular component. It is concluded that the structure of the nucleolus in rat embryos corresponds to the level of transcriptional activity and could be used for studying structure-function relationships during early development.

MEASUREMENT OF CHLORIDE TRANSPORT IN THE RAT BLASTOCYST USING THE FLUORESCENT DYE SPQ

D. R. Brison & H. J. Leese, Department of Biology, University of York, York YO1 5DD, U.K.

The chloride-sensitive fluorescent dye 6-methoxy-N-(3-sulfopropyl) quinolium (SPQ) has been used to measure the flux of chloride ions across the trophectoderm of single day 5 rat blastocysts. Dye loading into the blastocoel cavity was achieved by a 2 hr incubation in 20 mM SPQ at 37°C. The fluorescence of SPQ trapped in the blastocoel cavity was measured using epifluorescence microscopy, at excitation wavelengths of 340-380 nm, and emission >450 nm. On replacement of external Cl$^-$ with gluconate, a continuous increase in fluorescence was monitored, as a result of Cl$^-$ efflux from the blastocoel cavity. In the presence of DIDS (4,4-diisothiocyanatostilbene-2,2-disulphonic acid: 0.1 mM) the increase in fluorescence was greatly reduced, suggesting the activity of a Cl$^-$/HCO$_3^-$ exchanger on the apical surface of the trophectoderm. The furosemide-inhibitable Na$^+$/K$^+$/2Cl$^-$ cotransporter may also play a role. The Cl$^-$ channel blocker A9C (Anthracene-9-carboxylic acid: 0.1 mM) appeared to have no significant effect on Cl$^-$ flux. A large component of the flux (~50%) appeared to be independent of these pathways, and may be paracellular as in the mouse (Manejwela et al., Dev. Biol., 133:210, 1989), since the trophectoderm is a "leaky" epithelium (Biggers et al., Am. J. Physiol., 255:C419, 1988). This work represents a novel application of an ion-sensitive fluorescent dye, to the study of trans-epithelial transport in the mammalian blastocyst.

We acknowledge support from the U.K. Medical Research Council.

THE MATURE DOMESTIC CAT OOCYTE DOES NOT EXPRESS A CORTICAL GRANULE-FREE DOMAIN

ANN P. BYERS AND DAVID E. WILDT
National Zoological Park, Smithsonian Institution, Washington, DC 20008

Cortical granules (CG) are secretory, membrane-bound organelles found in the cortex of unfertilized oocytes. A premature release of CG, resulting in the formation of a CG-free domain (CGFD) over the metaphase II (MII) spindle, is associated with nuclear maturation in the mouse and hamster. CG release at fertilization is associated with the block to polyspermy in these same species. Using the domestic cat, the objectives of this study were to: 1) compare the distribution of CG in immature, in vitro-matured and in vivo-matured oocytes; 2) determine the size, location and density of CG in in vivo-matured oocytes; and 3) establish whether a CGFD is formed in mature cat oocytes which may be useful as a marker for stage and normalcy of maturation. Immature oocytes (n=110) were collected from ovaries obtained from local veterinary clinics. A portion of these oocytes (n=65) were matured in vitro using our laboratory's previously described methods (Gamete Res. 24: 343, 1989). In vivo-matured (MII) oocytes (n=38) were flushed from the oviducts of gonadotropin-treated, ovariohysterectomized cats. CG were visualized microscopically by Lens culinaris agglutinin-biotin and Texas red-strepavidin fluorescence (Ducibella et al., Dev. Bio. 130: 184, 1988). Immature, in vitro-matured and in vivo-matured oocytes had similar, uniform distributions of CG throughout the cortical region as measured by fluorescence microscopy. Transmission electron microscopy confirmed that CG remained distributed throughout the entire cortical region of the in vivo-matured oocyte. All oocytes contained CG which had a mean diameter of 0.28 ± 0.03 μ at a mean density of 51.5 ± 13 CG/100 μ of plasma membrane. No CGFD was detected in domestic cat oocytes, indicating that premature CG release does not occur in this species. Additional evidence for the lack of premature CG release comes from parallel, on-going studies demonstrating that "zona hardening" does not occur in the cat. This phenomenon occurs when the contents of prematurely released CG interact with the zona of mouse oocytes matured in serum free medium and results in an inability of sperm to penetrate the zona. These results describe, for the first time, the physical and functional characteristics of CG of domestic cat oocytes and illustrate that CG localization cannot be used as a marker for maturation in this species. These results also may be extrapolated to mean that the cat has a different mechanism for preventing polyspermic fertilization from that of the mouse and hamster. (Funded by Smithsonian Institution's Scholarly Studies Program)

INVESTIGATION OF THE TRANSCRIPTIONAL REGULATION OF A TROPHOBLAST SPECIFIC GENE
TERESA CALZONETTI AND JANET ROSSANT
Samuel Lunenfeld Research Institute, Mount Sinai Hospital, Toronto, Canada

The trophectoderm is the first cell type to differentiate in the developing mouse embryo. Subsequent to its formation at the blastocyst stage the trophectoderm gives rise to three cell types that along with maternal and fetal blood vessels form the placenta. This lab has previously cloned a novel trophoblast-specific cDNA termed 4311. In an effort to elucidate the molecular mechanisms responsible for the development of the trophoblast lineage in the mouse we are in the process of identifying the cis-acting regulatory elements that are required for the restricted expression of this gene. We have determined that the gene spans three kilobases and is composed of five exons. Primer extension analysis has shown that the transcription initiation site is ten nucleotides upstream of the 5' end of the cDNA. Preliminary analysis of sequences upstream of the transcription initiation site has revealed several putative transcription factor binding sites including a non canonical TATA box, four CAAT boxes, a CRE consensus site and an octamer sequence. The functional characterization of the upstream sequences will be carried out using two approaches. *lacZ* reporter constructs containing these upstream sequences will be tested in transgenic mice to define trophoblast specific regulatory elements and bandshift assays will be used to identify trophoblast specific DNA binding factors. (supported by MRC Canada)

EFFECT OF MICRODROP SIZE AND NUMBER OF EMBRYOS PER MICRODROP ON EARLY MURINE EMBRYO DEVELOPMENT

R. S. CANSECO, A. E. T. SPARKS, R. E. PEARSON, and F. C. GWAZDAUSKAS. Department of Dairy Science, Virginia Polytechnic Institute and State University, Blacksburg.

Our objective was to determine microdrop size and number of embryos per microdrop for optimum development of 1-cell murine embryos. Embryos (640) were obtained from immature C57BL6 female mice superovulated with 5 i.u. PMSG and 5 i.u. hCG 48 h apart and mated by CD1 males. Groups of 1, 5, 10, or 20 embryos were cultured in 5, 10, 20, or 40 μl drops of CZB under silicon oil at 37.5°C in a humidified atmosphere of 5% CO_2 and 95% air. Embryos were evaluated every 24 h for 134 h and scored 0 to 9 (0=degenerate; 1=2-cell; 2=4-cell; 3=8-cell; 4=compact morula; 5=early blastocyst; 6=blastocyst; 7=expanded blastocyst; 8=hatching blastocyst; 9=hatched blastocyst). At 134 h mean developmental score (MDS) for embryos cultured in 10 μl drops was higher than that of embryos cultured in 20 or 40 μl (5.8 ± .3 vs 4.0 ± .3 and 3.8 ± .3; P< .01). MDS for embryos cultured in 5 μl drops (5.0 ± .3) was not different from that of embryos cultured in 10 or 20 μl but was higher (P< .01) than that of embryos cultured in 40 μl drops. MDS for embryos cultured in groups of 5, 10, or 20, were higher than that of embryos cultured singly (4.7 ± .3, 5.5 ± .3 and 5.2 ± .3 vs 3.1 ± .3; P< .01). The highest MDS was with 20 embryos/10 μl drop (7.0 ± .6). The lowest MDS was with 1 embryo/20 μl drop (2.3 ± .6). The percentage of live embryos (PLE) cultured in 20 or 40 μl was lower than that of embryos cultured in 10 μl (47.5, 46.2 vs 68.7; P< .05). PLE cultured in 5 μl (60.6) was not different from the other treatments. PLE cultured singly was lower than that of embryos cultured in groups of 10 or 20/drop (38.7 vs 65.6 and 61.8; (P< .05). PLE cultured in groups of 5 (56.8) was intermediate and not different from groups of 1, 10, or 20. Our results suggest optimum embryo development occurs with culture of multiple embryos in ratios of 1:1 or 2:1 embryos per microliter. However, ratios of 1:4 or lower appear to be diluting substance necessary for embryo development.

THE HOMOLOGY BETWEEN A RABBIT BLASTOCYST PEPTIDE AND OVINE TROPHOBLAST PROTEIN-1 (oTP1)

CAO YONG QING, CHEN YOU ZHEN AND ZHENG MING
Laboratory of Reproductive Biology, Institute of Zoology, Academia Sinica, Beijing, China 100080

In a previous study we have isolated and characterized specific peptides from rabbit blastocyst fluid at preimplantation stage. These peptides are involved in 1) endometrial immunosuppression; 2) inhibition of endometrial $PGF_{2\alpha}$ secretion; 3) promotion of endometrial protein synthesis.

In the present investigation blastocysts were obtained from the uteri of rabbits on days 4 and 6 of gestation. A peptide with MW about 4,500 was purified by preparative isoelectrofocusing and HPLC. The amino-acid sequence of the peptide was analyzed by ABI 900A sequence analyzer. The results indicated that the peptide contains 54 amino acids with double residues such as: Leu-Leu, Asp-Asp and Met-Met, which occur throughout the amino acid sequence of oTP1; and Glu-Asn, Val-Ser, Tyr-Glu, Thr-Glu and Glu-Asn-Leu residues, which also exist in the sequence of oTP1. In addition, several amino acid residues which are found in most IFN_αs so far characterized are conserved in the rabbit blastocyst peptide. These include Pro, Leu, Val at positions 4, 15, 19 and Arg-Pro-Asp sequence (33-35) which also exists in the sequence (49-51) of rabbit blastocyst peptide.

In conclusion our sequence data reveal a significant homology between the rabbit blastocyst peptide and oTP1 as well as IFN_αs.

Expression of insulin-like growth factor-I (IGF-I) mRNA in the rat fallopian tube

Björn Carlsson, Anders Nilsson, Jan Törnell, Håkan Billig, Torbjörn Hillensjö
Department of Physiology and Department of Obstetrics and Gynecology*, University of Göteborg, Sweden.

Growth factors are now recognized as important regulators of embryonic development. Comparison between growth of early embryos under *in vivo* compared to *in vitro* conditions has indicated that factors present in the fallopian tube are beneficial for early development. This suggests that the fallopian tube may secrete factors which influence the embryo. For instance there are indications that IGF-I and insulin are of critical importance for embryo development. The present investigation was done to study presence and regulation of IGF-I mRNA in the fallopian tube.

RNA extracted from rat fallopian tube was analyzed on a Northern blot with a mouse IGF-I RNA probe. Three major bands were detected with estimated sizes of 7.0, 1.7 and 1.2-0.8 Kb. The levels of IGF-I mRNA in the fallopian tube appeared similar to that present in the uterus and ovary. The level of IGF-I transcripts was then measured with a solution hybridization RNase protection assay. The levels of IGF-I mRNA decreased after hypophysectomy. Administration of estradiol (0.5 µg/100 g BW) to hypophysectomized rats significantly increased the abundance of IGF-I transcripts.

The present study demonstrate that IGF-I mRNA is present in the fallopian tube and that it is regulated by estradiol. IGF-I may therefore modulate or mediate estradiol regulation of the fallopian tube and may perhaps also be secreted into the lumen as suggested by the presence of IGF-I in porcine tubal fluid. Future studies will clarify if the fallopian tube is to be regarded as a paracrine gland through which the embryo is transported and whether it participates in the regulation of early embryonic development through IGF-I or other growth factors.

p34^{cdc2} PROTEIN KINASE IN MOUSE OOCYTES AND EMBRYOS

T. CHOI[1], F. AOKI[1], M. YAMASHITA[2], Y. NAGAHAMA[2], K. KOHMOTO[1].
[1]Department of Animal Breeding, Faculty of Agriculture, University of Tokyo, Yayoi 1-1-1, Bunkyo-Ku, Tokyo, Japan. [2]Department of Development Biology, National Insititute for Basic Biology, Okazaki, Japan.

To investigate the regulation of mouse oocyte maturation and embryo development, we examined the changes in phosphorylation states of p34^{cdc2} and its histone H1 kinase activity during meiotic and mitotic cell cycles. Recently, p34^{cdc2} protein kinase has been shown to be a key regulator of M-phase in the eukaryotic cell cycle. The activation of p34^{cdc2} protein kinase requires its dephosphorylation and association with cyclin.

When the embryos obtained 10 h after insemination were examined by immunoblotting following SDS-PAGE with antibody against the peptide EGVPSTAIREISLLKE, conserved sequence of p34^{cdc2}, three different migrating bands (upper, middle and lower bands) were detected. The upper and middle bands are phosphorylated forms since these two bands shifted to the lower one by the alkaline phosphatase treatment. In oocyte maturation, only oocytes in germinal vesicle (GV) stage had three forms. The phosphorylated forms decreased gradually in oocytes up to 2 h after isolation from follicles, and thereafter the phosphorylated states did not change appreciablly until metaphase II. On the other hand, histone H1 kinase activity oscillated in this period. It was activated at the first and second metaphase and inactivated at the time of the first polar body extrusion. These results suggest that changes in phosphorylation states of p34^{cdc2} triggered its activation at the first metaphase, but not inactivation and reactivation at the first and second metaphase, respectively. In the first mitotic cell cycle, phosphorylated forms appeared at 4 h after insemination, increased greatly short time after metaphase, and were dephosphorylated in metaphase. Histone H1 kinase activity was high at only metaphase. This kinase activation is probably triggered by dephosphorylation of p34^{cdc2}.

CHROMATIN STRUCTURAL CHANGES DURING MOUSE EMBRYOGENESIS: HISTONE H1 IS FIRST DETECTED AT THE 4-CELL STAGE.

H. J. Clarke[1,2,3], C. Oblin[1] & M. Bustin[4] Depts. Ob/Gyn[1] & Biol[2] & Cntr Study Reprod[3], McGill University, Montreal, Quebec H3A 1A1; [4]LMC, NCI, NIH, Bethesda, MD 20892

Early mouse embryogenesis is characterized by major changes in chromatin activity, including initiation of DNA replication, activation of embryonic transcription, and loss of nuclear totipotency. The molecular basis of these changes is poorly understood. To identify changes in chromatin structure during early embryogenesis, we examined the distribution of histone H1 in oocytes and preimplantation embryos by immunofluorescence and immunoblotting using affinity-purified anti-H1 antibodies. Fluorescent signal was observed in the nuclei of germinal vesicle (GV)-stage oocytes previously injected with histone H1, as well as in the nuclei of granulosa cells. In contrast, no nuclear fluorescence was observed in uninjected GV-stage oocytes. As well, histone H1 was not detectable in immunoblotted egg lysates, although it was detected in lysates of 8-cell embryos. To determine when immunoreactivity appeared during embryogenesis, embryos were examined at different stages of development. In one- and 2-cell embryos, nuclear fluorescence was never observed. At the early 4-cell stage (54-56 h post-hCG), 5 of 52 embryos displayed nuclear fluorescence. At the late 4-cell stage (66-68 h post-hCG), however, 58 of 62 embryos displayed nuclear fluorescence. The intensity of the signal varied among embryos and often among nuclei within an embryo. In 8-cell embryos, morulae, and blastocysts, all nuclei were fluorescent in every case. When the transcriptional inhibitor, α-amanitin, was added to late 2-cell embryos (48 h post-hCG), only 5 of 61 displayed nuclear fluorescence at 68-74 h post-hCG. When the drug was added to early 4-cell embryos, 28 of 40 embryos showed nuclear fluorescence at 68-74 h post-hCG. We conclude that oocytes and early embryos contain little or no immunoreactive histone H1, that it appears during the 4-cell stage, and that its appearance requires embryonic transcription. These results provide evidence of molecular differentiation of chromatin during early mouse embryogenesis. Supported by MRC (Canada).

STEROIDOGENIC ENZYME EXPRESSION IN PORCINE CONCEPTUS AND PLACENTA

A.J. CONLEY, R.K. CHRISTENSON, S.P. FORD, R.D. GEISERT AND J.I. MASON
Univ. Texas Southwestern Medical Ctr., Dallas, TX. 75235; USDA-ARS, RLHUSMARC, Clay Center, NE. 68933; Iowa State Univ., Ames, IA. 50011; Oklahoma State Univ., Stillwater, OK. 74074

Estrogen secretion by pig blastocysts reaches a peak at day 11 or 12 post-mating, coincident with blastocyst elongation, then declines precipitously within 24 hours. Uterine luminal estrogen concentration also peaks at day 12 and declines on day 13, but is reported to rise again by day 15. Except for aromatase activity, little has been done to characterize steroidogenic competence of pig blastocysts or the developing trophoblast. Therefore, the following studies were performed to investigate steroidogenic enzyme expression in pig conceptuses between days 12 and 21 post-mating by Western immunoblot and Northern RNA analysis. In the first experiment, blastocyst diameter or length was recorded for blastocysts from litters collected on day 12 and frozen individually for Western analysis. The second experiment examined conceptus tissue pooled from each uterine horn of white crossbred and Chinese gilts on days 12, 14, 16 and 21 post-mating. Western and Northern analyses utilized antibodies raised against, and cDNAs encoding for, porcine 17α-hydroxylase cytochrome P450 (P450$17\alpha$), bovine cholesterol side chain cleavage cytochrome P450 (P450scc), human 3β-hydroxysteroid dehydrogenase (3βHSD) and human aromatase cytochrome P450 (P450arom). Within litters P450$17\alpha$ protein was detectable in 3mm blastocysts with highest levels apparent in 10-15mm (tubular) blastocysts. In some instances, filamentous blastocysts appeared to have less P450$17\alpha$ than did littermate blastocysts that were tubular. P450scc and P450arom followed a pattern similar to that seen for P450$17\alpha$. 3βHSD was undetectable by Western analysis in blastocysts of the stages examined although detectable in placenta from mid and late gestation. In pooled conceptus tissue, P450$17\alpha$ protein and mRNA were very high in day 12 tissues compared to tissues from all other days which did not differ. Neither P450scc nor 3βHSD were detectable by Northern analysis of conceptus RNA at any stage. These data suggest that the decrease in blastocyst estrogen secretion on day 13 in pigs may be due to a decrease in P450$17\alpha$ expression. The relatively low level of 3βHSD in embryos compared to placenta suggests that this enzyme may limit embryonic Δ^4-steroid hormone synthesis to the utilization of maternal progesterone rather than cholesterol during these stages of development.

HIGH IN VITRO FERTILIZATION RATE OF OOCYTES AND DEVELOPMENT TO 2-CELL EMBRYOS DOES NOT DEPEND ON GONADOTROPIN PRIMING IN IMMATURE RATS.

S.A.J. DANIEL and R.E. GORE-LANGTON
Dept. of Gynaecology, University Hospital and Dept. of Physiology, University of Western Ontario, London, Canada

Gonadotropins are not required for oocyte growth or acquisition of meiotic competence in cultured rat preantral follicles. However, a low fertilization rate suggested cytoplasmic maturation might depend on gonadotropins (Gamete Res. 24:108, 1989). Since we have been unable to show such a requirement for exogenous gonadotropins in vitro (unpublished), we have attempted to confirm evidence that fertilization of fully grown oocytes from immature rats requires gonadotropin priming (Zhang and Armstrong, Gamete Res. 23:267, 1989). Twenty-five day-old rats were primed with 10 IU pregnant mares' serum gonadotropin (PMS) or injected with saline vehicle. Oocyte-cumulus complexes were isolated from the largest ovarian follicles 28-30 h later and cultured in MEM supplemented with 1 mM pyruvate, 0.25 mM uridine, 50 ug/ml gentamicin and 2.5% fetal bovine serum for 12-14 h to allow for resumption of meiotic maturation. Oocytes were stripped of adherent cumulus cells and all oocytes showing germinal vesicle breakdown were inseminated with ejaculated sperm. Approximately 18 h later, oocytes were examined for fertilization (2 pronuclei). Development to 2-cell embryos was assessed following a further 24 h culture period. There were no significant differences between fertilization rates (69.8% (134/192) vs. 68.8% (88/128)) or development to 2-cell embryos (80.6% (108/134) vs 81.8% (72/88)) for PMS-primed and unprimed rats, respectively. Results are the combined data from three experiments. This data indicates that oocytes from the largest follicles in immature rats are able to undergo fertilization and cleavage to 2-cells without additional beneficial effects of exogenous gonadotropin priming. Whether endogenous gonadotropins are sufficient or gonadotropins are not essential for cytoplasmic maturation in rat is unresolved. (R.E.G.-L is supported by the Ontario Ministry of Health).

THE EPIDERMAL GROWTH FACTOR COMPONENT OF HUMAN FOLLICULAR FLUID IS A STIMULATORY FACTOR FOR MOUSE OOCYTE MATURATION IN VITRO

KAMALINI DAS, WILLIAM R. PHIPPS, HUGH C. HENSLEIGH, GEORGE E. TAGATZ;
Division of Reproductive Endocrinology, Department of Obstetrics and Gynecology, University of Minnesota, Minneapolis, Minnesota 55455

Human follicular fluid (FF), although generally considered to inhibit the onset of meiosis, is known to contain epidermal growth factor (EGF), which in concentrations as low as 0.1 ng/mL has been shown to stimulate the in vitro maturation of mouse oocytes. We studied the effect of human FF and the specific contribution of its EGF component on mouse oocyte maturation. Specimens of human FF surrounding mature and immature oocytes were obtained at the time of oocyte retrieval for in vitro fertilization, and then tested for their effects on incubated cumulus-enclosed mouse oocytes. The endpoints assessed were the percentage of oocytes undergoing germinal vesicle breakdown (GVBD) and polar body one (PB1) formation at different intervals over a 24-hour period, and the final degree of cumulus expansion achieved. A concentration-related stimulatory effect of mature FF was noted when compared to the spontaneous increase of GVBD and PB1 formation observed in the EGF-free control medium (Ham's F-10 with 7.5 % albumin from which EGF had been extracted by immunoprecipitation). In contrast, the overall effect of immature FF was inhibitory. After extraction of EGF from FF by immunoprecipitation from both immature and mature FF, the rates of GVBD and PB1 formation were decreased in both groups; this effect was reversed by the addition of EGF at a concentration of 5 ng/mL. Epidermal growth factor at this concentration was still not able to overcome the inhibitory effect of immature FF. Cumulus expansion was maximal for oocytes incubated with mature FF, and minimal for those incubated with the EGF-free media; however, the changes in cumulus expansion did not directly parallel the changes in nuclear maturation. As measured by RIA (Biomedical Technologies, MA), FF concentrations of EGF in the mature and immature FF specimens were similar (3.2 vs 3.0 ng/mL). In summary, the stimulatory effect of mature human FF on mouse oocyte maturation and cumulus expansion can be largely attributed to the presence of EGF. However, this stimulatory effect of EGF on oocyte maturation appears to depend upon lower FF concentrations of inhibitory substances present in mature FF as compared to immature FF.

INSULIN AND GLUCOSE MODULATE RAT BLASTOCYST DEVELOPMENT IN VITRO

R. DE HERTOGH, I. VANDERHEYDEN, E. DUFRASNE, D. ROBIN and J. DELCOURT
Physiology of Human Reproduction Unit, University of Louvain, Brussels, Belgium

The effect of glucose and insulin on rat blastocyst development in vitro was studied by incubating day 5 embryos for 24 or 48h in Ham's F10 medium. A differential cell staining method allowed the separate counting of inner cell mass (ICM) and trophectoderm (TE) cells at the end of the incubation time. Low insulin concentrations (3pM) in the incubation medium stimulated ICM and TE development in the presence of 1.1 or 6mM glucose. Higher insulin levels (30 to 600 pM) in a 6 mM glucose medium resulted in a dose dependent inhibition of ICM and to a lesser extent, TE development. In the absence of glucose, insulin was neither stimulatory nor inhibitory on ICM growth. Glucose consumption was studied in day 5 blastocysts after 24h incubation. ^3H_2O production from 5-^{3}H-glucose, as a function of glucose concentration, was curvilinear, with a steep active uptake (K: 0.15mM, Vmax: 20pmoles/embryo/h), followed by a slow linear increase. $^{14}CO_2$ production from U-^{14}C-glucose accounted for less than 3% of total glucose consumption and was due mainly to the pentose shunt. Insulin added to the culture did not change the glucose consumption kinetics through the different pathways. These results suggest that insulin and glucose interplay to modulate rat blastocyst development, as a function of their respective concentrations, in a permissive mode of action independent of glucose metabolism.

INFLUENCE OF CULTURE MEDIA ON THE HATCHING AND IMPLANTATION OF IN VITRO-PRODUCED BOVINE BLASTOCYSTS

F. DELHAISE, P. MERMILLOD, C. BOCCART, V. BRALION AND F. DESSY
Catholic University of Louvain, Domestic Animals Physiology,
Place Croix du Sud, 2, B-1348 Louvain la Neuve, Belgium

Bovine blastocysts produced by IVM and IVF were used. At day 8 post fertilization (PF) they were transferred on a mitotic-arrested mouse fibroblast feeder layer. Three culture media were used: (A): TCM199 containing 10% FCS, 10% calf serum (CS) and 10^{-4} M β-mercaptoethanol; (B): DMEM/F12 containing the same supplements; (C): DMEM containing the same supplements + 1% non-essential amino acids. The numbers of hatched and implanted embryos were recorded on days 11,12,14 and 15-16 PF. The diameters of the hatched not yet implanted embryos were recorded on days 11 and 12 PF; their volumes were calculated. Results are shown in tables 1 and 2.

Table 1 : rates of hatching and implantation

Medium	N.hatched/N.embryos (%)	N.implant./N.hatched (%)
DMEM	15/31 (48)	14/15 (93)
DMEM/F12	14/28 (50)	14/14 (100)
TCM 199	18/28 (64)	16/18 (89)

Table 2: Mean volume (nL) of the hatched non yet implanted blastocysts.

Medium	Volume + SD (n) Day 11 PF	Volume + SD (n) Day 12 PF	+%
DMEM	10.8 ± 8.3 (9)	10.6 ± 5.2 (11)	0%
DMEM/F12	25.4 ± 22.7 (10)	51.9 ± 50.2 (6)	104%
TCM 199	16.8 ± 11.1 (11)	26.4 ± 16.7 (10)	57%

Hatching occured from day 8 to day 15 PF and reached 48% to 64% (table 1); no significant difference in these rates was observed between the media. Nearly all the hatched blastocysts (89% to 100%) did implant between day 11 and 15 PF; no significant difference was observed between (A) (B) and (C). Blastocyst volumes of the hatched not yet implanted embryos at days 11 and 12 PF were significantly larger in DMEM/F12 and TCM199 than in DMEM (table 2). In DMEM there was no increase of volume; in TCM199 and DMEM/F12 the increases were respectively +57% and +104%.
As a conclusion, the rates of hatching and implantation appear similar for the tested media. However, DMEM/F12 and TCM199 promote the growth of blastocyst.
This research was supported by a grant of "I.R.S.I.A", Belgium.

REPROGRAMMING OF PROTEIN SYNTHESIS DURING FERTILIZATION IN PIG OOCYTES

J DING, N CLARKE, T NAGAI & RM MOOR.   Dept. Molecular Embryology, AFRC Institute of Anim. Physiol. and Genetics, Babraham, Cambridge CB2 4AT

Complexes of oocyte-cumulus-granulosa cells (oocyte complexes) were obtained from 3 - 6 mm non-atretic follicles dissected from the ovaries of slaughtered gilts. The oocyte complexes were cultured in 2 ml medium 199 with gonadotrophins and two opened follicle shells in a non-static incubation system at 39°C, and in a 5% CO_2 atmosphere. After 47 h culture, oocytes were fertilized in vitro with in vitro capacitated boar spermatozoa.
Changes in protein synthesis and protein phosphorylation during fertilization were analyzed by 1- and 2-dimensional SDS-polyacrylamide gel electrophoresis after labelling with [^{35}S]-methionine and/or [^{32}P]-orthophosphate. Post-translational modification of existing proteins during fertilization was also analyzed by gel electrophoresis after radio labelling of proteins, but the radio labelling of the proteins was made before in vitro insemination and during fertilization protein synthesis in oocytes was inhibited by cycloheximide.
A variety of protein changes were triggered by sperm penetration of oocytes. New protein bands observed during fertilization were mainly derived from post-translational modification and/or degradation of existing proteins. The most striking post-translational protein event identified at fertilization was the transition from a group of 25 kD phosphoproteins which have isoelectric points (pI) ranging from 6.7 to 6.0, to a single positively charged (pI=7.6) 22 kD non-phosphorylated protein species. This post-translational modification is not dependent on new protein synthesis.

HUMAN TROPHECTODERM BIOPSY AND CHORIONIC GONADOTROPIN (hCG) SECRETION

A. DOKRAS, I.L. SARGENT, R.L. GARDNER, D.H. BARLOW.
University of Oxford, IVF Unit, Nuffield Department of Obstetrics and Gynaecology, John Radcliffe Hospital, Oxford, OX3 9DU, U.K.

We have developed a technique of trophectoderm biopsy to obtain cells from human blastocysts for preimplantation genetic diagnosis. To determine whether this technique affects the subsequent development of the blastocyst, 45 manipulated blastocysts were observed from day 3-14 in culture, the amount of hCG secreted by each embryo measured and these results were compared with those of 26 non-manipulated controls.

A slit was made in the zona pellucida opposite the inner cell mass in 18 of the 45 blastocysts. This increased the rate of hatching but the other morphological changes up to day 14 were similar to those seen in the non-manipulated controls. There was no difference in the mean cumulative hCG secretion by these zona-slit controls (149.8±45.7mIU/ml) compared to the non-manipulated controls (146.2±23.7mIU/ml).

A slit was made in the zona of the other 27 blastocysts. Approximately 12-18 hours later, in 18 blastocysts, a biopsy of the herniating trophectoderm cells (5-30) was performed. The rate of hatching and adherence to the culture dish was similar to the non-manipulated controls. The mean cumulative hCG secretion decreased significantly (57.5±16.2mIU/ml, p<0.01) after the biopsy procedure. However, if a small biopsy was performed (<10 cells removed) the decrease in hCG secretion (87.6±24.8mIU/ml) was not significant, whereas after a large biopsy was performed (>10 cells) hCG levels fell to 19.9±9.1mIU/ml.

The amount of hCG secreted by the blastocyst after biopsy was inversely proportional to the number of trophectoderm cells removed. Moreover, the removal of less than 10 cells did not appear to impair the further development of the blastocyst. This biopsy method can therefore provide sufficient cells to allow replicate analysis necessary for reliable genetic diagnosis.

TWO-CELL BLOCK IN MOUSE EMBRYOS MAINTAINED BY HYPOXANTHINE AND cAMP-ELEVATING AGENTS: EFFECTS OF GLUCOSE, EDTA AND TIME OF EXPOSURE

SM DOWNS, Biology Department, Marquette University, Milwaukee, WI

The present study was carried out to compare the effects of hypoxanthine and cAMP-elevating agents on the early development of preimplantation mouse embryos under a variety of experimental conditions. (C57BL/6J X SJL/J)F_1 female mice, 20-23-days-old, were primed for 2 d with PMSG, followed by hCG injection and mating overnight with an F_1 male. Pronuclear-stage embryos were isolated the next morning and placed in Whitten's medium containing a number of different supplements. Embryos were assessed after 2 d of culture for the extent of development. Dose response experiments utilizing the cAMP analogs, dbcAMP and Sp-cAMPS, and the phosphodiesterase inhibitor, 3-isobutyl-1-methylxanthine (IBMX), revealed a dose dependent increase in the number of embryos arrested at the two cell stage, with an additional block at the 1-cell stage at higher concentrations. Hypoxanthine produced a two-cell block but did not suppress first cleavage. Removal of D-glucose from Whitten's medium reduced the inhibitory effect of hypoxanthine but had no effect on dbcAMP- or IBMX-arrested embryos. If the exposure of embryos to these agents was delayed, hypoxanthine was able to prevent second cleavage as long as the embryos were in the two-cell stage upon transfer to hypoxanthine-containing medium (up to 47-48 h post-hCG), but dbcAMP and IBMX were much less inhibitory at the later times. Addition of 100 mM EDTA eliminated the two-cell block produced by dbcAMP, IBMX, and hypoxanthine. These data demonstrate a greater sensitivity of the two-cell embryo than the one-cell embryo to cAMP and hypoxanthine and suggest that hypoxanthine may not be exerting its effects solely by maintaining elevated cAMP levels. Furthermore, it has been demonstrated that EDTA can prevent the inhibitory action of these agents on preimplantation embryo development, suggesting a role for trace metals in mediation of this effect. Supported by funds from the NIH.

HISTAMINE CONCENTRATION AND MAST CELL ULTRASTRUCTURE IN THE NORMAL HUMAN ENDOMETRIUM AT THE PERIOVULATORY STAGE OF THE MENSTRUAL CYCLE.

L. DRUDY. RCSI DEPARTMENT OF OBSTETRICS AND GYNAECOLOGY, ROTUNDA HOSPITAL, DUBLIN 1. IRELAND.

The successful implantation of the blastocyst into the endometrium and the development of a decidual response involves many complex maternal and embryonic reactions. Histamine is derived mainly from mast cells which are present in large numbers in the human uterus. The transformation of uterine stromal cells into decidual cells is an essential step in natural implantation and it has been proposed that uterine histamine of mast cell origin is involved in the initiation of implantation and the decidual cell response in the rat. In the present study biopsies incorporating endometrium and myometrium were taken from the anterior uterine wall following hysterectomy for patients throughout the menstrual cycle. Extracted histamine was condensed with 0-phthaldialdehyde to form a fluorophore and its fluorescence was measured at 450 mu using a spectrofluorometer. Higher mean values for histamine concentration are observed in the late proliferative stage of the menstrual cycle than in the early secretory phase. Transmission electron microscopy studies showed considerable variability in mast cell ultrastructure not only at the various phases of the menstrual cycle but also from patient to patient particularly in the endometrium. However, some specific morphological features were observed for phases of the menstrual cycle. In the late proliferative stage of the menstrual cycle, slight invagination of the mast cell membrane and vacuolation were observed in mast cells containing particulate granules; slight invagination of the mast cell membrane was also observed in the endometrial/myometrial border. Very electron-dense granules were also observed in the cytoplasm of these mast cells, with few vacuoles. In the early secretory phase mast cells in the endometrium contained granules with a very dense particulate structure. Constriction of the particulate structures was a common finding. The present observations have considerable relevance in investigating uterine homeostasis per patient at mid-cycle.

Stimulation of the Cortical Reaction by Protein Kinase C Agonists in the Mouse Egg.

T. Ducibella[1], P. Duffy[1], S. Kurasawa[2], G.S. Kopf[2], and R.M. Schultz[3], [1]Depts. of OB/GYN (Div. Reprod. Endo.) and Anat./Cellular Biology, Tufts U. School of Medicine, Boston, MA 02111, [2]Div. of Reprod. Biol., and [3]Dept. of Biology, U. of Pennsylvania, Philadelphia, PA 19104.

Mammalian egg activation at fertilization leads to cortical granule (CG) release, resulting in modifications of the *zona pellucida* (ZP) that are responsible for the polyspermy block. An investigation of the potential involvement of protein kinase C (PKC) in the activation of CG exocytosis was conducted with the PKC agonist 12-O-tetradecanoyl phorbol 13-acetate (TPA), and the biologically inactive phorbol diester control, 4α-phorbol 12,13 didecanoate (PDD). Germinal vesicle (GV) and ovulated metaphase II (MII) stages were coincubated in the presence of either 10ng/ml of TPA or PDD. PKC-dependent CG loss and ZP2 modification were determined by comparing the relative mean CG densities in the cortex or mean ZP2 to ZP2f conversion, respectively, in TPA- vs PDD-treated eggs. CG densities were determined after staining with *Lens culinaris* agglutinin and quantification by image analysis. ZP2 and ZP2f were assayed by ZP iodination and PAGE. After 60 and 120 min., TPA treatment (0-60 min.) resulted in greater than 25% and 50% CG loss, respectively, at both stages. At both time points, GV oocyte ZP2f levels were similar to those in fertilized eggs. MII egg ZP2 underwent 50% conversion by 60 min. and reached fertilization levels by 120 min. Complete CG release was not required for the high % of ZP2 to ZP2f conversion observed at fertilization. In contrast to these PKC results, agonists causing calcium-mediated activation stimulated only MII eggs. MII eggs underwent the ZP2f conversion after microinjection of inositol trisphosphate (IP_3, 1uM final concentration), whereas GV oocytes did not. These results suggest that the activation of the CG and ZP reactions may utilize a PKC-dependent mechanism. This mechanism appears to be present in fully grown GV oocytes, unlike the IP_3-dependent pathway that is detected after the resumption of meiotic maturation.

ISOLATION OF EMBRYONIC CELL LINES FROM PORCINE BLASTOCYSTS

RAYMOND W. GERFEN, CAROL J. FAJFAR-WHETSTONE AND MATTHEW B. WHEELER
Laboratory of Molecular Embryology, Dept. of Animal Sciences, University of Illinois, Urbana, IL 61801

Several embryonic cell lines have been established from transplanted, hatched porcine blastocysts with embryonic stem cell-like characteristics. Porcine embryos were cultured in a Whitten's based medium until hatching from the zona pellucida. Blastocysts were then transferred to a mitotically-inactivated mouse embryonic fibroblast (STO) monolayer containing Dulbecco's modified Eagle's medium with 20% fetal bovine serum. Blastocyst attachment occurred within 24 hr. After 7-10 days of culture at 39 C in 5% CO_2 in air, large, flat cell colonies appeared. Whole colonies were disaggregated with 0.25% trypsin in 0.03% EDTA and replated onto feeder layers. Initially two cell types appeared, one being large in size with a trophoblastic-like cell morphology that disappeared by the 3rd passage. The other cell type was epithelial-like containing a large prominent nucleus with several nucleoli. Cell size and rate of growth varied between isolates but was similar to murine ES cells. However, porcine ES-like cells appeared translucent and formed extensive monolayers. The porcine ES-like cells have demonstrated the ability to differentiate into epithelial, muscle and neuronal-like cells, indicating a pluripotential cell line. Cystic embryoid bodies were formed when these cells were placed in suspension culture with cells from all three embryonic germ layers being present. To date we have collected 150 total embryos from eight Duroc, twelve Meishan and one Yorkshire gilt(s). We have isolated eleven putative stem cell colonies and are in the process of sub-culturing, expanding and freezing these cell lines for ES-cell characterization. The efficiency for producing stem cells from Duroc, Meishan and Yorkshire embryos was 14.8, 17.2 and 0%, respectively. As controls for our procedures, we have also isolated ES-cells from four strains of mice, ICR, C57/B6, 129J and hybrid C57/B6 X C3H. The efficiency rates for producing ES-cells from the mouse strains were 25%, 43%, 66% and 15%, respectively. The efficiency rates for producing putative stem cell colonies from swine blastocysts appeared to be lower than that of mice. However, the efficiency of producing stem cells from both mouse and swine embryos may be affected by strain and cell feeder layer. These differences in efficiency of isolation are some aspects that we are currently investigating.

OBSERVED DIFFERENCES IN THE DEVELOPMENT OF HUMAN ZYGOTES IN VITRO.

GORDON AC., DRUDY L., KONDAVEETI U., BARRY-KINSELLA C. HENNELLY B. HARRISON R.F. HARI, ROTUNDA HOSPITAL, DUBLIN 1. IRELAND.

Mammalian zygotes are known to develop in vitro at similar rates to those observed in vivo. The human zygote resulting from in vitro fertilization (IVF) can cleave beyond the four cell stage within 42-46 hrs post-insemination (pi).
This study examined the present IVF system by classifying cleavage 42-46 hrs pi as either one-, two-, three-, four- and >four-cell. The cleavage index was determined by calculating the number of zygotes showing more than two cells 42-46 hrs pi and dividing by the total number of zygotes cleaving at that time.
Five hundred and eighteen human zygotes, showing two pronuclei 16-19 hrs pi, were studied. Two ovarian stimulation regimes were employed. Protocol one involved the administration of a Gonadotrophin Releasing Hormone analogue prior to commencing follicular recruitment using human Menopausal Gonadotrophin (hMG). Protocol two utilized a combination of Clomiphene Citrate and hMG.
Overall results indicated that 5.6 % of the total zygotes failed to develop beyond the one-cell stage. Two-, three-, four- and >four cells represented 26.6 %, 12.0 %, 46.7 % and 9.1 % of the total zygotes respectively. Comparison of zygotes from the two stimulation regimes indicated significant differences between the proportion of zygotes at the two-cell (32.6 % vs. 19.5 %) ($p<0.001$) and at the four-cell stage (42.6 % vs. 51.7 %) ($p<0.05$) 42-46 hrs pi, for Protocol one and two respectively. The overall cleavage index for the zygotes studied gave a value of 0.72. The cleavage index for Protocol one was 0.66 and for Protocol two was 0.79.
The developmental rate of human zygotes in vitro shows a variety of cleavage stages 42-46 hrs pi. The use of a cleavage index may be of relevance in monitoring the IVF system and the effects of alterations within eg. ovarian stimulation or culture conditions.

MURINE ANDROGENONES DO NOT SHOW CELLULAR POLARIZATION

L.J. HAGEMANN, C. NAVARA and N.L. FIRST
University of Wisconsin, Madison, WI 53706

Blastomere polarization occurs in the outer cells of murine embryos at the late 8-cell stage. This leads to blastocysts containing 2 distinct cell types -- apolar inner cell mass cells and polarized trophectoderm cells. One method of visualizing polarization is by localized binding of lectins to microvilli-coated apical cell membranes. Because androgenones (paternal genome) undergo trophoblast hypertrophy with no fetus, we expected androgenetic blastocysts to contain only polarized blastomeres. Zonae pellucidae of normal and androgenetic blastocysts cultured for 5 days were removed with acid Tyrode's. The blastocysts were treated for 30 min with Ca^{2+}/Mg^{2+}-free medium (CMF) and attached to poly-lysine coated coverslips. After fixing in 3.7% formaldehyde and staining with 700 µg/ml fluorescence Con-A and 5 µg/ml DAPI to highlight nuclei, the blastocysts were viewed under light microscopy. All blastocoeles had collapsed due to CMF treatment. Although both control and androgenetic blastocysts appeared to compact and cavitate normally, only control blastocysts bound Con-A just on the apical surface. Androgenetic blastomeres showed binding over the entire surface of each blastomere. Likewise, compared to controls, androgenones contained fewer cells (average of 55 vs 27 cells, respectively) with nuclei that appeared enlarged. These data question the importance of polarization in cavitation and differentiation.

EFFECT OF BIOPSY ON SUBSEQUENT VIABILITY OF MOUSE AND HUMAN EMBRYOS IN VITRO

G. M. HARTSHORNE* AND S. AVERY[+]
* BOURN HALL CLINIC, BOURN, CAMBRIDGE, CB3 7TR, UK.
[+] HALLAM MEDICAL CENTRE, 112 HARLEY STREET, LONDON, W1N 1AF, UK.

The authors have examined the viability of human and mouse embryos using fluorescein diacetate (FDA) (1) and a Fluovert microscope and automatic exposure photographic apparatus (Leitz Instruments, UK). Fluorescence was quantified as high, medium, low or undetectable using the automatic exposure meter. High fluorescence reflected high viability.

The necrotic appearance under light microscopy of human embryos too poor for cryopreservation correlated well with FDA viability, except at the morula stage. Tripronucleate embryos usually showed good viability and further cleavage was observed. FDA viability also correlated well with the appearance of three human blastocysts after freezing and thawing, clearly demarking the viable areas.

In mouse blastocysts (F_1, female MF1 x male CFLP), repeated FDA analysis was associated with reduced viability as assessed by morphology and intensity of fluorescence. Mouse blastocysts were biopsied by aspirating the blastocoelic cavity. Control groups comprised blastocysts from the same batch which were not biopsied, or which were subjected to a sham biopsy consisting of puncturing the trophectoderm. Similar rates of subsequent development and hatching in vitro were observed, but viability assessed by FDA was lower in both sham and biopsied blastocysts in comparison with controls. The human cryopreserved blastocysts were also biopsied but no conclusions could be drawn owing to the small numbers.

The authors conclude that viability assessed by light microscopy usually correlates well with results obtained using FDA. However, human morulae which appear abnormal may retain high FDA viability and biopsied mouse blastocysts which show reduced FDA viability may subsequently develop normally in vitro.

(1) Mohr LR, Trounson AO, (1980). J. Reprod. Fert. 58, 189-196.

PHYSIOLOGICAL LEVELS OF IGF-1 STIMULATE MOUSE EMBRYOS

MARK B. HARVEY and PETER L. KAYE
Department of Physiology and Pharmacology, The University of
Queensland, Australia 4072

The insulin growth factor family has recently been implicated in early development. We have investigated the effects of IGF-1 on the development of mouse embryos *in vitro*. When added to the medium for culture of 2-cell embryos, 170 pM (1 ng/ml) IGF-1 maximally stimulated the number of cells in the resultant blastocysts after 54 h (47.4 ± 1.4 v 42.7 ± 1.2 cells/blastocyst), entirely by increasing the number of cells in the ICM (16.0 ± 0.5 v 12.6 ± 0.5 cells/ICM; EC_{50} = 60pM). This stimulation was also achieved when ICMs were isolated from blastocysts prior to culture for 24 h with 170 pM IGF-1 (22.3 ± 1.0 v 17.5 ± 0.8 cells/ICM). There was no effect of IGF-1 on TE cell proliferation in blastocysts. In morphology studies, IGF-1 increased the proportion of blastocysts (62 ± 3% v 49 ± 4%) after 54 h culture from the 2-cell stage with an EC_{50} around 60pM IGF-1. These levels of IGF-1 are in the range for IGF-1 receptor mediation, but cross reaction with other receptors is not excluded. Nonetheless, the results show that physiological concentrations of IGF-1 (17-170 pM; 0.1-1 ng/ml) which has been observed in the reproductive tract, affect the early embryo, suggesting a normal role for this factor in the regulation of growth of the developing conceptus before implantation.

RESPONSES TO GROWTH HORMONE THERAPY IN A WOMAN WITH ABNORMAL FOLLICULAR DEVELOPMENT

Jamieson ME, Fleming R, Shaker AG, Yates RWS and Coutts JRT,
Department of Obstetrics and Gynaecology, Glasgow Royal Infirmary, Scotland.

Growth hormone (GH) stimulates production of insulin-like growth factor-1 which may augment some FSH dependent ovarian cell processes (1). We present a case study of a patient with a history of abnormal follicular and oocyte development after stimulation for IVF, who was treated with concurrent GH, GnRH-analog and hMG therapy. Previous cycles (n=6) using a variety of stimulation regimes failed to attain standard criteria for administration of hCG (3 follicles ≥17mm), regardless of duration (max 20 days) and doses of gonadotrophins (max 80 amps). A high proportion (17/24; 71%) of oocytes showed incomplete cumulus expansion with a mean score of 2.0 (scoring system: 1-3, no cumulus expansion-fully expanded). Subsequent fertilisation was low (9/24; 37.5%) and embryo quality poor (mean 5.4, scoring system 1-10).

Adjuvant GH treatment did not affect the pattern of follicular development (only 1 mature follicle). Four of the 9 oocytes recovered were mature (mean score 2.4), 6/8 fertilised (1 damaged) and subsequent embryo development was normal (mean score 7.5). This supports evidence that cumulus expansion after luteinisation may be FSH dependent (2) and growth factor mediated and is important for oocyte maturation. Transfer of three embryos resulted in an ongoing singleton pregnancy.

1. Adashi et al (1988) in 'Growth factors and the ovary'
   Ed A N Hirshfield pp 95-106. Plenum Press NY.
2. Eppig J J (1980) Biol Reprod 23; 545.

PREIMPLANTATION GENETIC ANALYSIS

KAUFMANN RA, TAKEUCHI K, MORSY M, HODGEN GD. The Jones Institute for Reproductive Medicine, Department of Obstetrics and Gynecology, Eastern Virginia Medical School, 855 West Brambleton Avenue, Suite B, Norfolk, VA 23510

Among the nascent assisted reproductive technologies at the forefront of preventive clinical care in the 1990's is preimplantation genetic analysis. We have conducted a number of pre-clinical investigations that are mandatory prior to initiation of human application. We have biopsied 4- and 8-cell mouse pre-embryos using enucleation, aspiration or extrusion of single blastomeres. Development to the blastocyst stage for the 4-cell pre-embryo were: 94.6% (control), 80.7%, 90.1% and 83.1%, respectively. For the 8-cell pre-embryos, the blastocyst formation rates were: 96.7% (control), 89.1%, 91.7% and 91.5%. After biopsy at the 4-cell stage and transfer to the oviducts, live birth rates were slightly lower in the enucleation group than in the blastomere aspiration and extrusion groups (49.2% vs 58.8% and 56.3%, respectively); for the 8-cell stage, there were no differences. Thus, both the *in vitro* development to expanded blastocyst formation and *in vivo* live birth rates were not different between biopsied and control groups. There were no developmental abnormalities in body or organ weights of neonates or at 3 weeks of age or in their subsequent ability to produce a second generation. The litter size from mated biopsied offspring for both the 4-cell and 8-cell biopsied groups was 8.20 ± 0.89 versus control 8.80 ± 0.75 ($\bar{x} \pm $ SD). We successfully amplified and diagnosed sickle cell disease and Tay Sachs disease from single lymphoblasts and the ornithine transcarbamylase deficiency mutation from single blastomeres of Sparse Fur mouse pre-embryos. In addition, we have performed amplification for the sex-determining region of the Y chromosome from genomic DNA (co-amplified with X chromosome or an autosome target sequence to prevent false negative results) or single cells in the mouse and human. These preclinical studies demonstrate that preimplantation embryo biopsy and genetic analysis can be undertaken with accuracy, reproducibility and reliability.

EFFECT OF INCUBATION VOLUME AND EMBRYO DENSITY ON THE DEVELOPMENT AND VIABILITY OF MOUSE EMBRYOS IN VITRO

MICHELLE LANE AND DAVID K. GARDNER
Centre for Early Human Development, Monash Medical Centre, 246 Clayton Road, Clayton, Melbourne, Victoria, 3168, Australia.

The cleavage rate and viability of embryos developed in vitro is reduced compared to embryos in vivo. The composition of embryo culture media does not reflect the environment that the embryos is exposed to in vivo. Furthermore, the physical conditions of the oviduct are not mimicked, eg. embryos are cultured in large volumes compared to the sub-microlitre volumes of fluid present in the lumen of the reproductive tract. Any autocrine factor(s) produced by the embryo(s) during development will therefore be greatly diluted out in culture and have minimal, if any, effect.
This study investigates the role of incubation volume and embryo density on development, cell number and viability. Two-cell embryos from random-bred Swiss mice were cultured individually (in 5, 10, 20, 40, 80, 160 and 320μl) or in groups (1, 2, 4, 8 and 16), in 5, 20 and 320μl of medium (MTF) (1), under paraffin oil. Cell number was determined at 113h post hCG, and embryo viability assessed by transfering morulae/ early blastocysts to pseudopregnant recipients.
Blastocyst formation increased significantly with embryo density in 5 and 320μl droplets (P<0.05). Cell number of embryos grown individually was maximum at 20μl (81.9+4.3), and minimum at 320μl (55.5+2.6). When embryos were grown in groups in 20μl, there was a significant increase in cell number when two embryos were present (98.1+7.6), and no further significant differences were observed with larger groups. In contrast, in 320μl, there was no increase in cell number until 4 embryos were present (92.7+7.1), and the maximum response required 8 embryos (116.8+9.8). Initial results from embryo transfers have revealed that embryos cultured singly in 20μl resulted in significantly more implantations (P<0.05) than those grown in 320μl.
These data suggest that preimplantation mouse embryos are capable of producing autocrine factor(s) which stimulate cleavage and increase viability. In most embryo culture systems, such factors will be diluted resulting in delayed cleavage and loss of viability. This study has implications for clinical IVF, where human embryos are cultured singly in large volumes, and embryo viability is poor.
(1) Gardner,D.K. and Leese,H.J. (1990) J. Reprod. Fert. **88**, 361.

THE ROLE OF LACTATE DEHYDROGENASE IN OOCYTE AND EMBRYO METABOLISM.

MICHAEL LEGGE.
Department of Biochemistry, University of Otago, P.O. BOX 56, Dunedin, NZ

Oocyte and preimplantation embryos have high lactate dehydrogenase (LD;EC 1.1.1.27) activity which is exclusively the LD1 isoenzyme (H or B isoenzyme). This LD isoenzyme is involved in aerobic metabolism and is the major LD isoenzyme of heart muscle, brain and erythrocytes. The function of the exceptionally high LD1 activity found in both oocytes and preimplantation embryos (five to six thousand times the amount for normal metabolic regulation), has puzzled investigators for some time. In this presentation data is shown to demonstrate the ubiquitous nature of LD1 in oocytes and preimplantation embryos with regards to substrate affinity and examines alternative metabolic pathways to account for the high LD1 activity. Oocytes and preimplantation embryos were obtained from superovulated inbred strains of mice. LD1 activity was determined spectrophotometrically using a range of α-keto aldehydes as substrates. The specific activity of one of the α-keto aldehydes tested, glyoxylate, has been found to be at least as high as that for conventional substrates. Further investigations relating to the significance of this result are currently underway.

VITRIFICATION OF BOVINE OOCYTES

E.N. LEVCHENKO

Internat. Lab. of Biotechnology, Inst. Anim. Product. 949 92 Nitra, CSFR

This study examined the viability of bovine oocytes at the germinal vesicle stage after vitrification. Two vitrification solutions were used: 25% glycerol + 25% propylene glycol (VS-1) and 40% glycerol + 10% propylene glycol (VS-2). Cumulus-intact oocytes (n=1488) were exposed in three steps (0.4-0.8-1.2 M) into 0.1 ml drop in 8-10 minutes of glycerol in PBS. After exposure, 3-5 oocytes were put into VS in 30 sec. and then they were vitrified using the dropping method. Dilution of VS was prepared using 1.0 M sucrose in PBS. Survivability was evaluated after culture. Survival rate of vitrified-warmed oocytes was more than 88% based primarily on morphological appearance of the cumulus. Seventy-six% of oocytes after vitrification began development, but 42% reached the M-I + M-II stages of maturation when VS-I was used. A different proportion was obtained when VS-2 was used: 82% oocytes began development and 61% reached the M-I + M-II stage (X^2=6.69; p<0.05). Data obtained suggested that the percentage correlation between the cryoprotectors glycerol - propylene glycol has considerable significance for preservation of oocytes by vitrification.

EFFECT OF OXYGEN CONCENTRATION ON RABBIT EMBRYO DEVELOPMENT

JIANMING LI AND R. H. FOOTE
Department of Animal Science, Cornell University, Ithaca, New York 14853

Embryos were collected from superovulated Dutch rabbits 19 hours after injection of LH and insemination. The embryos are in the one-cell stage at that time and those judged to be normal by absence of granular cytoplasm and regular shape were distributed randomly into culture dishes containing RD Medium as described by Carney and Foote (J. Reprod. Fertil. 91:113-123. 1991). However, low glucose RD medium was used in the present experiments. In Experiment 1, O_2 concentrations of 5, 10 and 15%, with 5% CO_2 plus 90, 85 and 80% N_2, respectively, were tested. In Experiment 2, O_2 concentrations of 1, 5 and 20% were combined with 5% CO_2 and the rest of N_2. Culture was at 39°C. At the end of 84 hr, embryos were stained with Hoechst 33342 stain to count the number of cells with the aid of a fluorescent microscope. In Experiment 1, the proportion of embryos reaching the hatching blastocyst stage after 84 hr in culture in 5, 10 and 15% O_2 was 48, 38 and 21%, ($P<0.01$) and the cell number per embryo averaged 258, 226 and 188, respectively ($P<0.01$). In Experiment 2, the proportion of hatching embryos after 84 hr in 1, 5 and 20% O_2 was 67, 72 and 29% ($P<0.01$), respectively, with cell numbers in the 1 and 5% O_2 levels higher than in the 20% O_2 level ($P<0.01$). These results indicate that reduction of O_2 below the frequently used level in 95% air is beneficial to rabbit embryo development.

IN VITRO HUMAN ENDOMETRIAL STROMAL CELLS ENHANCE THE DEVELOPMENT OF MOUSE EMBRYO

H.C. LIU[1], J.K. TSENG[2], X.Y.TANG[1], J. MAZELA[2], Z. ROSENWAKS[1], L. TSENG[2]. Dept. of Ob/Gyn., Center for Reproductive Medicine, Cornell University, NY NY 10021[1] and Dept. of OB/Gyn., School of Medicine. State University of New York at Stony Brook, Stony Brook, NY 11794[2].

Human endometrial stromal cells synthesize a set of secretory proteins. Many of these proteins are induced by progestin and relaxin (JCEM 71: 889-899, 1990, 72:1014-1024, 1991). The present study was under taken to investigate whether these secretory proteins would enhance the embryonic development. Mouse embryos at 2-cell stage obtained from 8 weeks mated mice (CB6F1 strain) after PMSG/HCG stimulation were isolated and plated on human endometrial stromal cells (late secretory endometrium) pretreated with (1) no hormone, the control; (2) 0.2 uM medroxyprogesterone acetate (MPA); (3) MPA and 20 ng/ml porcine relaxin (RLX) for 19 days. The co-cultures were carried out in serum free RPMI 1640 for 13 days. In all three conditions, mouse embryos co-cultured with endometrial stromal cells grew rapidly and 80-90% of embryos reached the stage of hatching on Day 3 of culture. The embryos grew continuously and stretched out after hatching. Cell to cell contact between stromal and embryonic cells was observed. Larger and healthier embryos were developed when they were co-cultured in endometrial stromal cells pretreated with MPA or MPA plus RLX as compared with the control. In contrast, 80-100% mouse embryos cultured in medium alone or supplemented with hormone(s) were remained at the 4-8 cells stage in 3 days of culture and subsequently degenerated. The enhancement of embryonic development was also observed when embryos were cultured in the conditioned medium of endometrial stromal cells. However, the embryos remained as a cluster and did not stretched out after hatching. This indicates that presence of endometrial cells is required for the embryos to stretch out. These results suggest that endometrial stromal cells secrete embryo trophic factor(s) which enhance the development of embryos. After 6 days of culture, the mouse embryos were labelled with [^{35}S]methionine in methionine free medium for 16 h. Six secretory proteins were shown on the autoradiograph of SDS-PAGE. Among which a major secretory protein (60K Dalton) was greatly enhanced in those embryos cultured with the conditioned medium from MPA and RLX stimulated stromal cells. The ability of endometrial stromal cells to induce prolactin, insulin-like growth factors binding proteins (IGFBP-1) and other secretory proteins was not altered in those cells co-cultured with embryos. (Supported in part by NIH grant 19247).

MEASUREMENT OF THE MAXIMAL ACTIVITY OF KEY ENZYMES OF ENERGY METABOLISM IN SINGLE HUMAN PREIMPLANTATION EMBRYOS

K.L.MARTIN, K.HARDY* & H.J.LEESE, Department of Biology, University of York, York YO1 5DD, U.K. *Institute of Obstetrics and Gynaecology, Royal Postgraduate Medical School, Hammersmith Hospital, Du Cane Road, London W12 0NN, U.K.

Non-invasive measurements of nutrient uptake in single human preimplantation embryos have shown that pyruvate consumption exceeds that of glucose during the early developmental stages (2.5 to 4.5 days post-insemination), before glucose becomes the predominant substrate in the blastocyst (5 to 6 days post-insemination; Hardy et al., Hum. Reprod., 4:188, 1989). In order to understand the biochemical nature of the switch from pyruvate to glucose, we have measured the maximal activities of key enzymes of energy metabolism in single human preimplantation embryos, using a microfluorescence technique. Mean activities [nmol/embryo ± s.e.m.(n)] for embryos from day 2 to day 6 were as follows: from the glycolytic pathway, hexokinase [0.030 ± 0.004 (23)], 6-phosphofructokinase [0.61 ± 0.12 (21)], aldolase [0.042 ± 0.0074 (18)], glucose phosphate isomerase [3.38 ± 0.40 (18)], pyruvate kinase [3.22 ± 0.35 (28)], lactate dehydrogenase [0.91 ± 0.10 (29)], from the pentose phosphate pathway, glucose-6-phosphate dehydrogenase [4.04 ± 0.31 (27)], and from the TCA cycle, α-ketoglutarate dehydrogenase [0.068 ± 0.02 (9)]. Glycogen phosphorylase was not detectable at the single embryo level. Hexokinase activity increased from day 3 to day 5, while that of G6PDH decreased from day 2 to day 6. The other enzymes showed no obvious pattern of activity with development. The results suggest that hexokinase plays an important role in the regulation of glucose metabolism in the early human embryo.

We acknowledge support from the U.K. Medical Research Council.

Full ethical permission for the work was obtained.

INTERLEUKIN−6 EXPRESSION IN PORCINE, BOVINE AND OVINE PREIMPLANTATION CONCEPTUSES

N. MATHIALAGAN and R.M. ROBERTS, Departments of Animal Sciences and Biochemistry, University of Missouri, Columbia MO 65211

Interleukin−6 (IL−6) is a cytokine with multiple biological effects. It can influence cellular proliferation and differentiation and a range of immune and inflammatory responses. Its synthesis is not confined to cells of the immune system but has also been noted in cells of epithelial and endothelial origin as well as in human placental tissue and mouse blastocysts. The latter results have suggested that the cytokine plays an active role in pregnancy. In the present study we cloned a pig IL−6 cDNA from a spleen library to provide a probe for studying IL−6 expression during pregnancy of several domestic *Artiodactyla*. The porcine library was screened under relatively non−stringent conditions with a human IL−6 probe that was eventually shown to possess 82% nucleotide sequence with the full−length porcine IL−6 cDNA we isolated. This cDNA was 1049 bp long and encoded a 212 amino acid polypeptide with a 25−residue signal sequence. It shared 61% and 43% amino acid sequence identity with human and mouse IL−6 respectively. Complete conservation was noted between all three proteins in the region Leu[60] to Leu[65] and between Leu[60] to Asn[82] when pig and human only were compared. In addition, a conserved alignment of the four cysteine (Cys[47,53,76,86]) residues in this region of the molecules was observed between all three proteins. PCR procedures with primers designed from conserved regions of sequence were used to identify IL−6 cDNA in cDNA libraries and reverse transcribed mRNA derived from elongating preimplantation ovine (days 13 to 25), bovine (days 18 to 19) and porcine (days 13 to 17) conceptuses. Identity of appropriately sized DNA bands on gels was further confirmed by Southern blotting procedures. Conceptuses from all three species expressed IL−6 mRNA. These observations suggest that IL−6 may be involved in evoking early maternal responses to the presence of a developing conceptus within the uterus. Supported by grants from the NIH (HD21980) and USDA (89−37240−4586).

WHAT IS THE MOST APPROPRIATE ENDPOINT FOR EVALUATING PRE-IMPLANTATION EMBRYONIC DEVELOPMENT *IN VITRO*?

SUSAN H. McKIERNAN and BARRY D. BAVISTER
Department of Veterinary Science, University of Wisconsin, Madison, WI 53706.

"Perhaps the most important aspect of culturing mammalian embryos in vitro is the accuracy with which the results are assessed" (Brinster, 1971, in: The Biology of the Blastocyst, RJ Blandau, ed.). Hamster blastocysts and 1-cell embryos were collected from naturally cycling females 74h and 10h post egg-activation (post-EA), respectively. Blastocysts were immediately fixed and 1-cell embryos were cultured in Hamster Embryo Culture Medium (HECM-4, containing no pyruvate and 20 amino acids) in 5% O_2 : 10% CO_2; one group for 64h, the other for 74h. Morphological and nuclear number data were compared.

No. of embryos	collection time (h post-EA)	culture time (h)	% ≤ 8-cells	% morulae	% blastocysts	$\bar{X}$ nuclear number
14	74	0	0	0	100	18±0.5
12	10	64	17	83	0	15±0.5
10	10	74	0	70	30	33±1.1

Although mean nuclear numbers in embryos that developed for 74h post-EA (entirely *in vivo* or 64h *in vitro*) were not significantly different (18 *vs.* 15, respectively), the percentages of blastocysts were highly different (100% *vs.* 0%, respectively). Additionally, when embryos were cultured for 74h (84h post-EA, approx. 10h beyond the time of normal implantation), the mean nuclear number was doubled (33) but development still lagged behind that of embryos grown *in vivo* (30 *vs.* 100% blastocysts, respectively). These data clearly show that the two standard endpoints used to assess development, i.e., morphology and nuclear number, are inconsistent. These inconsistencies need to be reconciled in evaluating the effectiveness of different culture media for supporting normal embryo development, otherwise erroneous conclusions may be drawn. Supported by NIH grant no. HD 22023.

EFFECTS OF ALCOHOL ON BLASTOCYST IMPLANTATION AND IMPLANTATION SITE BLOOD FLOW

J.A. MITCHELL, H. GOLDMAN, B. VAN KAINEN AND A. MILSTONE
Department of Anatomy & Cell Biology and the Fetal Alcohol Research Center,
Wayne State University, Detroit, MI.

The effects of daily doses of alcohol during the first four days of pregnancy on blastocyst implantation were determined. Pregnant rats (Sprague-Dawley) were randomly assigned to control (saline) or treated (alcohol : 2g/kg body wt : 15% v/v EtOH/water) groups. Rats were dosed daily by gastric gavage between 10:00 -12:00 h on day 0 (sperm in vaginal lavage) through day 3. On day 4 the progress of implantation was determined by counting the number of implantation sites that are rendered visible by the Blue reaction 15 minutes after iv. injection of 0.25 ml Evans Blue dye (1%). Alcohol treatment accelerated implantation. The percentages of blastocysts implanting at selected times in alcohol vs. saline-treated rats were as follows (7 implantations/cornu = 100%) : 11:00 a.m. 36 vs 17%, 12:00 p.m. 65 vs 19%, 2:00 p.m. 86 vs 50%, 3:00 p.m. 100 vs 53%. One hundred % implantation was not achieved in controls until 4:00 h.

The effects of alcohol on uterine blood flow at sites of blastocyst implantation versus adjacent non-sites was measured in conscious unrestrained rats on day 5 pregnancy. Implantation sites were rendered visible by iv. injection of Evans Blue dye. Alcohol (0.75 ml 100% ethanol diluted 1:2 v/v in saline/rat) was injected into the femoral vein; controls received an equivalent volume of vehicle (5 rats/group: mean wt 235g). Blood flow was measured at 15 minutes post-injection by the rubidium 86 fractionation method. In controls, blood flow was significantly greater at implantation sites that at non-sites (0.77 ± 0.01 vs. 0.47 ± 0.01 ml/min/g, p<0.01). Alcohol markedly increased perfusion at both sites and non sites (2.10 ± 0.05 and 0.77 ± 0.4 ml/min/g, p<0.01) vs. control non-sites (0.47 ± 0.01). The magnitude of alcohol-induced increase in blood flow was much greater at implantation sites (173%) than at non-sites (64%). Cardiac output did not differ between control and alcohol treated animals.

In conclusion alcohol advances the time of blastocyst implantation and enhances blood flow to the implantation site. (Supported by NIAAA Grant No. P50-AA-07606-04).

ENHANCEMENT OF NUCLEAR FORMATION IN METAPHASE II
MOUSE OOCYTES BY THE SYNERGISTIC ACTION OF THE
CALCIUM IONOPHORE AND PROTEIN SYNTHESIS INHIBITION

MOSES, RUTH M. and MASUI, YOSHIO
Department of Zoology, University of Toronto
Toronto, Ontario , Canada M5S 1A1

The calcium ionophore, A23187 and the protein synthesis inhibitor, cycloheximide, are each known to cause parthenogenetic activation and induce nuclear formation in mature mouse oocytes arrested at metaphase of the second meiotic division. We found that when immature oocytes are matured in vitro to metaphase II, responsiveness to either of these drugs gradually develops. Oocytes matured in vitro for 14-16 hours show little response to either drug alone, but are induced to form nuclei when both drugs are used simultaneously. Oocytes matured in vitro for 23-24 hours are fully responsive to either drug, but if they are treated with the microtubular inhibitor, colcemid, nuclear formation after treatment with either A23187 or cycloheximide is inhibited. However, if oocytes are exposed to all three drugs simultaneously, nuclei form. These results suggest that both increased intracellular calcium and inhibition of synthesis of some protein(s) are required for nuclear formation to occur. For only one activation stimulus to be effective, cytoplasmic maturity and microtubular polymerization are required. Presumably, in fully mature oocytes, increased intracellular calcium results in inhibition of synthesis of some protein(s) and inhibition of protein synthesis leads to increased intracellular calcium during oocyte activation.

PIG MEMBRANA GRANULOSA CELLS PREVENT RESUMPTION OF MEIOSIS
IN CATTLE AND MOUSE OOCYTES

JAN MOTLIK, JAROSLAV KALOUS, PETR SUTOVSKY, ZORA RIMKEVICOVA
Institute of Animal Physiology and Genetics, 277 21 Libechov,
Czechoslovakia

Pig oocyte-cumulus complexes (OCC) isolated with an attached piece of membrana granulosa do not resume meiosis in vitro. The maturation inhibiting activity of pig membrana granulosa (PMG) was tested in interspecies combinations with cattle and mouse OCC. The large pieces of PMG containing about $3-5 \times 10^5$ granulosa cells were isolated with the basement membrane. The pieces of granulosa tended to roll up with an inside-out orientation. Cattle and mouse OCC were placed on the basement membrane inside the PMG rolls. Germinal vesicle breakdown (GVBD) was observed only in 16, 17 and 21% of cattle oocytes placed in the rolled pieces of PMG for 8, 16 and 24h, respectively. After 6h of culture, the one third of mouse OCC placed on PMG underwent GVBD. The inhibitory effect of PMG upon resumption of meiosis in cattle and mouse oocytes was fully reversible. Cattle cumulus cells were closely associated with the pig basement membrane, however, no gap junctions were formed among heterologous granulosa cells. Thus the inhibitory factor secreted by PMG is not species-specific and it can act, at least in vitro, without the mediation of gap junctions.

IN VITRO AND IN VIVO DEVELOPMENT OF ZONA PELLUCIDA FREE GOAT DEMI-EMBRYOS

M. A. NOWSHARI* and W. HOLTZ

Institute of Animal Husbandry and Genetics, University of Goettingen, Albrecht Thear Weg-1, 3400-Goettingen, FRG.

Although a protective role for the zona pellucida during pre-implantation embryo development is well established, the developmental potential of zona-free embryos subjected to manipulation (splitting) or freezing has not been examined in detail for species other than mouse and cattle. The aim of the present investigation therefore was to investigate 1) the in vitro and in vivo development of zona-free goat demi-embryos (ZFDE) 2) the feasibility of freezing demi-embryos without zona pellucida. Embryos were collected from superovulated goats on day 5 or 6 after estrus and bisected using a simple splitting technique. Of 103 morulae and 77 blastocysts that were split, 200 and 151 demi-embryos, respectively, with no major morphological aberrations, were obtained. ZFDE obtained from blastocysts were cultured in vitro for two hours and those obtained from morulae for 24 hours. Demi- embryos were transferred to recipients directly after culture or after freezing. A high percentage (91%) of all ZFDE studied developed further during in vitro culture, 76% of these developed from morulae to blastocysts after 24 h. Subsequent transfer of 11 pairs of ZFDE resulted in 9 pregnancies. 79% of frozen ZFDE were morphologically intact after thawing. Subsequent transfer of 11 pairs of these resulted in 6 pregnancies. The zona pellucida does not seem to play an essential role in the development of goat demi-embryos *in vitro* and *in vivo* after transfer to recipients. Furthermore, it does not appear to be necessary for protection during freezing of demi-embryos.

* Present address: Deutsches Primatenzentrum GmbH, Kellnerweg 4, Goettingen, FRG.

A STUDY TO ESTABLISH THE OBJECTIVE CRITERIA FOR THE SELECTION OF EXCELLENT OVARIAN OOCYTES.

TANEAKI OIKAWA(1), KEI-ICHIRO YOGO(2) AND SHICHRO SUGAWARA(2): (1) RESEARCH INSTITUTE FOR THE FUNCTIONAL PEPTIDES, YAMAGATA 990, JAPAN. (2) DEPARTMENT OF AGRICULTURE, TOHOKU UNIVERSITY, SENDAI, MIYAGI 980, JAPAN.

In this study, for the purpose of utilizing effectively immature oocytes recovered from the ovaries of cows slaughtered, we performed some experiments to establish the objective criteria for the selection of excellent ovarian oocytes as follows; To exmine 1) the lightmicroscopical structural changes of ooplasma, 2) the changes of the polarising properties (P.P.) of ooplasmic membrane, 3) the meiotic nuclear status of oocytes before and after in vitro maturation and 4) the developmental potency of the oocytes recovered from four different cumulus oocytes complexes (COCs) and matured, fertilized and developed in vitro. As the result, it became clear that there are four types of the COCs (Type 1, 2, 3 and 4), that the oocytes recovered from four different types of the COCs are classified into two groupes; one of which its ooplasmic membrane shows the P.P. (Type 1 and 4) and another one of which its ooplasmic membrane does not show the P.P. (Type 2 and 3), that the in vitro meiotic maturation of oocyte is possible in the Type 2, 3 and 4 and that, in in vitro fertilization system at least used in this study, only the oocyte of which the ooplasmic membrane does not show the P.P. Type 2 and 3) has more high developmental potency to form the blastcyst. This study was partially supported by a "Grant-in-Aid for Scientific Research on Priority Areas", No. 63640004, from the Ministry of Education of Japan to T.O.

EVIDENCE THAT THE EQUINE EMBRYONIC CAPSULE IS A DEVELOPMENTALLY REGULATED MUCIN-LIKE GLYCOPROTEIN.

ORIOL[1], J. G., SHAROM[2], F. J., MEYERS[3], P. J. and BETTERIDGE[1], K. J.
Departments of Biomedical Sciences[1], Chemistry and Biochemistry[2], Equine Research Centre[3], University of Guelph, Guelph, Ontario, Canada N1G 2W1

The capsule, which envelops the horse conceptus during the second and third week of pregnancy, consists of a glycoprotein that is O-glycosylated mainly on threonine residues and, like mucins, is resistant to chemical and enzymatic solubilization. In the present study, we have further investigated the glycosylation sites. When capsular glycoproteins ^{3}H-labelled on sialic acid residues were passed through a jacalin lectin column (specific for O-linked chains), >80% of the material bound tightly and was eluted with 20 mM α-methylgalactopyranoside, showing that most of the glycosylation sites are via O-linkages, as are those of mucin-like glycoproteins. Determination of the capsule's sugar composition also showed it to have mucin-like characteristics; amino sugars and sialic acid make up a high proportion (60%) of the carbohydrate-rich capsule, while sulphated and neutral sugars (except mannose, undetectable by lectin histochemistry) are minor components. Uronic acid, characteristic of proteoglycans, was undetectable by colorometric assay. Histochemical analysis using peroxidase-labelled peanut lectin showed the presence of ß-D-Gal-(1-3)-D-GalNAc, again characteristic of mucins. The biosynthesis of the capsule during development was studied by weighing 35 accurately aged capsules (including 10 kindly supplied by Dr. Katrin Hinrichs, Tufts University, MA). The dry mass of the capsule increased rapidly with blastocyst expansion between 10.5 and 18.5 days after ovulation (from 0.46 mg to 4.46 mg), but then may decrease until disappearance of the capsule shortly after Day 22. The role of the capsule in pregnancy establishment remains unclear, but O-linked oligosaccharides are known to have cell recognition functions.
(Supported by NSERC and the Equine Research Centre).

SODIUM TRANSPORT PROPERTIES OF THE BOVINE AND EQUINE BLASTOCYST

OVERSTROM[1], E.W., DUBY[2], R.T., DOBRINSKY[2], J., HINRICHS[1], K., ROCHE[3], J.F., and M.P. BOLAND[3]. Tufts Univer. Sch. Veterinary Med[1], N. Grafton, MA, Univer. Massachusetts[2], Amherst, MA, Univer. College Dublin[3], Dublin, Ireland.

The trophectoderm of the mouse, rabbit and pig blastocyst acquires solute-specific transport properties that are stage- and species-specific. The functional expression of apical (channels) and basal-lateral (Na$^+$/K$^+$ ATPase) sodium transport mechanisms are thought to play key roles in the prerequisite processes of cavitation and blastocyst expansion prior to implantation in all mammals, including primates. This study was initiated to characterize the trans-trophectodermal [^{22}Na] transport properties of the preimplantation cow and horse blastocyst. Cow blastocysts were recovered at slaughter from uterine flushings of superovulated cattle on Day 8 of pregnancy; horse blastocysts were collected non-surgically from the uterus of cycling mares on Days 8-11 of pregnancy. Blastocysts were placed in flux medium (CR2-cow, F12-horse; 10% FBS, and 2.5 or 25 uCi/ml [^{22}Na]Cl) and were incubated for 1 h (37°C, 5% CO$_2$ in humidified air). After washing in cold, tracer-free medium, labeled samples of blastocoel fluid were quantified by scintillation spectroscopy. Additionally, [^{22}Na] flux experiments were conducted with cow blastocysts in the presence of amiloride (10^{-4}M, apical sodium channel blocker), or ouabain (10^{-4}M, Na$^+$/K$^+$ ATPase inhibitor, micro- injected into the blastocoel). Between Days 8 and 10 in the horse, the net [^{22}Na] transport rate increased from 0.85 ± 0.19 to 1.91 ± 0.45 pmoles/cm^2/h, while blastocoel volume increased 14-fold (1.74 to 24.39 ul). However, on Day 11, the [^{22}Na] transport rate decreased 35% even as blastocoel fluid volume expanded 4-fold. Sodium transport was diminished 31.6% in ouabain treated cow blastocysts compared with controls (2.95 vs. 4.31 pmole/cm^2/h), whereas amiloride appeared to stimulate net [^{22}Na] influx. These data provide evidence that Na$^+$ transport processes are functional, and developmentally expressed in the preimplantation cow and horse blastocyst. Supported by BioResearch Ireland, and the Wellcome Foundation.

EXPRESSION OF PDGF A CHAIN AND PDGF α RECEPTOR DURING MOUSE EMBRYOGENESIS SUGGESTS POSSIBLE ROLE IN CELL CYCLE CONTROL.

SL PALMIERI, JD BIGGERS and M MERCOLA.
Dept. of Cellular & Molecular Physiol. & Lab. of Human Reprod. & Reprod. Biol., Harvard Medical School

The action of growth factors and growth factor receptors is postulated to have a role in the regulation of early embryogenesis. As mouse preimplantation development can occur in culture in the absence of exogenous growth factors, it is of interest to determine whether the embryo makes growth factors to stimulate differentiation within the embryo itself. One candidate for an embryo derived growth factor that acts on embryo tissue is platelet derived growth factor (PDGF). Previous work has shown that PDGF A, which binds to the α receptor, is the predominant form of early embryonic PDGF in both *Xenopus* and mouse.

We have used ^{35}S-labeled sense or antisense RNA probes with *in situ* techniques to map the expression of PDGF A chain and PDGF α receptor mRNA. Antisense probes for both genes show label in all cells of preimplantation stage embryos. In contrast, in early postimplantation (d7.5), mRNA for PDGF α receptor is localized to the mesoderm layer of both the embryo and the extra-embryonic membranes while the message for PDGF A chain is found in the embryonic ectoderm and in the ectoderm lining the ectoplacental cavity. Sense probes show no hybridization to any of the developmental stages examined. Using antibodies specific for mouse PDGF A chain and PDGF α receptor, we have shown that both proteins are present in all cells of preimplantation stage embryos. Preimmune serum for each antibody confirms that the binding is specific for PDGF A chain and α receptor proteins. We have also determined that the receptor is active in d7.5 embryos by measuring its ability to autophosphorylate using a tyrosine kinase assay.

Demonstration of the expression of both PDGF A chain and its receptor at the same time in development is an initial step in confirming the hypothesis that early embryos produce PDGF in order to stimulate events within the embryo, in a manner similar to autocrine control of growth in v-*sis* transformed cells. To examine this hypothesis, we are creating PDGF-deficient embryos by the injection of synthetic mRNA encoding mutant PDGFs capable of suppressing endogenous PDGF. A single blastomere of two-cell stage embryos injected with dominant-negative RNA ceases further division, whereas its uninjected sister cell develops normally to the blastocyst stage in culture.

Taken together, our results suggest that in cleavage stage embryos PDGF A promotes cell division by an autostimulatory pathway. Following implantation, however, PDGF produced in embryonic ectoderm may act in a paracrine fashion by stimulating receptors on embryonic mesoderm, probably to influence early mesoderm development.

ANEUPLOIDY AND ARRESTED PRE-IMPLANTATION DEVELOPMENT IN HAMSTERS EXPOSED ACUTELY TO THE FUNGICIDE METHYL BENZIMIDAZOLE CARBAMATE (MBC)

S PERREAULT, S JEFFAY, R BARBEE and B LIBBUS
US EPA-HERL, METI, Inc. and Genetic Research, Inc. RTP, NC.

MBC, a fungicide and microtubule poison, causes early pregnancy loss in female hamsters when exposure occurs late on proestrus (i.e. coinciding with the first meiotic division of the oocyte). The purpose of this study was to characterize the site and mechanism of this loss through direct assessments of oocytes and pre-implantation embryos. Groups of female hamsters (n = 10 treated and 10 control per experiment) were given a single high dose of MBC (1000 mg/kg, p.o.) or corn oil at 3:30 PM on proestrus (with lights on at 7 AM and off at 9 PM). In one experiment they were killed shortly after ovulation (d.1) to recover oocytes for chromosome analysis. In other experiments they were bred overnight and killed on d. 1 - 5 of pregnancy to evaluate embryos and implantation sites. Chromosome analysis in unfertilized oocytes revealed an MBC-induced increase in aneuploidy (30.3% vs. 6.2% in controls). When animals were bred after dosing, MBC had no effect on the number of eggs recovered or fertilized on d. 1, but significantly decreased the percentage of embryos reaching the 8- cell stage on d. 3 PM (51% vs. 79% in controls), and increased the percentage of immature (less than blastocyst) embryos recovered on d. 4 PM (48% vs. 12% in controls). The mean number of implantation sites, revealed by Evans Blue staining, was also significantly lower on d. 4 PM (1.6 vs. 8.1 in controls) and d. 5 AM (7.1 vs. 14.3 in controls). These simple direct assessments elucidated a mechanism of MBC-induced early pregnancy loss, namely chromosome non-disjunction, and demonstrated a subsequent arrest of pre-implantation embryonic development accompanied by a decrease in the likelihood of implantation.

INSULIN-LIKE GROWTH FACTOR BINDING PROTEIN SECRETION BY PREIMPLANTATION SHEEP EMBRYOS

A.J. PETERSON, A. LEDGARD AND S.C. HODGKINSON
Ruakura Agricultural Centre, Hamilton, New Zealand

The secretion of insulin-like growth factor (IGF) binding proteins (IGFBPs) by preimplantation sheep embryos was determined from days 9-19 after conception using specific Western ligand blotting procedures. No IGFBPs were detected until d15 but by d19 the pattern of IGFBP was similar to that of fetal sheep plasma with bands observed at 18, 24-32, 40-50 and >200 kDa. Competitive tracer binding using IGF-I and II revealed 18 and >200 kDa bands to be specific for IGF-II.
There was a different pattern in the appearance of the IGFBP for IGF-I and -II during development. On d15 only the species of ~24-29 kDa was present for IGF-I whereas this plus another at 32 kDa were detected using IGF-II tracer. A similar pattern occurred at d16 with the addition of both a ~40 and a fainter 50 kDa BP for both IGF-I and -II. By d19 there was a marked increase in the binding in the ~24-29 kDa and 40-50 kDa areas. The 32 kDa area bound both IGF-I and -II and an additional IGFBP was observed ~38 kDa. The IGF-II specific BP of 18 kDa and >200 kDa also appeared on d19.
The role of the IGFBPs during preimplantation development remains to be elucidated, however their appearance during the rapid expansion of trophoblast and allantois suggests that they may be involved in the local interaction between IGFs present in the uterine fluid and these extraembryonic membranes. This may ensure the appropriate control, not only of embryonic metabolism but also of the requisite neovascularization of the extraembryonic membranes at this stage of rapid fetal growth before placentation.

TROPHOBLAST TISSUE CULTURE OF HUMAN INTRAUTERINE AND ECTOPIC PREGNANCIES AND TREATMENT WITH METHOTREXATE

Pfuhl,J.-P.;Schäfer,D.;Baumann,R.
Dept OB/GYN University Frankfurt, Annastr.26 b, D-6100 Darmstadt, Germany

We describe the effects of methotrexate (MTX) on trophoblast tissue cultures (TTC) derived from intrauterine (IUP) and ectopic pregnancies (EUP).
After trypsination the Chang medium was used for propagation and changed every second day. Human chorionic gonadotropin (HCG) levels were measured. MTX in concentrations from 1.0×10^{-2} mol/l to 3.8×10^{-5} mol/l was administered either directly at or six days after tissue culture onset.
All TTC showed a secretion of HCG within the first 16 to 20 days. MTX concentrations lower than 3.8×10^{-4} mol/l generally had no effect on HCG secretion. TTC of EUPs needed a concentration about ten times higher to lead to an equivalent reduction of HCG levels compared to IUPs. In cultures treated on day 6 a reduction of HCG values occurred only about 8 days after treatment. Some few IUPs and EUPs showed no reduction of HCG levels after treatment.
Our study shows that the established culture system can be used to test substances proposed for conservative treatment of EUP. Furthermore it shows that one has to be careful when extrapolating from data derived by IUPs to EUPs. In some regimens it takes about 8 days until the effect becomes visible. In combination with our clinical data we therefore recommend not to repeat an MTX dose too rashly in treatment of patients with EUP. The possibility that among patients non responders could exist should be borne in mind.

GONADOTROPINS INFLUENCE THE EXTENT OF MEIOTIC PROGRESSION DURING IN VITRO MATURATION OF HAMSTER OOCYTES.
CARLOS E. PLANCHA[1,2] AND DAVID F. ALBERTINI.[2]
Inst. Histology and Embryology, Lisbon Medical School, Lisbon, Portugal[1] and Dept. of Anatomy/Cellular Biology, Tufts Univ. Health Science Schools, Boston, MA.[2]

To define better the optimal conditions required for completion of meiotic maturation in cultured mammalian oocytes, we examined the effects of gonadotropins on meiotic progression in "in vitro" matured hamster oocytes. Cumulus oocyte complexes obtained from PMSG-primed 6-9 week old golden hamsters were cultured for 14-18 hrs in medium 199/10% newborn calf serum lacking gonadotropin supplements (control) or in media supplemented with 5 μg/ml rat FSH, 5 μg/ml hCG, or both FSH and hCG. Cultures were evaluated for the extent of cumulus expansion and meiotic progression using triple label fluorescence microscopy to assess, simultaneously, chromatin, microtubule, and actin filament organization. Cumulus expansion failed to occur in both control and hCG treatment groups whereas full expansion was observed in FSH and FSH/hCG groups. With respect to the extent of meiotic progression, only 36% of oocytes cultured in the absence of gonadotropin proceeded past metaphase of meiosis-1 (MI) whereas 94% of oocytes cultured with FSH progressed to metaphase of meiosis-2 (MII). In hCG treated cultures, only 32% of oocytes progressed past MI and in FSH/hCG cultures 52% passed the MI stage. Moreover, oocytes exposed to hCG, either alone or in combination with FSH, exhibited a higher incidence of degeneration and parthenogenic activation. These data suggest that FSH is required for the normal completion of maturation whereas hCG impairs meiotic progression under the "in vitro" conditions employed in this study. (Supported by NIH HD 20068 and by Instituto Nacional de Investigacao Cientifica and Fundac Luso-Americana para o Desenvolvimento.)

EFFECTS OF α-AMANITIN ON EARLY BOVINE EMBRYO DEVELOPMENT

L. PLANTE, D. L. SHEPHERD AND W. A. KING
Department of Biomedical Sciences, OVC, University of Guelph, Guelph, Ontario, Canada, N1G 2W1

Alpha-amanitin binds to polymerase II and blocks mRNA synthesis. We report here its effects at the zygote and morula stage on incorporation of ^{3}H-uridine into RNA and on the ultrastructure (EM) of bovine embryos produced by in vitro oocyte maturation and fertilization (IVMF). Zygotes were co-cultured with oviductal cells in B2 medium + 10% estrus cow serum at 39°C, 5% CO_2. Some zygotes (18 h post-fertilization=hpi) were cultured in the presence of α-amanitin (0, 10, 50, 100 or 500 μg/ml in the same medium conditioned and supplemented with oviductal cells) for 44, 72 or 90 hpi to provide 60 embryos. IVMF-morulae (n=32) were similarly incubated with α-amanitin for 24h. Embryos were then fixed directly for EM or were incubated first with ^{3}H-uridine (20 μM in Ham's-F10, 2h), then with excess unlabelled uridine before fixation (methanol:acetic acid) on glass slides for autoradiography. ^{3}H-uridine was detected in embryos 72 hpi (4- & 8-cell) in the absence of α-amanitin, but no incorporation above background was seen in embryos of the same stage following the 10 to 100 μg/ml treatments. However, when morulae were incubated with ^{3}H-uridine, all cells were heavily labelled at 0 μg/ml of α-amanitin, a few cells were heavily labelled at 50 to 500 μg/ml and with 10 μg/ml, labelling was intermediate. EM studies on embryos at 44 hpi (4- & 8-cell) showed dose-related nuclear degeneration following treatment with 10 to 100 μg/ml α-amanitin. At 90 hpi (8- & 16-cell), all treated embryos (10 to 500 μg/ml) revealed nuclei in advanced stages of degeneration with very dark, dense heterochromatin at the nuclear periphery. In proportion to α-amanitin concentrations (10 to 500 μg/ml) morulae also showed EM degeneration with condensation of heterochromatin, disorganization of nucleoli, disappearance of rough endoplasmic reticulum and polysomes. The external cells were severely affected by the drug. However, at all concentrations, there were always 1-2 internal cells that appeared relatively normal. In conclusion, α-amanitin is an effective inhibitor of RNA synthesis in early bovine embryos, but induces nuclear degeneration. In morulae, internal cells were less affected and some of them were capable of RNA synthesis even after exposure to 100-500 μg/ml α-amanitin. (Supported by NSERC, CAAB and SEMEX).

IMMUNE INTERFERON (IFN-γ) INHIBITS LATERAL MOBILITY OF A PLASMA MEMBRANE PROTEIN IN EARLY STAGE MOUSE EMBRYOS MEASURED BY FLUORESCENCE PHOTOBLEACHING RECOVERY.
KATALIN POLGAR, PATRICK YAKONO, DEBORAH ANDERSON, DAVID GOLAN, JOSEPH HILL. Depts. of Obstetrics and Gynecology, Medicine and Biological Chemistry and Molecular Pharmacology, Harvard Medical School, Boston, MA 02115.

The T-lymphocyte cytokine IFN-γ blocks mouse embryonic development (Hill et al., 1988) and has a dramatic effect on the ultrastructural morphology of one- and two-cell mouse embryos (Polgár et al., 1990). In other systems, IFN-γ induces striking phenotypic alterations including cytoskeletal rearrangements (Stolpen et el., 1988). We hypothesized that IFN-γ could exert its effect on embryos through cytoskeletal rearrangements that might also lead to altered membrane protein mobility. To test this hypothesis we used the fluorescence photobleaching recovery technique to measure the lateral mobility of an embryonic surface glycoprotein recognized by the monoclonal antibody S75. Using this technique it has been found previously that the lateral diffusion of proteins in the plasma membrane is often much slower than that of lipids, an effect attributed to interactions between transmembrane proteins and cytoskeletal components. In our study there was a significant (40-80%) decrease in the fractional mobility of FITC-S75 in IFN-γ-treated as compared to control one-cell or two-cell embryos. The mobility of FITC-S75 in IFN-γ-treated embryos was distinct from that in non-viable embryos. These data provide evidence that IFN-γ may induce rearrangements of membrane skeletal and cytoskeletal components or of plasma membrane lipid domains.

EXPRESSION OF SPERM BINDING AND EMBRYOTROPHIC ACTIVITY OF THE BOVINE OVIDUCT THROUGH A XENOPUS OOCYTE IN VITRO TRANSLATION SYSTEM

J.W. Pollard, H.W. Davey, J.M. Scodras, W.A. King, A.G. Wildeman and K.J. Betteridge

Bovine oviduct epithelial cells (BOEC) in co-culture maintain both the motility and fertilizing capacity of bovine sperm, and enable bovine embryos to pass the 8-16 cell block. We have used a Xenopus oocyte in-vitro translation and secretion system to determine whether these activities are encoded in BOEC mRNA. Four peri-ovulatory oviducts, ipsilateral to the active ovaries, were collected from 4 synchronized cows at slaughter to prepare 4 BOEC samples. These were either frozen immediately in liquid nitrogen (LN_2), or cultured for 48 h in Menezo B2 medium + 10% estrous cow serum (ECS) to obtain control conditioned medium (BOEC-CM), which was also stored in liquid LN_2 until used for embryo culture. Poly (A)+ RNA was isolated from each frozen BOEC sample. Stage 5 and 6 Xenopus oocytes obtained surgically were divided into two equal groups. Those in the first group (control, XCT) were microsurgically injected with ~20-50 nl of buffer alone. Those in the second group (XmRNA) were injected with buffer containing ~50 ng of BOEC mRNA. After 24 h, surviving injected oocytes were selected for culture. Groups of 10 XSCT and 10 XmRNA oocytes were cultured separately in 200 μl of B2+ECS for 96 h to provide respective conditioned media (XCT-CM, or XmRNA-CM), then removed for use in a bovine sperm binding assay. The effects of medium conditioning were compared in a co-culture bioassay in which groups of 4-cell bovine embryos (n=20) were cultured in 50 μl droplets of either XCT-CM, XmRNA-CM, or BOEC-CM for 122h at 39°C, then evaluated for development beyond the 8-16 cell block to either the compact morula or blastocyst stage. Each bioassay was replicated 4 times. The sperm binding assay indicated that bovine sperm bound in high numbers to the vitelline membranes of Xenopus oocytes injected with BOEC mRNA, but in very low frequency to controls. Results of the embryo culture bioassays are tabulated below.

Culture Treatment	No. Compact Morulae and Blastocysts/ No. 4-cell Embryos Cultured			
	Oviduct 1	Oviduct 2	Oviduct 3	Oviduct 4
XCT-CM	6/80[a]*	9/80[a]	7/80[a]	9/80[a]
XmRNA-CM	20/80[b]	23/80[b]	12/80[a]	13/80[a]
BOEC-CM	39/80[c]	47/80[c]	40/80[b]	53/80[b]

* within columns different superscripts represent significant differences (P< .05)
Results overall suggest that BOEC sperm binding and embryotrophic activity may be at least partially encoded in BOEC mRNA, and can be expressed in a Xenopus oocyte in-vitro translation and secretion system.

This work was supported by the CAAB. NSERC. SEMEX-Canada. Agriculture Canada and the Ontario Premiers Council Technology Fund

COOPERATIVE INTERACTION AMONG MOUSE ZYGOTES CULTURED IN PROTEIN-FREE
MEDIUM: BLASTCYST DEVELOPMENT AND HATCHING

PATRICK QUINN, TOSHIO HIRAYAMA AND RICHARD P. MARRS
Institute for Reproductive Research, Los Angeles, CA 90017

Since the report of Cholewa & Whitten (JRF 22:553 1970) of the growth of 2-cell mouse embryos in vitro in the absence of fixed nitrogen and the authors' comment that mouse zygotes however would not develop to the blastocyst stage in the absence of exogenous protein, it has generally been assumed that in vitro, murine zygotes require some form of added protein to develop to the expanded blastocyst stage. It has recently been found however, that the presence of EDTA and/or aminoacids can overcome this block to development in the absence of exogenous protein (Fissore et al. Biol Reprod 41:835 1989) and that the embryonic density used during culture influences rates of development (Wiley et al. F&S 45:111 1986; Paria & Dey PNAS 87:4756 1990). Because of variable results obtained when attempting to culture zygotes for 24 hours in medium containing BSA and then transferring them to protein-free medium, a more thorough investigation of the role of BSA and embryonic density on the development of mouse zygotes in protein-free medium was undertaken.

Effect of embryonic density and BSA on development						
	% Blastocysts			% Hatched		
Zygote density (#/10ul)	1	10	20	1	10	20
+ BSA (5 mg/ml)	53	87	91	27	76	64
- BSA	6	54	65	0	11	45

The results show that mouse zygotes can develop to fully expanded blastocysts and hatch from their zona pellucida in the absence of added BSA in the medium if a sufficient density of embryos is used. The addition of EDTA enhanced the development of single zygotes cultured in 10 ul of medium but did not stimulate the hatching rate as did the density of 10 to 20 zygotes in 10 ul of medium. These results confirm the previously reported effects of embryonic density on the in vitro development of murine embryos, and for the first time show that mouse zygotes can develop into fully expanded hatched blastocysts in medium devoid of exogenous protein.

EFFECTS OF EPIDERMAL GROWTH FACTOR, INSULIN-LIKE GROWTH FACTOR-I, AND PORCINE FOLLICULAR FLUID ON PORCINE OOCYTE NUCLEAR MATURATION IN VITRO

MICHAEL L. REED, JOSE L. ESTRADA, AND ROBERT M. PETTERS
DEPT. OF ANIMAL SCIENCE, N. C. STATE UNIVERSITY, RALEIGH, NC 27695-7621.

Undefined follicular factors are required during in vitro maturation of pig oocytes. Epidermal growth factor (EGF), insulin-like growth factor-I (IGF-I), and dialysed porcine follicular fluid (pFF) were evaluated for their effects on pig oocyte nuclear maturation in vitro. Eight different maturation media were made in a split-plot factorial design with pFF (0% vs. 10% v/v pFF) as the whole plot component, and EGF (0.0 vs. 50 ng/ml) and/or IGF-I (0.0 vs. 100 ng/ml) as the factorial subplot component. Pig follicular granulosa-cumulus-oocyte complexes (GCOC) were obtained from slaughterhouse ovaries, washed and cultured at 38.5°C in a humidified incubator with 5% CO_2 and 95% air for 42 h. Following culture, GCOC were mechanically stripped of granulosa-cumulus cells and evaluated for nuclear maturation by light microscopy. The percentage of Metaphase II oocytes for control, IGF-I, EGF, and IGF-I + EGF treatments without pFF were 50.7%, 52.6%, 80.9% and 84.3% (control & IGF-I groups significantly less, $p<.001$). The same treatments in the presence of pFF were similar and high (84.2, 84.9, 82.1 and 86.8%, respectively). These results demonstrate that EGF, in the absence of pFF, promotes a similar level of oocyte nuclear maturation as does pFF alone or pFF with EGF and/or IGF-I. IGF-I does not appear to influence nuclear maturation of GCOC. Funded by NICHHD, NIH, through cooperative agreement HD 21937.

PERIODIC Ca^{2+} OSCILLATIONS INDUCED IN THE MURINE OOCYTE AFTER ELECTROPORATION PULSE IN MEDIUM CONTAINING INOSITOL 1,4,5-TRIPHOSPHATE

L.F. RICKORDS and K.L. WHITE
Department of Animal Science, Louisiana State University, Baton Rouge, LA 70803

The objective of this study was to monitor alterations in intracellular Ca^{2+} concentration after electroporation of inositol 1,4,5,-triphosphate (IP3) into the cytosol of the murine oocyte. Electroporation media consisted of phosphate buffered saline containing 0.0 mM Ca^{2+} (PBS-) or 0.9 mM Ca^{2+} (PBS+), PBS- with 25 µM IP3 or PBS+ with 25 µM IP3 (PBS-, PBS+, PBS- w/IP3, PBS+ w/IP3, respectively). Cumulus-free oocytes were incubated in 100 µl drops of PBS+ containing 2 µM of calcium indicator fluo-3/AM for 60 min at 37°C. Fluo-3/AM loaded oocytes were equilibrated for 7 min in assigned treatment media prior to application of a 3 V, 5 sec AC pulse followed by a 1.56 KV/cm DC pulse. Alterations in fluorescence intensity were monitored for 6.5 to 20 min after DC pulse by photon counting spectrofluorometry. Fluorometric analysis demonstrated an immediate and dramatic rise in intracellular free Ca^{2+} (Ca$^{2+}_i$) following DC pulse in PBS+, PBS- w/IP3, PBS+ w/IP3. Ca$^{2+}_i$ levels remained elevated for the duration of measurement in PBS+ and PBS+ w/IP3. In contrast, PBS- w/IP3 exhibited an immediate rise in Ca$^{2+}_i$ for 3-4 min followed by a dramatic drop in Ca$^{2+}_i$. A series of transient oscillations in Ca$^{2+}_i$ was then observed. Transient Ca$^{2+}_i$ rises were measured for 20 min after electrical pulse. Significantly lower Ca$^{2+}_i$ levels were observed in PBS- (P<0.01) with no secondary peaks observed. There was no difference (P>0.05) between PBS+ and PBS+ w/IP3 due to the sustained elevated levels of Ca$^{2+}_i$ resulting from the influx of extracellular Ca^{2+} that masked any IP3 induced release of stored Ca^{2+}. The transient oscillations in Ca$^{2+}_i$ reported here, appear similar to those that normally occur at fertilization. Studies are currently underway to determine if IP3 can enhance the rate of electrical pulse induced activation and subsequent development.

CHANGES IN GLUCOSE AND GLUTAMINE METABOLISM DURING DEVELOPMENT OF BOVINE EMBRYOS PRODUCED *IN VITRO*

D. RIEGER, N.M. LOSKUTOFF and K.J. BETTERIDGE
Dept of Biomedical Sciences, University of Guelph, Guelph, Ont., Canada, N1G 2W1

To determine the patterns of energy metabolism in bovine embryos, oocytes were isolated from ovaries collected at slaughter, matured and fertilized *in vitro*, and then co-cultured with bovine oviduct epithelial cells in Ménézo's B2 medium. At various developmental stages (2-cell to expanded blastocyst), embryos were removed from co-culture, washed and cultured individually for 3h at 39°C in a 4 µl droplet of B2 containing ^{14}C- or ^{3}H-labelled glucose or glutamine, suspended over 1.5 ml of 25 mM NaHCO$_3$ within a sealed 2.5 ml vial. After culture, the NaHCO$_3$ was mixed with scintillation fluid and counted to determine the amounts of ^{14}CO$_2$ and ^{3}H$_2$O produced, and hence the amounts of substrates metabolized. As shown below, the metabolism of [1-^{14}C]-glucose (pentose-phosphate pathway activity) increased significantly between the 8- and 16-cell stages (time of activation of the embryonic genome) and continued to increase to the expanded blastocyst stage. The pattern of [5-^{3}H]-glucose (total glucose) metabolism, was similar except that the first marked increase occurred between the 16-cell and morula stages. Conversely, the metabolism of [^{14}C(U)]- or [3,4-^{3}H(N)]-glutamine (Krebs cycle) was high at the 2-cell stage, decreased to a minimum at the morula to blastocyst stage and then increased in the expanded blastocyst. The high level of glutamine metabolism in the early stages may be related to its importance for development, as has been shown for hamster and mouse embryos.

Substrate	2-cell	4-cell	8-cell	16-cell	Compacted morula	Early to mid-blastocyst	Expanded blastocyst
[1-^{14}C]-glucose	0.40 ± 0.10^a	0.46 ± 0.10^{ab}	0.98 ± 0.21^b	2.16 ± 0.20^c	2.52 ± 0.19^c	4.50 ± 0.74^d	5.86 ± 1.36^e
[5-^{3}H]-glucose	1.14 ± 0.34^a	2.95 ± 0.63^b	1.50 ± 0.58^a	2.30 ± 0.57^{ab}	7.75 ± 1.60^c	23.90 ± 2.33^d	37.00 ± 10.3^d
Glutamine	6.72 ± 0.57^a	7.21 ± 0.84^a	3.96 ± 0.49^b	2.63 ± 0.43^c	1.74 ± 0.15^c	1.89 ± 0.36^c	5.04 ± 1.57^b

Each mean (pmoles/embryo/3h ± s.e.m.) is for 13 to 15 embryos, except for the expanded blastocysts (N = 4 or 5). Within each line, values with no superscripts in common are significantly different (P ≤ 0.05).

MATURATION AND FERTILIZATION OF MOUSE OOCYTES IN SERUM-FREE MEDIA

R. T. Serta, M.M. Seibel and A. A. Kiessling,
Faulkner Centre for Reproductive Medicine, Department of Surgery, Harvard Medical School, Boston, Ma., 02130.

The ability to mature human oocytes in vitro would be immensely useful to assisted reproductive technology programs. Studies of mouse oocytes have shown it is possible to achieve offspring from oocytes matured in vitro under carefully controlled conditions[1]. One essential condition was continual exposure to follicular fluid (FF) or serum to prevent the "zona hardening" which appeared to block sperm entry.[2,3] To begin to understand the factors in human FF essential for normal oocyte maturation, we have studied the effects of fetal calf serum (FCS), human preovulatory serum (HPOS), FF and bovine serum albumin (BSA) on the developmental potential of mouse oocytes matured in vitro. Prepubertal B6SJL females were injected with gonadotropin and 48 hours later, oocytes were harvested from antral follicles. They were incubated in each maturation condition for 15 hours and then inseminated. Six hours later eggs were scored for polar body extrusion and transferred into fresh culture dishes containing Ham's F-10, modified to support development to blastocysts of B6SJL zygotes. The Table lists the results of these experiments.

Condition	#of eggs	PB	Percent Development Two cell	Blastocysts
Modified Ham's F-10	129	40 (31%)	13 (32%)	5 (38%)
" +FCS (10%)	97	41 (42%)	14 (34%)	1 (7%)
" + HPOS (10%)	77	32 (41%)	17 (53%)	7 (41%)
" +FF (10%)	237	127 (53%)	75 (59%)	45 (60%)
" +BSA (4mg/ml)	98	65 (66%)	47 (72%)	15 (31%)

These results demonstrate that under these culture conditions, mouse oocytes can mature and fertilize in the absence of serum or follicular fluid. This suggests that the previously reported "zona hardening" was not simply due to lack of a serum factor. It is possible that maintaining or developing the integrity of sperm zona receptors is an egg function that may be inhibited in suboptimal culture conditions. In addition, our results support previous reports of the inhibitory effects of fetal calf serum on mouse embryo developmental potential. The modified Ham's F-10 employed in these studies provides a protein-free medium for studies of the effects of hormones and other growth factors on egg maturation.

[1]A. Schroeder, S. Downs, J. Eppig. (1988). Annals N.Y. Acad. Sciences 541:197-204.
[2]M. DeFelici, A. Salustri, G. Siracusa. (1985). Gamete Research 12:227-235.
[3]S. Downs, A Schroeder, J. Eppig. (1986). Gamete Research 15:115-122.

EVALUATION OF PROTEIN-FREE CULTURE MEDIA TO SUPPORT THE DEVELOPMENT IN VITRO OF MORULAE RECOVERED BY NON-SURGICAL UTERINE FLUSHING FROM NATURALLY-BRED RHESUS MONKEYS.

POLANI B. SESHAGIRI AND JOHN P. HEARN
Wisconsin Regional Primate Research Center, Univ. of Wisconsin, Madison, WI 53715.

The development in vitro of preimplantation embryos is regulated by epigenetic factors, including the medium used for culture. Inadequate medium is a cause of early embryonic mortality in vitro. Moreover, most of the media used so far are complex, containing various sera. To optimize a medium, we evaluated several protein-free chemically defined media known to support rodent embryo development. In vivo developed zona-intact normal morulae, recovered from naturally-bred rhesus monkey (Macaca mulatta) females on day 5 (day 0: the day following LH peak) of pregnancy by employing a non-surgical uterine flushing technique, were cultured in 3 media: i) CMRL-1066 (control); ii) hamster embryo culture medium-1 (HECM-1): supports hamster 2-cell embryo development (Seshagiri and Bavister, J. Exp. Zool., 257: 51); and iii) simplex optimization medium (SOM): supports mouse 1-cells (Lawitts and Biggers, J. Reprod. Fert., 91: 543). For each treatment, three embryos were cultured in separate dishes in 1 ml medium for 48-96h in a humidified atmosphere of 5% CO_2 in air at 37°C. After culture, at least one embryo from each group was subjected to Hoechst-staining. The cell numbers (nuclei/embryo) for freshly recovered early and hatching blastocysts were 94 (n 1) and 502 (n 1), respectively. In 36-48h, CMRL-1066 (531, n 1) and HECM-1 (394, n 1) both containing 20% bovine fetal serum readily supported embryo development to hatched blastocysts (100%). In contrast, embryos cultured in serum-free media, CMRL-1066 (199, n 1) or HECM-1 (109, n 1) or SOM (122, n 1) developed to fully expanded blastocysts (100%) and none of them hatched (0%), despite providing an additional 24-48h. Similarly, embryos cultured in HECM-1 supplemented with either Ham's F-10 vitamins (87, n 3) or 1% insulin/transferrin/selenium (126, n 2) developed to fully expanded blastocysts (both 100%) and did not hatch. These results show that rhesus monkey morulae can develop to fully expanded blastocysts in a protein-free medium, but hatching-promoting factors appear to be necessary for hatching. If identified and characterized, such factors should be added to protein-free media for use in primate embryo culture. Supported by NIH Grant RR00167 and a grant (to JPH) from The Graduate School, University of Wisconsin-Madison.

OOCYTE QUALITY MAY BE COMPROMISED IN IVF CYCLES BY PRE-EXISTING OVARIAN CYSTS.

LEON SHEEAN, DONNA WESOLOWSKI, CYNTHIA AUSTIN, JAMES GOLDFARB, University MacDonald Womens Hospital, Department of OB/GYN, Cleveland, Ohio 44106.

A retrospective analysis of 83 consecutive cycles was performed in patients seeking treatment in our IVF program. None of the couples in this study demonstrated male factor infertility. Endocrine suppression with GNRH-a (Lupron) was initiated in the luteal phase of the previous cycle. After establishing baseline conditions (estradiol <100 pg/ml), controlled ovarian hyperstimulation was achieved with concurrent hMG (Pergonal). Standard IVF procedures were performed on all patients. Results shown below were tabulated by the absence or presence of ovarian cysts 10mm in average diameter, at initiation of hMG stimulation:

	No Cysts	Cysts ≤ 10mm	Cysts > 10mm
Number of Cycles	33	26	20
Oocytes Recovered	334	259	229
Oocytes/Cycle	10.1	10.0	11.5
Fertilized	225 (67%)	174 (67%)	108 (47%) *
Polyspermic	16 (4.8%)	19 (7.3%)	14 (6.1%)
Embryos Transferred/Cycle	4.7	4.7	4.0
Pregnancies: Chemical	0	3	1
Clinical	10	8	3
Ectopic	0	1	0
Ongoing/Del'd	9 (27%)	7 (27%)	3 (15%)

These data suggest a less than optimal outcome may result in IVF cycles initiated when non-estrogenic ovarian cysts are found to exceed 10mm in average diameter. We believe that the potential for fertilization and development of mature oocytes may be impaired by these cystic structures although "apparent" oocyte maturity at aspiration is unaffected.

ENHANCED IN VITRO EMBRYO DEVELOPMENT USING OVIDUCTAL CELL CO-CULTURE FOLLOWED BY UTERINE CELL CO-CULTURE FOR EARLY BOVINE EMBRYOS DERIVED FROM IN VITRO MATURATION AND FERTILIZATION

GURPREET SINGH* and N.L. FIRST
University of Wisconsin, Madison, WI 53706

Oviductal cell co-cultures (O) or their conditioned media have been used to overcome the developmental block in vitro for bovine early embryos. However, the efficiency of development to blastocysts/morulae (Bl+M) in such cases is approximately 30%. In vivo bovine embryos are normally found in the uterus 3-4 days after fertilization. In the present study early bovine embryos were obtained by in vitro maturation (IVM) and fertilization (IVF). In Group I (O-U) - bovine oocytes were co-cultured for the first 3 days post insemination (PI) with O and then switched to uterine cell co-cultures (U). At 48 h PI the embryos were separated as 2, 4 and 8 cell embryos and cultured separately throughout. Only uniformly and evenly divided embryos were kept for the study. Development to Bl+M was observed on day 7 PI. As controls, Group II (O-O) - embryos were transferred to O at the same time as in Group I. In Group III (U-U) embryos were cultured directly in U and shifted to U again as in Group I. 94.7% of early cleaving 8 cell embryos at 48 h PI developed to Bl+M in Group I whereas in Groups II and III 58.6 and 54.7% were Bl+M. 71.0, 60.0 and 40.5% of 4 cell embryos were at Bl+M stage in Groups I, II and III, respectively. Similarly 40.5, 22.2 and 54.5 % of 2 cell embryos developed to Bl+M in group I, II and III, respectively. These experiments suggest that enhanced embryo development can be obtained utilizing a two-step embryo culture using oviductal and uterine cells. Work supported by W.R. Grace and Co.-Conn. and *Department of Biotechnology Associateship, DBT, Government of India.

INVOLVEMENT OF cAMP- AND Ca^{2+} - DEPENDENT INTRACELLULAR MECHANISMS IN THE REGULATION OF BOVINE OOCYTE MATURATION IN VITRO

A.V. SIROTKIN, C.G. MIKHAELJAN, A.K. GOLUBEV

Genetical Inst., Leningrad, 189620, USSR, Inst. Anim. Product. 949 92 Nitra, CSFR

The role of cAMP- and Ca^{2+} -dependent intracellular mechanisms in the regulation of bovine oocytes nuclear and cytoplasmic maturation was investigated through in vitro experiments. It was found that dbcAMP or theophylline (inhibitors of intracellular cAMP metabolism) or heparin (which stimulates cAMP production) inhibit both meiosis, reinitiation and completion in cumulus-enclosed oocytes in a dose-dependent manner. After removal of these drugs, the oocytes became mature and capable of cleaving following insemination. In denuded oocytes these treatments inhibited reinitiation, but stimulated completion of meiotic maturation. Ca^{2+} ionophore A23187 and Ca^{2+} channel blockers verapamil, diltiazem or PN200-110 had no marked influence on nuclear maturation of cumulus-enclosed oocytes, but these oocytes were incapable of cleaving after insemination. Data obtained suggest that: (1) reinitiation, completion of nuclear maturation and cytoplasmic maturation are relatively independent processes, (2) bovine oocyte nuclear maturation is under inhibitory control of cAMP-dependent intracellular mechanisms; cytoplasmic maturation is regulated mainly by Ca^{2+} -dependent mechanisms; (3) cAMP-dependent structures may control nuclear maturation by (a) direct action on the oocyte and (b) through the surrounding cumulus.

EVALUATION OF BOVINE MORULA BISECTION AND BIOPSY ON SUBSEQUENT EMBRYONIC DEVELOPMENT IN VITRO.

A.E.T. SPARKS, R.S. CANSECO, AND F.C. GWAZDAUSKAS, Department of Dairy Science, Virginia Polytechnic Institute and State University, Blacksburg

Bovine morulae (n=60;d 6-6.5) were collected nonsurgically from superovulated dairy cows to compare the effects of bisection and biopsy on subsequent embryo development. Excellent, good, and fair quality embryos were selected for microsurgery. Bisections (n=34) and biopsies (n=26) were performed by holding embryos with a 30 μm fire polished, holding pipette, passing a microsurgery blade through the zona pellucida, cutting embryonic tissue against the holding pipette, and removing the biopsy or demi-embryo with the microsurgery blade. Biopsies removed approximately 1/4 of the embryonic tissue while bisections removed approximately 1/2 of the embryo. Following microsurgery embryos were cultured in 25 μl microdrops of Ham's F-10 supplemented with 10% new born calf serum in an atmosphere of 5% CO_2 in air at 37.0 C for 48 h. Embryos were morphologically evaluated at 24 and 48 h of culture and assigned the following scores: 1=degenerate, 2=morula, 3=early blastocyst, 4=blastocysts, 5=expanded blastocyst. Nuclei were stained at 48 h with Hoechst 33342 DNA stain for determination of cell number. Mean development scores at 24 and 48 h and mean cell number for each treatment were analyzed by analysis of variance. Least square means for biopsied embryo development scores at 24 and 48 h were 2.5 ± .3 and 3.0 ± .5 ($\bar{x}$ ± S.E.) and did not differ from 24 h (2.7 ± .2) or 48 h (3.1 ± .4) development scores of bisected embryos. Mean cell number after 48 h of culture did not differ between treatments with 43.3 ± 4.5 (n=23) for biopsied embryos and 38.8 ± 4.5 (n=11) for bisected embryos. These results indicate that recovery of 1/4 of embryonic tissue has no advantage over removal of 1/2 of the embryo in regards to subsequent embryo development in vitro. Therefore, embryo bisection and biopsy may both be suitable methods for recovery of tissue samples for genetic analysis with bisection yielding the larger sample.

EARLY DEVELOPMENT OF GUINEA PIG EMBRYOS *IN VITRO* IN TWO SERUM-FREE MEDIA.

O. SUZUKI, T. ASANO, A. OGURA, M. OHIKE, Y. YAMAMOTO and Y. NOGUCHI
Department of Veterinary Science, National Institute of Health, Tokyo 141, JAPAN.

Preimplantation embryos in various developmental stages from Hartley guinea pigs were collected and cultured *in vitro* in two chemically-defined media. Collection schedule of embryos were as follows(plug=Day 1 of pregnancy): one-cell on Day 1, two-cell on Day 2, four-cell on Day 3 and eight-cell on Day 4. In Whitten's medium(WM), one-cell embryos arrested at the four-cell stage(13/22). Two-cell and four-cell embryos developed into morulae and then degenerated(9 from 18 two-cell and 8 from 13 four-cell, respectively). A part of eight-cell embryos(11/21) reached the blastocyst stage, but further development were not observed in WM. In Kane's medium(KM) containing amino acids and vitamins of Ham's F10, one-cell embryos arrested at the two-cell stage(3/9). Two-cell embryos were not yet cultured in KM. Some embryos at four-cell or later developed into blastocysts(8 from 28 four-cell and 6 from 13 eight-cell embryos). In contrast to the culture in WM, blastocysts which had developed *in vitro* exhibited further morphological changes in KM, such as the trophoblast outgrowth through a small hole of the zona pellucida, the attachment to a non-coated plastic culture dish and the multilayered vesicle formation of trophoblasts outside the zona. But only a few blastocysts completely hatched out from the zona. Small portion of embryonal cells remained in the zona until the end of culture. These observations suggest that 1) guinea pig embryos might change in the requirement of nutrients at the 4-cell stage, 2) the proliferation and differentiation of trophoblasts might not need any growth factors, and 3) trophoblasts in the guinea pig might actively produce cell-attachment factors to adhere the endometrium of the uterus.

GLUCOSE METABOLISM IN OVIDUCTAL AND NON-OVIDUCTAL CELL COCULTURES IS ASSOCIATED WITH ENHANCEMENT OF MOUSE PRE-EMBRYONIC DEVELOPMENT *IN VITRO*.

TAKEUCHI K, KAUFMANN RA, HODGEN GD, SANDOW B. The Jones Institute for Reproductive Medicine, Department of Obstetrics and Gynecology, Eastern Virginia Medical School, 855 West Brambleton Avenue, Suite B, Norfolk, VA 23510

Pre-embryo development can be stimulated by coculture with a variety of cell types, but the mechanism of this stimulatory effect is unclear. The present study was designed to evaluate the development of mouse pre-embryos in coculture with oviductal or non-oviductal cells, and to determine whether glucose metabolism in the cocultures is associated with pre-embryonic development. One-cell Swiss Webster mouse pre-embryos were cocultured in Ham's F-10 plus 10% FBS (control) with monolayers of the following cell types: mouse oviductal epithelial cells, mouse oviductal fibroblasts, monkey oviductal epithelial cells, CHO cells, 3T3 cells, and L-929 cells. Blastocyst formation was evaluated after 5 days of culture; glucose, lactate, and pyruvate levels in the culture media were measured after 24 hours of culture using a UV method. Blastocyst formation rates were as follows: control (n = 165), 13%; mouse oviductal epithelial cells (n = 153), 80%; mouse oviductal fibroblasts (n = 92), 13%; monkey oviductal epithelial cells (n = 154), 74%; CHO cells (n = 145), 66%; 3T3 cells (n = 122), 12%; L-929 cells (n = 89), 55%. Glucose, lactate, and pyruvate levels (mM), respectively, were as follows: control, 7.30, 1.85, 0.23; mouse oviductal epithelial cells, 4.99, 3.74, 0.14; mouse oviductal fibroblasts, 5.70, 1.84, 0.17; monkey oviductal epithelial cells, 4.35, 4.91, 0.16; CHO cells, 4.01, 6.38, 0.16; 3T3 cells, 6.62, 3.01, 0.20; L-929 cells, 4.42, 3.75, 0.17. Thus, blastocyst development was significantly enhanced by coculture with mouse or monkey oviductal epithelial cells, CHO cells, or L-929 cells, indicating that the coculture effect lacks species- or tissue-specificity. Furthermore, decreased levels of glucose and increased levels of lactate were seen in the culture medium of the groups that exhibited enhanced pre-embryo development, suggesting that glucose metabolism may be a critical factor in early pre-embryo development.

EFFICIENT ISOLATION OF HOMOGENOUS CELLS FROM THE INNER CELL MASS OF PIG AND SHEEP BLASTOCYSTS.

NEIL TALBOT, ANNE POWELL, VERNON PURSEL, AND CAIRD E. REXROAD, JR. USDA, Agricultural Research Service, Reproduction Laboratory, Beltsville, MD 20705.

To establish embryonic stem cell lines from blastocysts, trophectodermal and primitive endodermal cells must be separated from the desired primitive ectodermal cells. Separation is possible, both efficiently and effectively, by treating the blastocyst with a combination of physical dissections and immunodissections. Sheep and pig blastocysts collected at 7-8 days post coitus were cultured in vitro until hatched. The inner cell mass (ICM) of each blastocyst was isolated by exposing the blastocyst to a rabbit anti-sheep splenocyte serum followed by exposure to guinea pig serum. This initial immunodissection removed most of the trophectoderm. The ICM was then cultured for 3-4 days on STO feeder cells. The colonizing ICM cells consisted of rapidly and radially emanating flat, vesicle-forming cells and a centrally located group of cells which grew up as an organized ball of cells. In the case of sheep ICMs, the ball of cells was physically dissected intact and then exposed to a second round of immunodissection to remove a surrounding outer layer of cells. For pig ICMs the central ball of cells could usually be obtained free of any other contaminating cell by simple physical dissection. The final dissections resulted in the isolation of a homogenous mass of cells which when grown on STO feeder cells produced tight colonies of small cells with large nuclei and prominent nucleoli resembling mouse ES cells. The continuous culture of these cells and their pluripotency are being studied.

AN EXAMINATION OF THE REQUIREMENT FOR GLUCOSE IN PREIMPLANTATION SHEEP EMBRYOS DURING IN VITRO CULTURE.

J.G. THOMPSON, A.C. SIMPSON, P.A. PUGH AND H.R. TERVIT
Ruakura Agricultural Centre, P.B., Hamilton, New Zealand.

Glycolysis readily occurs in preimplantation sheep embryos, but entry of glycolytic products into the TCA cycle is limited in the presence of exogenous pyruvate+lactate (PL) (1). This has led us to examine if in vitro development can occur when glucose is the only carbohydrate present. Estimates for maximal glycolytic rates were obtained for 8-cell and blastocyst stages by measuring conversion of [5-^{3}H]-glucose to ^3H_2O for 3h over a range of glucose concentrations (0.1mM - 5.0mM). Ninety percent maximal glycolytic activity occurred at approximately 4.5mM for 8-cell embryos and 1.0mM for blastocysts. One- and 2-cell sheep embryos were then incubated for 120 h at 39 C under humidified $5\%CO_2$, $5\%O_2$, $90\%N_2$ in SOF medium supplemented with 5mg/ml human serum albumin and MEM non-essential amino acids in one of 4 glucose concentrations (0, 1.5, 3 or 6mM) $\pm$ 0.33mM P + 3.3mM L. Development (%) to blastocyst stage was:

Pyruvate+Lactate	Glucose (mM)	0	1.5	3.0	6.0
+		37	59	35	19
-		19	24	0	0

Overall, the presence of glucose was detrimental (P<0.001) to embryonic development. In contrast, the presence of PL was beneficial (P<0.001) to development. Moreover, the effect of glucose is interactive with the presence of PL (P<0.05). An optimum level of glucose occurs between 0-3mM in the presence of PL (P<0.1). We suggest that glucose metabolism is not critical for embryonic development, but is beneficial at low concentrations. However, higher concentrations can inhibit development, possibly by inhibiting the TCA cycle (2).
(1) Thompson, J.G. et al (1991) Reprod.Fert.Dev. 3:in press.
(2) Schini, S.A. & Bavister, B.D. (1988) Biol. Reprod. 39:1183-1192.

MEASUREMENT OF PYRUVATE UPTAKE BY HUMAN EMBRYOS IN THE NATURAL CYCLE
USING A NON-INVASIVE TECHNIQUE.

K.TURNER, K.MARTIN*, B.WOODWARD, H.ROBERTS, N.MONKS, H.LEESE*, E.LENTON
Sheffield Fertility Centre, 26, Glen Rd, Sheffield, S7 1RA, England.
*Department of Biology, University of York, York, YO1 5DD, England.

The uptake of pyruvate by 32 human embryos derived from natural cycles was measured non-invasively using an ultra-fluorometric technique. Embryos with two pronuclei at fertilization check were cultured in microdroplets of culture medium for a period of approximately 24 hours, after which the microdroplets were collected and assayed for pyruvate uptake. Pyruvate consumption by individual embryos was then correlated with cycle outcome following embryo transfer to patient.

The results showed that those embryos culminating in a clinical pregnancy tended to have a pyruvate uptake of between 16 and 26 pmol/hour. Embryos with a pyruvate uptake above or below this defined region appeared to have a smaller chance of resulting in a clinical pregnancy.

In addition, the results were examined with respect to morphological grading of embryos and patient infertility. Embryos graded as good, reasonable or poor showed some correlation with pyruvate consumption but emphasize the point that morphological grading of embryos is an inadequate method for assessing the quality of embryos. With regard to patient infertility, it appeared that tubal damage patients were more likely to have embryos with a pyruvate uptake within the above defined region. Patients with unexplained infertility that did not conceive, however, tended to produce embryos with a pyruvate uptake outside this range. These preliminary observations may help to discriminate between embryos with the potential to implant and those not likely to result in a pregnancy.

CELLULAR INTERACTIONS IN AN IN VITRO MODEL OF IMPLANTATION.

R.F.WATERHOUSE AND S.J.KIMBER.
Cell and Structural Biology,Manchester University,Manchester,U.K.

Successful pregnancy is dependant upon initial attachment of the blastocyst to the endometrial epithelium and subsequent implantation. The use of in vitro models has allowed examination of the complex cellular interactions occurring during implantation. By utilising some of the cell- and stage-specific markers of uterine cells the degree of differentiation and resemblance to the natural state in utero can be determined.

We have developed an in vitro system based on that of McCormack and Glasser (1980) in which murine luminal uterine epithelial (UE) cells are grown on semi-permeable collagen membranes. This provides the necessary conditions for expression of a polarised phenotype which resembles that found in vivo.

Using this system,we have examined how UE cell growth in vitro compares with that seen in vivo. Preliminary observations will be presented on the interaction between UE cells and embryos,and how this is affected by the presence of reproductive steroids in the culture medium.

Ref. McCormack,S.A. and Glasser,S.R. (1980). Endocrinology. 106. (5),1634.

EXPRESSION OF INTERFERON-Γ RECEPTOR mRNA DURING MOUSE EMBRYOGENESIS

Tsung-Chieh Jackson Wu, Lai Wang and Ju-Jui Yvonne Wan
Department of Obstetrics and Gynecology, UCLA, Los Angeles, CA; and
Department of Pathology, Harbor UCLA Medical Center, Torrance, CA.

Interferon Γ (IFN-Γ) modulates a variety of gene expressions including major histocompatibility complex (MHC), antiviral enzymes and a large number of proteins of unknown function. It accelerates differentiation of human promyelocytic cells and keratinocytes, and upregulates MHC and c-fos gene expression in embryonal carcinoma F9 cells. IFN-Γ exerts its effects through specific cell surface IFN-Γ receptor whose cDNA has been cloned. To elucidate the potential role of IFN-Γ during development, we examined the ontogeny of IFN-Γ receptor mRNA, and semi-quantitated its levels in mouse oocytes and embryos at various stages of gestation, using highly sensitive reverse transcriptase-polymerase chain reaction. Total RNA extracted from ovulated oocytes, 2-cell embryos, morulae and blastocysts (n=40 each) was reverse transcribed into cDNA. A pair of primers flanking 627 bp of cDNA was used for amplification of 30 cycles. The identity of the amplified product was confirmed by sizing and Southern blot analysis. β-Actin mRNA served as control. The results indicate that IFN-Γ receptor mRNA was expressed in the unfertilized oocytes, but the amount decreased in 2-cell embryos. Morulae and blastocysts expressed increasing levels of IFN-Γ receptor transcripts as the embryonic genome became activated. Postimplantation embryos expressed IFN-Γ receptor mRNA throughout gestation. Thus, IFN-Γ receptor is one of the very early genes expressed in embryos even before implantation. The detection of IFN-Γ receptor mRNA in mouse oocytes and embryos suggests that IFN-Γ may be involved in the MHC expression and differentiation process of oocytes and embryos.

EXPRESSION OF THE PROTO-ONCOGENE C-FOS BY THE OVINE TROPHOBLAST: IN VITRO AND IN VIVO ANALYSIS

F. XAVIER[1], M. GUILLOMOT[1], J. MARTAL[1] and P. GAYE[2], I.N.R.A. [1] Unité Endocrinologie de l'Embryon and [2] Unité Endocrinologie moléculaire, 78352 Jouy-en-Josas cédex, France

Expression of the c-fos proto-oncogene by ovine trophoblastic cells was analyzed during the peri-implantation period by Northern and slot blots and indirect immunohistofluorescence. The expression of the c-fos proto-oncogene was transient, occurred at a maximal level at the beginning of implantation (days 14-15) and decreased thereafter. Indirect immunofluorescence technique confirmed the temporal expression of c-fos. On days 12 and 13 of gestation, no significant staining appeared in the trophoblast layer, whereas on day 14 the entire trophoblast was reactive. The c-fos protein was observed either exclusively into the nucleus or both in the cytoplasm and the nucleus. By day 15, the periembryonic trophoblast was negative whereas the distal trophoblast was still strongly positive. On day 17, c-fos was expressed only in non-adherent trophoblastic cells. Ovine trophoblastic cells expressed no more c-fos when they established cellular contact with the uterine epithelium during the implantation process. We have developed primary cultures of trophoblastic cells from day 14-15 tissues. The subconfluence state was reached the 4th day after seeding. The trophoblastic origin of the cells in culture was confirmed by immunofluorescence technique using various antibodies (anti cytokeratine, anti oTP and anti trophectoderm). In our culture conditions, the trophoblastic cells retained some of their specific characteristics. Interestingly, the c-fos proto-oncogene was still expressed in 24h serum-starved trophoblastic cells. An autocrine regulation of the c-fos expression might exist in ovine trophoblast. Studies are in progress to analyze the mechanisms which control the transient expression of the c-fos proto-oncogene by trophoblastic cells in vitro.

EFFECT OF ANTI-SENSE OLIGONUCLEOTIDES TO MAJOR HISTOCOMPATIBILITY COMPLEX (MHC) mRNA ON PREIMPLANTATION EMBRYO DEVELOPMENT

YUANXIN XU AND CAROL M. WARNER
DEPARTMENT OF BIOLOGY, NORTHEASTERN UNIVERSITY, BOSTON, MA 02115

A gene has been described, Ped (Preimplantation embryo development), that influences the rate of cleavage division of preimplantation mouse embryos. Formal linkage studies have shown that the Ped gene is located in the Q region of the H-2 complex, the major histocompatibility complex (MHC) of the mouse. The Q region encodes ten genes: Q1 - Q10. Analysis of the congenic strains B6.K1 and B6.K2 has suggested that the Ped gene is encoded by Q3, Q5, Q6, Q7, Q8, and/or Q9. In order to test which of these genes encodes the Ped gene product, mouse embryos were treated with anti-sense oligonucleotides. Anti-sense oligonucleotides which bind specifically to the leader sequence, $\alpha 1$, or transmembrane domains of Q7/Q9 mRNA were used to block Qa-2 antigen expression. Corresponding sense oligonucleotides and and oligonucleotide against globin mRNA (not expressed by mouse embryos) were used as controls. Two-cell embryos from C57BL/6 (Ped fast, Qa-2 positive) were cultured with the oligonucleotides. Qa-2 antigen expression was blocked with the Q7/Q9 anti-sense oligonucleotides. H-2D expression was not influenced by these oligonucleotides. Studies on Ped gene phenotype showed that embryos cultured with the anti-sense oligonucleotides had fewer cells per embryo than the ones cultured with the sense oligonucleotides ($P<0.01$).

Human oviductal cells improve the hatching rate of human embryos in vitro

W.S.B. Yeung, P.C. Ho and *S.T.H. Chan   Department of Obstetrics and Gynaecology, *Department of Zoology, University of Hong Kong, Hong Kong.

It is well known that the current in vitro embryo culture systems for the assisted human reproduction programs are suboptimal. In this study, we demonstrated that the developmental capacity of human embryos in vitro was improved when co-cultured with oviductal cells. Human oviductal tissue was obtained from patients admitted for tubal ligation. The cells were enzymatically dispersed and were cultured in DMEM/F12 medium supplemented with 20% maternal serum, insulin, transferrin and epidermal growth factor. Subcultured oviductal cells were used for co-culture with the embryos. 102 spare human embryos from our assisted reproduction program were randomly divided into 3 groups. The embryos in Group A (n=40) were only cultured in Earle's balanced salt solution supplemented with 15% maternal serum (sEBSS). The embryos in Group B (n=42) were cultured in sEBSS for the first 2 days post-insemination, they were subsequently co-cultured with oviductal cells in sEBSS. The embryos in Group C (n=20) were co-cultured with oviductal cells at the pronucleated or 2-cell stage (1 day post-insemination) in sEBSS. The degree of fragmentation of the developing embryos, the rates in the formation of compacted embryos, cavitated embryos, expanded blastocysts and hatching blastocysts of the 3 groups were assessed. Among these parameters, only the blastocyst hatching rate between Group A (5.6%) and Group C (40.0%) showed significant statistical difference (Fisher exact test; $p<0.05$). The corresponding rate for Group B was 16.2%. We concluded that co-culture with human oviductal cell culture could enhance the hatching rate of human embryos in vitro. The degree of improvement was greater when the co-culture started at the pronucleated stage or 2-cell stage than at later stages.

IN VITRO DEVELOPMENT OF CHINESE MEISHAN PIG EMBRYOS

C. R. YOUNGS, L. K. MCGINNIS, and S. P. FORD
Department of Animal Science, Iowa State University, Ames, IA 50011 U.S.A.

The Meishan breed of pig is noted for its prolificacy, with some litters containing 30 or more piglets. The biological mechanisms responsible for this high level of prolificacy are unknown, although uniformity of embryo development leading to an increased embryo survival has been proposed (Bazer et al., 1988, J. Reprod. Fert. 84:37). A study was initiated to compare the in vitro development of Meishan and Yorkshire embryos. Five Meishan and five Yorkshire naturally cycling gilts were used for embryo donors. One- to three-cell embryos (n=51 and 44 for Meishan and Yorkshire, respectively) were cultured at 39°C in 5% CO_2 in humidified air in modified Whitten's medium containing no glucose and reduced sodium lactate (12.93 mM). Observations of embryo development were made daily. At the end of the 144 hour culture period embryos were stained with the DNA-specific fluorochrome Hoechst 33342 to enable counting of cell nuclei. Embryos from an additional two Meishan gilts were transferred to assess viability following 96 hours of in vitro culture. Analysis of data revealed no breed differences (P>.10) in number of ovulations, number of embryos recovered, or final embryo cell counts. However, Meishan embryos were slower (P<.01) to reach the compact morula and blastocyst stage than were Yorkshire embryos. Seventy-seven percent of Yorkshire embryos and 73% of Meishan embryos developed to blastocysts. Recipient gilts were slaughtered between days 26-29. One recipient contained no fetuses but had evident fetal resorption sites. The other recipient contained 6 fetuses (43% embryo survival). These data indicate that Meishan embryos have a different in vitro developmental pattern than Yorkshire embryos. In addition, Meishan embryos cultured in vitro were capable of establishing a pregnancy following embryo transfer.

EFFECT OF MATURATION MEDIA ON MALE PRONUCLEUS FORMATION OF PIG OOCYTES MATURED AND FERTILIZED IN VITRO

MITSUTOSHI YOSHIDA AND KOJI ISHIGAKI
FACULTY OF AGRICULTURE, SHIZUOKA UNIVERSITY, SHIZUOKA 422, JAPAN

The present study was carried out to examine the effect of maturation media on male pronucleus formation of pig oocytes matured and fertilized in vitro. Follicular oocytes collected from prepubertal gilts at a local slaughter house were cultured (36 h) in three different media (mTCM-199, Waymouth MB 752/1 and mTLP-PVA), fertilized in vitro and assessed for nuclear maturation and male pronucleus formation. The addition of 10% (v/v) pig follicular fluid (pFF) to maturation media significantly increased the rate of nuclear maturation of pig oocytes (88-94%) compared with the control (64-70%, P<0.01). The rate of male pronucleus formation of pig oocytes was significantly increased in Waymouth medium with (91%) or without pFF (88%) compared with the others (45-63%, P<0.01). Furthermore, the addition of cysteine (the same concentration in Waymouth medium, 0.57 mM) to mTLP-PVA significantly increased the rate of male pronucleus formation of pig oocytes (94%) compared with the control (22%, P<0.01). In addition, the inhibition of glutathione (GSH) synthesis of pig oocytes matured in Waymouth medium by GSH inhibitor, buthionine sulfoximine, resulted in reducing the ability to form male pronucleus after in-vitro fertilization. The results indicate that the composition of maturation medium affects the ability of pig oocytes to form male pronucleus after fertilization and the medium containing the high concentration of cysteine (possibly as a substrata of GSH), such as Waymouth medium, can remarkably promote this ability.

Author Index

A

Albertini, D.F., 3

B

Baltz, J.M., 97
Bavister, B.D., 57
Belin, D., 38
Biggers, J.D., 97

D

Dey, S.K., 264

E

Eppig, J.J., 43

F

Fahy, M.M., 169
Ferguson-Smith, A.C., 144
Fischer, B., 83

G

Gardiner, C.S., 200
Gocza, E., 157

H

Hahnel, A., 115
Hardy, K., 184
Hogan, A., 115
Holdener-Kenny, B., 131
Huarte, J., 38

I

Ivanyi, E., 157

K

Kane, M.T., 169
Kimber, S.J., 244

L

Lechene, C., 97
Leese, H.J., 73
Lindenberg, S., 244
Ling, P., 229
Lopata, A., 276

M

Magnuson, T., 131
Mathialagan, N., 229
Mattson, B.A., 3
McKiernan, S.H., 57
Menino, A.R., Jr., 200
Merentes-Diaz, E., 157
Messinger, S., 3

N

Nagy, A., 157

O

O'Connell, M.L., 38
Oliva, K., 276

P

Paria, B.C., 264
Pedersen, R.A., 212
Plancha, C.E., 3

R

Racowsky, C., 22
Rappolee, D.A., 212
Roberts, R.M., 229
Rossant, J., 157

S

Schultz, G.A., 115
Sharan, S.K., 131

Stallings-Mann, M., 229
Strickland, S., 38
Sturm, K.S., 212
Surani, M.A., 144

T

Trout, W.E., 229

V

Vassalli, J.-D., 38

W

Waterhouse, R., 244
Watson, A.J., 115
Werb, Z., 212
Wickramasinghe, D., 3

Subject Index

Acetyl-CoA carboxylase, 179
Acetylation, 10–11, 13
Acid load/acidification, 66, 99–104, 106–107
Acidosis, 97–98, 104, 106
Actin, 4, 6, 115, 123, 216, 326, 336
Adenine, 28
Adenylate cyclase, 27, 202, 300
Adhesion molecule, glia (AMOG), 204, 207
Alanine, 59, 62
Albino gene (mouse), 134–135, 138, 150
Albino-deletion complex (c-region), 131–136
Albino-specific-repeat probe, 136
Alkaline load/alkalinization, 99, 101, 104–106
Alkalosis, 98–99, 106
Allantois, 215, 219, 234, 325
α-amanitin, 115–117, 188, 307, 326
Amiloride sensitivity, 97, 101–102, 104
Amino acids, 75, 77–78, 97, 169–170, 175–176, 178–179. See also specific acid
  requirements for, 57–63, 65, 69, 75
  sequence, 236–237
Ammonia/ammonium ion, 99, 101, 104–106
AMP, cyclic, 22, 24–29, 31–32, 49–50, 300, 311, 332
Amphibians, 74, 212, 221
Anaphase, 4–5, 8–9, 11
Androgenotes, 144, 148–149, 153–154, 212–213, 220, 222–223, 314
Aneuploidy, 161–162, 324

Angelman syndrome (AS), 153, 213
Antibodies, 8, 13–15, 254–255, 265, 299, 301, 307, 308, 334, 336
Antigens, 254–256, 299
Antileukoproteinase, 237
Antiport, Na^+/H^+, 66, 97–99, 101–103, 107–108
Antisense signals, 121
Antrum, 6, 8, 16, 23–24, 43, 50
Aprotinin, 236
Areolae (pig chorion), 231, 233
Arginine, 59, 61–62
Arrest, cleavage stage, 185–186, 188, 190–192, 196, 272. See also Block; Meiosis, arrest
Arrester, follicular, 23–25, 27
Asparagine, 62–63
Aspartic acid, 62–63
Asters, 10–11, 17
Astrocyte, 204
Ataxia, 153
ATP, 73, 76–79, 97, 103, 190, 267–269
ATPase, 76–77, 97, 103, 201–207, 323
  subunits of, 203–207
Autocrine factors, 83, 264, 273, 277, 280, 291, 316, 324, 336
Autoradiography, 91, 117, 119, 121, 172–174, 178, 266–268, 270, 318, 326
Axial differentiation, 212, 215, 218
Azaserine, 300

Baboon blastocyst, 194, 279
Basement membrane, 245, 247–249, 255

341

Beckwith-Wiedemann syndrome (BWS), 153
Bicarbonate/chloride (HCO_3^-/Cl^-) exchanger, 97-98, 102-106, 108
Bicarbonate ion (HCO_3^-), 102-108
Birds, 212, 221
Blastocoel, 77, 79, 97, 99, 108, 190, 192, 314
 fluid, 200-202, 207, 306, 323
 formation, 200-202, 205-207. *See also* Cavitation
Blastocyst, 44-51, 58-64, 73-74, 76-78, 80, 84-89, 91-92, 100, 108, 116-117, 119-124, 150-151, 157, 169-179, 184-188, 190-197, 202, 205-207, 212, 214-215, 217, 230, 232, 244, 246, 252-253, 264, 266-273, 277-279
 expansion, 88, 172-173, 175-178, 185, 190, 192-195, 200-202, 206-207, 215, 289, 291, 323, 330, 337
 hatching, 58, 85, 124, 192-193, 201-202, 214, 244, 256, 272-273, 278-279, 289, 310, 311, 313, 328, 330, 333, 337
 injection, 150-151, 157, 159
 morphology, 90-92
 secretion of hCG, 277-292
Blastokinin, 229
Blastomere, 84, 87, 101, 120, 122, 192-193, 196, 202, 205, 314, 316
Block
 culture, 116-117
 developmental, 78, 117, 124, 230, 300, 301, 311
 two-cell, 78, 116, 133, 311
Blood system, 215, 231, 251, 255, 264, 305, 320
Bovine serum albumin (BSA), 58-59, 75, 170, 176-177, 190, 253, 286-287, 328, 330
Bovine. *See* Cow
Brachyury (T) gene, 140
Brain, 147-149, 212-213, 215, 317
Breakpoints, chromosomal, 133-139, 145
Brinster's medium, 176
5-bromodeoxyuridine, 80

C-region, mouse chromosome 7, 131-136
Caenorhabditis elegans, 39
Calcium, 97, 170, 172, 178, 314, 329, 332
Calcium channels, 97, 332
Carbohydrate energy substrates, 57
Carbon atoms, 78
Carbonic anhydrase, 67-68
Carcinoma, 78, 91, 152-153, 159, 213, 249, 254-256, 284. *See also* Oncogenes
β-carotene, 238
Cartilage, 148-149
Cat, 44, 230, 304
Cathepsins, 230, 237
Cavitation, 116, 194, 202, 314, 323, 337. *See also* Blastocoel
Cell
 adhesion, 247-248, 251-253, 255-256, 264
 cycle, 4-5, 15-17, 19, 59, 186
 differentiation, 83, 88-89, 92
 number, 63, 118, 124, 171, 191, 193-197, 214, 315, 316, 320, 330, 332, 338
 proliferation, 83, 85-90, 92, 172, 175, 177, 179, 217, 219
 proliferation rate, 118, 124
Cellagen, 249-250, 253-254, 257
Centrioles, 10, 17
Centrosome, 5, 10, 12, 16-19
Cesarean section, 159-160
Chelation, 66, 178
Chemical saturation mutagenesis screen, 139
Chimera, 132, 148-152, 157-163, 212-213, 219
Chloride/bicarbonate exchanger, 202
Chloride, 98, 100, 103-106, 202, 304
Choleratoxin, 25
Choline, 25
Chondroitin sulphate proteoglycan, 245
Choriocarcinoma cells, 284
Chorion, 231
Chorionic gonadotropin, human (hCG), 18, 29, 79, 118, 185, 197, 250, 277-292, 311, 318, 325, 326
Choroid plexus, 147, 213

Chromatin, 4–9, 16, 22, 307, 326
Chromatography, 174–175, 177, 237, 254–255, 300
Chromosome 2, 146, 221–222
Chromosome 6, 146, 221–222
Chromosome 7, 131–138, 145–147, 149–153, 220–222
Chromosome 8, 221
Chromosome 9, 131, 139
Chromosome 11 (human), 153
Chromosome 11, 146, 221–222
Chromosome 17, 131, 140, 146, 220–222
Chromosome abnormalities, 186
Chromosome breakpoints, 133, 135–138
Chromosome walk, 133, 136–138
Chromosome X, 220–221, 316
Chromosome Y, 302, 316
Chromosomes
  marking of, 144
  maternal, 144–147, 154
  meiotic, 10, 17
  paternal, 144–147, 151, 154, 212, 222
Cis-acting proteins, 132
Citrate, 79, 177–179
Clam oocytes, 39
Cleavage, 3, 43–45, 51, 61, 74–75, 84, 117, 122, 169, 184–186, 188–189, 191–192, 194, 200, 212, 214, 277–278, 300, 314, 337
Clomiphene citrate, 277, 303, 313
Cloning
  cDNA, 120, 229, 234, 236–237
  genomic, 131–133, 136–138
Coat color, 158
Cognitive disability, 153
Collagen, 215, 245, 247–249, 255, 257, 335
Colonocytes, 78
Compaction (of embryo), 84, 116, 133, 185–186, 190, 202, 205, 251, 300, 314, 337
Competence. *See* Embryogenesis, competence for; Meiosis, competence; Oocyte, competence
Complement component C-3, 230
Complementation analysis, 132, 135, 138, 145–146, 221–222

Controlled pooling, 61
Corpus luteum, 234, 277, 302
Cortical granules, 304, 312
Cotransport, 102–103, 201–202, 174, 304
Cow, 118, 236, 322, 327, 329
  embryo, 74, 78, 90–91, 116–123, 206, 310, 323, 326, 329, 331
  implantation, 256, 258
  oocyte, 24, 27–28, 44, 118, 300, 317, 321, 322, 329, 332
Cryopreservation of embryos, 278, 301, 314, 317, 322
Culture block, 116–117
Cumulus cells, 15, 18, 23–26, 28, 59, 118, 278, 300, 309, 315. *See also* Oocyte-cumulus complex
Cyanide, 73, 103
Cyanoketone, 30
Cysteine, 59, 61–63, 338
Cytocalyx, 277
Cytochalasin, 201
Cytokine, 255, 319, 327
Cytoplasm, meiotic maturation, 3, 44
Cytoplasmic polyadenylation element (CPE), 39–40
Cytoskeleton, 204, 256, 327
Cytotrophoblast, 153, 277, 284, 287–288, 291

Dead cell index, 193–196
Death
  of cells, 193–196, 218, 257
  of embryo, 132–133, 145–147, 149, 151–152, 160, 207, 212, 222, 238, 305, 330
  fetal, 145–146, 222
  perinatal, 148, 158–163, 213
Decidual reaction, 214, 245
Deciduum, 230, 245, 247
Deficiency, 144, 221–222
Deletion-breakpoint-fusion fragment, 136–139
Deletions (in genes), 131–137, 144
Developmental delay, human, 153
Diacylglycerol (DAG), 170, 175
Diakinesis, 8–9
Diapause, embryonic, 268–269, 303

Dictyate stage. *See* Germinal vesicle stage
Diethylene glycol distearate, 119
Differences, interspecific, 10-11, 17, 19, 24-28, 74-75, 77, 79, 89-90, 92, 115-117
Differentiation, 118, 190, 192, 200, 202-203, 207, 216, 218-219, 336. *See also* Blastocyst; Inner cell mass
4,4'-diisothiocyanostilbene 2,2'-disulfonic acid (DIDS), 103-106, 304
Dilute-short ear-deletion complex, 131, 139
Dimethyl sulphoxide (DMSO), 281-282
2,4-Dinitriphenol, 73
Diploid androgenotes. *See* Paternal disomy
Diplotene, 22, 43
Disomy, 145-147, 149-153
Disuccinimidyl suberate, 267
DNA
  cloned, 132, 134-135, 234, 319
  complementary (cDNA), 89, 118, 120, 123, 140, 229, 234, 236-237, 305, 308, 319
  deletion in, 133, 136-137
  foreign, 131
  primer pairs, 120
  size-selected, 136
  synthesis of, 78, 86, 88, 189, 301, 307
Dosage (gene), 144-145
Drosophila spermatocytes, 39
Dulbecco's modification, 279, 313
Duplication (of gene), 144, 221-222
Dynein, 5

Eagle's medium, 279, 313
Earle's medium, 25, 188, 337
Early pregnancy factor, 277
Eco-restriction enzyme, 136-138
Ectoderm, 133-135, 138, 215-216, 218-219, 324, 334
Ectoplacental cone, 215-216, 218
Edema, 245
EDTA, 178, 311, 313, 328
Eed gene, 132-135, 138-139
Egg. *See* Oocyte
Elastase, 237
Electroencephalograph, 153
Electron microscopy, 249-250, 254, 256-258, 304, 312
Electron probe X-ray analysis, 100, 105, 107-108
Electrophoresis, 235, 237, 303, 310
Electroporation, 329
Embryo. *See also* Blastocyst; Cavitation; Compaction; Death; Human embryo; Morula; Zygote
  1-cell. *See* Zygote
  2-cell, 44-51, 57, 59-60, 65, 75-77, 79, 85-87, 99-107, 115-118, 120, 122, 133, 171, 173-174, 177, 185, 187-188, 204-206, 214, 217, 219, 269, 271, 278
  4-cell, 57, 60-61, 65, 75, 85-86, 116-118, 185, 187-189, 202
  6-cell, 133, 278
  8-cell, 57-65, 73, 75-77, 80, 85-87, 116-117, 121, 124, 159-160, 171, 173-174, 178, 185-189, 191-192, 200, 202, 205, 251, 267-270, 272
  16-cell, 85, 87, 97, 116-117, 120, 123, 185, 187-189, 194-195, 202
  32-cell, 192, 195
  abnormal, 276
  antigenicity, 229-230, 237, 239
  biopsy, 311, 314, 316, 332
  cryopreservation, 278, 301, 314, 317, 322
  culture, 246-259
  donation, 186
  ectodermic defect, 138
  flushed, 186, 193
  hypotaurine and, 63-64
  isoparental. *See* Disomy
  postimplantation, 57, 83-84
  protein content, 169-171, 173, 179, 206-207
  resorption, 159-160
  size of, 146-148, 152, 215, 308, 310
  somite stage, 116, 148-149, 212, 215
  splitting, 322, 332
  surplus, 184-185, 256, 278
  tetraploid, 158
  transfer (human), 58, 65, 75, 185, 196, 276-278, 302, 315, 335
Embryogenesis, competence, 43-47

Embryonic genome activation, 133
Embryonic stem cells (ES), 124, 131–133, 140, 154, 157–163, 213, 219, 313, 334
En-2 gene, 132
Endocytosis, 5, 86, 231, 233, 257–258
Endoderm, 215–216, 218–219, 334
Endometrium, 88, 207, 229–231, 233–235, 237–238, 244, 246–258, 276–277, 302, 303, 306, 312, 330, 335
Endoplasmic reticulum, 203, 250, 258, 326
Endothelium, 251
Energy reserves, 74, 78
Enhancer binding site, 204
Enhancer-trap construct, 132
Entactin, 245
Enterocytes, 78
Epiblast, 219
Epidermal growth factor (EGF), 83–85, 87–92, 122–124, 176, 264–273, 289–291, 309, 328, 337
Epidermal growth factor receptor (EGF-R), 265, 267–268, 273, 289, 291
Epigenetic factors, 58, 69, 144, 154, 218, 330
Epithelium, 190, 192, 204, 230–231, 233, 244–259, 264–265, 273, 277, 304, 329, 335
Epitope, 251
Estrogen/estradiol, 24, 27–33, 91, 118, 229–230, 233–234, 237–238, 245, 251, 264, 268–269, 273, 276, 308, 331
Estrus, 234, 238, 247, 324
Ethanol, 320
Ethidium bromide, 120
Ethylisopropylamiloride (EIPA), 102, 108
N-ethylmaleimide (NEM), 103
Exed gene, 133–135, 138–139
Exocrine mechanisms, 229
Exocytosis, 234, 312
Exonuclease, 38
Extracellular matrix (ECM), 245, 247, 249, 255–256
Extraembryonic tissues, 133, 138–139, 147–148, 153, 158, 192, 200, 212, 214–216, 218–219

Fatty acid metabolism, 179
Fertilization, 10, 15, 18, 43–44, 46, 51, 57, 59, 63–64, 99, 184, 186, 207, 279, 282, 285–286, 292
  in vitro (IVF), 19, 64–65, 75, 118–119, 184–187, 256, 276–280, 308, 310, 313, 322, 329, 330, 331, 338. *See also* Success rates
Fetal calf serum, 118, 249, 279, 308, 326, 330, 332
Fetus, 58, 145–146, 158–159, 185, 192, 222
Fibroblast, 78, 230, 333
Fibroblast growth factor (FGF), 83, 85, 87, 90, 124, 239
Fibronectin, 245, 247, 255
Fish, 212
Fitness 1 gene, 139
Fluorescent staining, 4, 6–9, 11, 13–15, 24, 100, 193, 235, 253, 304, 307, 312, 314, 318, 319, 326, 327, 329
Fodrin, 204
Follicle. *See also* Cumulus cells; Granulosa cells; Oocyte-cumulus complex
  antrum, 6, 8, 16, 23–24, 43, 50
  aspiration, 276, 278
Follicle stimulating hormone (FSH), 28–29, 46–50, 118, 315, 326
Follicular arrester, 23–25, 27
Follicular fluid, 23, 27, 29–30, 76, 309, 328, 330, 338
Folliculogenesis, 6, 15–16, 18, 48, 250
Forskolin, 25
Fos gene, 87, 89, 336
Freeze-fracture, 24–25
Frog oocyte, 24, 27
Fucose, 255–256
Fucosylation, 251
Furosemide sensitivity, 97, 304

G-proteins, 5
G_2 phase, 4, 16, 301
G418 selection, 158
Galactose, 247
β-galactosidase, 158
Galactosyltransferase, 247
Gap junctions, 18, 23–25, 31–33, 250, 321
Gastrulation, 132

Gene expression, 15, 85, 88-89. *See also* Oncogenes
Gene for
 albino complex, 131-136, 138, 150
 ATPase subunits, 204, 206-207
 brachyury (T), 140
 c-*myb*, 132
 c-*src*, 132
 dilute-short ear-deletion complex, 131, 139
 eed, 132-135, 138-139
 En-2, 132
 exed, 133-135, 138-139
 fitness 1, 139
 fos, 87, 89, 336
 growth factors, 85, 87, 91-92, 118, 122-124, 145
 growth factor receptors, 91-92, 118, 122-123
 H19, 145, 147, 149, 220, 222
 hepatocyte-specific developmental regulator (*hsdr*-1), 133-135
 hox-1.5, 132
 IGF-II, 145, 149, 152
 IGF-II-R, 146-147, 149, 152
 immunoglobulin μ-chain, 132
 insulin, 124
 insulin-like growth factor, 132, 284
 insulin receptor, 85
 int-1, 132
 int-2, 147
 jdf, 134-135
 lacZ reporter, 132, 157, 305
 major histocompatibility complex (MHC), 336, 337
 mesoderm (msd), 132, 134-135
 β_2-microglobulin, 132
 Mod-2, 134-135
 *mst*87F, 39
 myc, 87, 89, 299
 Oct-3, 43
 Pid, 337
 Shaker, 134-135, 138
 snRNA, 118, 120-122
 Superoxide dismutase-2 (Sod-2), 146
 T maternal effect (Tme), 146, 152, 222
 tp, 134-135
 uteroglobin, 229
 Wnt-1, 132
Gene targeting, 124, 132, 147, 157
Gene-trap constructs, 131-132
Genetic diagnosis, 316, 332
Genetic engineering, 132
Genome
 activation, 79, 115, 122, 186, 188-190
 library, 133, 136-137, 139
 map, physical, 146
 parental, 144-154
Gerbil embryo, 117
Germinal vesicle breakdown (GVBD), 16-17, 22, 25, 27, 29-31, 43-44, 48, 308, 309, 321
Germinal vesicle (GV) stage, 7-9, 17, 22-23, 28, 45, 47-50, 120, 122, 307, 312, 317
Germline, 144, 146, 148, 157-159, 163, 213
Gigantism, neonatal, 153
Glucocorticoids, 229
Glucose, 59, 75-76, 79, 173, 178, 188-191, 299, 309, 319, 329, 333, 334, 338
Glucose 6-phosphate, 78
Glutamic acid, 62
Glutamine, 58-59, 61-63, 69, 78, 329
Glutaminolysis, 78-79
Glycan, 170
Glycerol phosphate, 189
Glycine, 57, 61-63, 66-69
Glycine/taurine shuttle, 67-68
Glycocalyx, 254
Glycogen, 74
Glycolipid, 254
Glycolysis, 73, 78-79, 319, 334
Glycoprotein, 234, 236, 253, 255, 277, 323, 327
Glycosylation, 203, 236, 247, 251, 287, 323
Glycosylphosphoinositides, 170
Goat
 embryo, 322
 oocyte, 11, 17
Golgi, 5, 204, 250
Gonadotropins. *See* specific gonadotropin
GPI isozyme analysis, 159-160
Granulocyte colony stimulating growth factor, 122

Granulosa cells, 18, 23, 25, 28, 31-33, 117, 300, 307, 321, 328
Growth enhancement, 148-149, 151
Growth factor receptors, 83-85, 87, 91-92, 118, 122-124, 176, 214, 216.
   See also specific growth factor receptor
Growth factors (GF), 83-84, 88, 92, 98, 118, 122, 170, 175-177, 179, 216, 230, 264, 269, 271, 273, 284, 324.
   See also Epidermal GF; Fibroblast GF; Granulocyte colony stimulating GF; Insulin; Insulin-like GF; Kaposi's sarcoma-type GF; Nerve GF; Platelet-derived GF; Transforming GF
Growth retardation, 145-149, 152-153, 160
Guanine, 38, 80
Guanosine, 27
Guinea pig embryo, 333
Gut, 215
Gwatkin's amino acids, 59
Gynogenote, 144, 148-149, 212

H-type 1 structure, 255
H19 gene, 145, 147, 149, 220, 222
Ham's F-10 medium, 58, 176, 279, 309, 326, 332, 333
Hammersmith Hospital, 185
Hamster
   embryo, 57-67, 69, 75, 90-91, 116-117, 172, 179, 320, 324
   embryo culture medium (HECM), 59-65, 320, 330
   oocyte, 17, 23-32, 324, 326
Hanks' salts, 25
Heart, 73, 148, 185, 213, 215, 219, 317
Hematopoiesis, 159, 235
Hemoglobin, 235
Heparan sulphate proteoglycan (HSPG), 245, 247, 253, 255
Heparin, 247, 332
Hepatocyte-specific developmental regulator gene, 133-135
Heterokaryon, 202
Hexokinase, 319
Histamine, 312

Histidine, 62
Histone, 4-5, 40, 115, 307
Histotroph, 232-234, 238
Hoechst stains, 6-7, 14-15, 61-62, 318, 330, 332, 338
Horse
   embryo, 90, 206
   embryonic capsule, 323
   uterus, 236
Hox-1.5 gene, 132
Human, 91, 230, 252, 278, 302, 312
   carcinoma, 91, 153
   cord serum, 278-283, 285-287, 289-292
   embryo, 78, 89, 116-117, 184-197, 256, 259, 278, 280, 284, 301, 311, 313, 314, 316, 319, 335, 337
   implantation, 244-247, 249, 251, 256, 259
Hyaluronate, 247, 255
Hybridization, in situ, 85, 119-123, 133, 137, 265
Hydatidiform mole, 153
Hydrogen ions. *See* Protons
Hydroxyproline, 59
3β-hydroxysteroid dehydrogenase, 30
Hyperactivity, 153
Hyperphagia, 153
Hyperstimulation, 276, 331
Hypoglycemia, 153
Hypogonadism, 153
Hypomethylation, 139
Hypophysectomy, 268, 306
Hypotaurine, 61, 63-66, 69
Hypoxanthine, 27-28, 300, 311
Hypoxanthine guanine phosphoribosyltransferase, 80
Hysterectomy, 184, 304, 312

IGF-II gene, 145, 149, 152
Immune response, 229, 237, 239, 251, 276-277, 319
Immunoblotting, 255, 301, 307, 308, 325
Immunocytochemistry, 86-87, 91, 202, 215, 254, 265
Immunofluorescence, 85, 122, 205, 251, 307, 336

Immunoglobulin, 251
Immunoglobulin μ-chain gene, 132
Immunoprecipitation, 247, 309
Immunosuppression, 277, 306
Immunosurgery, 150
Implantation, 79, 88, 116, 124, 132, 135, 148, 179, 185, 192, 196–197, 207, 214–215, 218, 230–231, 237, 244–259, 264, 268, 273, 276, 280, 320, 336
  delay, 268–269, 273
  in vitro, 245–259, 335
  window, 245–246, 264, 276
Imprinting
  evolution of, 220–221
  genomic, 144–154, 161–162, 212–217, 219–223
  human, 153, 213
  reciprocal, 148, 152
Infertility treatment, 184, 335
Inheritance, non-Mendelian, 144, 154. See also Maternal imprinting; Paternal imprinting
Inner cell mass (ICM), 84–87, 92, 120–121, 150–151, 157–161, 163, 176, 184, 190–197, 200, 212–213, 215, 217–218, 244, 248, 266, 270, 309, 311, 314, 315, 334
Inositol, 170, 172–176, 179
Inositol phosphates, 170, 172, 174–175
Insemination, 184, 187, 191–192, 278
Insulin, 83–88, 90–91, 170, 216–217, 219–222, 284–289, 291–292, 299, 306, 309, 330, 337
  gene, 124
  receptor, 86–87, 122, 124, 216–217, 219, 221–222, 299
Insulin-like growth factor receptors, 118, 122–124, 146, 149, 152, 284, 287
Insulin-like growth factor binding protein, 325
Insulin-like growth factors (IGF), 83–87, 90, 122–124, 132, 146–147, 149, 152–153, 176, 213, 230, 238, 273, 306, 315, 325, 328
Int-1 gene, 132

Int-2 gene, 147
Integrin-like molecules, 251
Integrins, 247, 255
Interferons, 277, 327, 336
Interleukins, 122, 277, 319
Intermediate filaments (IF), 5
Ion exchangers, 76–77, 97–98, 102–106, 108
Ionic gradient, 200–203, 207
Iron, 178, 233–235, 238, 287
Isobutyl methylxanthine, 50, 311
Isocitrate, 179
Isoleucine, 58–59, 61–63
Isoparental embryo. See Disomy

Jdf gene, 134–135

K_2, 91
Kaposi's sarcoma-type growth factor, 122–123
K_I, 105
Kidney, 213
Kinetics, 65, 105
Kinetochore, 5, 10–11
K_m, 105
Kunitz inhibitor, 236–237

α-lactalbumin, 247
Lactate, 58–59, 61, 74–75, 78, 188, 333, 334, 338
Lactate dehydrogenase, 75, 80, 317
Lacto-N-fucopentaose 1, 251–255
Lactoferrin, 230
β-lactoglobulin-like protein, 230
LacZ reporter gene, 132, 157, 305
Lagomorpha, 90
Lamina, nuclear, 8–10
Laminin, 215, 245, 247–248, 255
Lamins, 4–5, 7–9
Lectin-carbohydrate interaction, 255
Lectin-like cell adhesion molecules (LECAM), 251, 255
Lectin-like receptors, 251
Leptomeninges, 213
Lethality. See Death
Leucine, 62–63, 77, 80
Leukemia inhibitory factor, 157

Lewis blood group antigens, 251
Liposomes, 258
Lithium, 175, 299
Liver, 159, 213
Lung, 159, 229
Luteinizing hormone (LH), 18, 24–25, 27–28, 30–32, 43, 186, 318
Lymphocytes, 239, 251, 255
Lysine, 59, 61–62
Lysosomes, 152, 234, 236
Lysozyme, 236, 239

M-phase, 4, 6, 8, 10, 16, 301, 307
M-phase kinase, 4–5, 10, 16, 19. *See also* Mitosis promoting factor
Macroglossia, 153
Magnesium ion, 178, 314
Major histocompatibility complex (MHC), 336, 337
Malate dehydrogenase, 80
Malic enzyme, 134–135
Manganese ion, 178
Mannose-6-phosphate, 152, 236
Mannose-6-phosphate receptor, 213, 217, 220–221
Marmoset blastocyst, 279
Marsupials, 221
Mast cells, 237, 312
Maternal
  control, 115–118, 120, 122
  disomy, 145–147, 149, 152–153
  imprinting, 145–147, 152–154, 213, 217, 219–220, 222
  inheritance, 3, 18, 217
  message, 38–39, 43–44, 79, 84, 115, 124
Matrigel, 249
Maturation promoting factor, 40
Meiosis. *See also* specific stages
  arrest, 22–24, 26–28, 30, 32–33, 43–44, 300, 302, 321, 332
  commitment, 26, 29–32
  competence, 12, 16, 18, 22, 43–44, 49–50, 308
  maturation, 3, 22, 120, 276, 300, 309, 322, 326, 328, 332
  nondisjunction, 145, 150
Menopausal gonadotropin, human (hMG), 277, 313, 331

Menstrual cycle, 312
Mental retardation, 153
Mesoderm, 132, 135, 147–149, 152, 213, 215–216, 218–219, 324
Metabolic coupling assays, 24–25
Metabolism, 73–79, 118, 124, 188, 190, 317, 319, 329, 333
Metaphase, 4–5, 8–13, 15–17, 22, 43, 50, 312
Metastasis, 255
Methionine, 58, 62, 85, 235, 318
Methotrexate, 325
Methylation, 154
Methylbenzimidazole carbamate, 324
Michaelis-Menten kinetics, 105
Microfilaments (SF), 5, 202
Microglobulin (B) gene, 132
Micrographs, 119, 192, 231, 250, 252, 254, 256–258, 265–266, 270
Microinjection, 150
Microtubule, 4–5, 10–12, 16, 19, 321, 324, 326
Microtubule associated protein (MAP), 10, 12
Microtubule organizing center, (MTOC), 10–11
Microvilli, 4–6, 8, 231, 249–250, 254, 257
Milk, 252
Minimum essential medium (MEM), 278–283, 285–290, 308, 310, 337
Mitochondria, 5, 73, 75, 135, 202
Mitogen, 91, 239, 288
Mitosis, 4–5, 301
Mitosis promoting factor, 4, 8, 15–17. *See also* M-phase kinase
Mitotic index, 193, 195
Mod-2 gene, 134–135
Monoclonal antibodies, 251, 255, 327
Monocytes, 251
Monolayer culture, 247–250, 252–254, 256, 313
Monosomies, 186
Mortality. *See* Death
Morula, 51, 58–59, 61–66, 77, 84–87, 117, 169, 171–174, 178, 185–189, 200, 202, 205, 214, 266–267

Mouse, 40, 91, 135, 213–214, 217, 249–250, 265, 273, 335
  embryo, 44–51, 57–58, 66, 73–76, 78–80, 84–87, 90–91, 97, 99–105, 115–122, 124, 131–132, 157, 169, 171, 173–175, 178–179, 185–189, 192–195, 202, 204–205, 220, 265–267, 270, 280, 299, 300, 301, 305, 307, 311, 313, 314, 315, 316, 318, 324, 327, 328, 336
  genetics, 89, 131–140, 144–152, 221, 229, 305
  implantation, 237, 244, 247–253, 256–257, 268, 273
  oocyte, 6–16, 24, 26–28, 32, 39–40, 44–51, 66, 85–86, 118, 120, 131, 200, 204, 212, 217, 307, 312, 317, 321, 329, 333, 336
  parthenogenetic, 212–221
  pseudopregnant, 150, 159, 316
  Quackenbush, 280
  transgenic, 132, 229, 305
Mouse colony stimulating factor 1, 230
Msd gene, 132, 134–135
Mucin, 170–171, 323
Mucopolysaccharides, 190
Mucoproteins, 190
Muscle, skeletal, 147–149, 152, 213
Mutagenesis, 131–132, 138, 147, 157
Mutations, 39, 131, 139. *See also* Mouse genetics
Myb gene, 132
Myc proto-oncogene, 87, 89, 299
Myo-inositol, 170, 173–175
Myometrium, 237, 312
Myosin light chain kinase, 4

Nerve growth factor, 85, 122–124
Neural tube, 212
Neuroectoderm, 148, 152
Neutrophils, 251, 255
Nidogen, 247
Nitrogen, 57, 78
Nondisjunction, 145, 150
Northern blot, 89, 140, 237, 265, 306, 308, 336
Nucleolin, 5
Nucleolus, 5–8, 117, 303, 326, 334

Nucleus, 3–5, 8, 22, 120
NuSerum, 249

Obesity, 153
Oct-3 gene, 43
Okadaic acid, 301
Oligonucleotide primers, 123
Oligosaccharides, 247, 251–254, 258–259, 323
Oncogenes, 83, 87–89, 92, 299, 336
Oocyte, 9–10, 18–19, 49–51. *See also* Cumulus cells; Germinal vesicle stage; Meiosis
  competence of, 4, 6–7, 14–17, 22, 43–44, 307, 315
  developmental potential, 18–19
  growth factors in, 85–87
  size, 45, 74, 79
  translational control by, 38–41
Oocyte-cumulus complex, 15, 24–26, 29–33, 46, 48–50, 118, 300, 310, 321, 322, 326, 328, 332
Oogenesis, 3, 6–7, 12, 15–18, 115
  polyadenylation during, 38–41
Oolemma, 23, 322
Ooplasm, 3, 120, 122, 322
Osmosis, 66, 201–203
Osteosarcoma, 213
Ouabain, 201, 205, 323
Ovalbulin, 236
Ovarian cysts, 331
Ovariectomy, 251, 268–269, 273, 304
Ovary, 79, 118, 276–278, 322, 328, 329
Oviduct
  cell co-culture, 117–119, 326, 327, 329, 331, 333, 337
  nutrients in, 59, 63–68, 73–74, 76, 108, 172, 176, 273, 306
  passage through, 79, 190
  transfer into, 214
Ovulation, 15–16, 18–19, 250, 323, 324
Oxoglutarate dehydrogenase, 79
Oxygen, 73, 76–77, 318

Pancreas, 213, 236
Paracrine mechanisms, 83, 176, 264, 273, 306, 324
Paraformaldehyde, 119

Parental imprinting, 144–154
Parthenogenesis, 144, 148, 151, 154, 302, 321
Parthenogenote, 212–223
Paternal
  disomy, 145–147, 149–151, 153, 212, 314
  imprinting, 145–148, 152–154, 216, 220–222
Ped gene, 337
Perchloric acid, 174
pH regulation, 66–69, 76, 97–108, 235
Phenylalanine, 58–59, 61–63
Phosphatases, 16, 19, 172, 234–235, 238, 301, 307
Phosphatidylinositol, 170
6-phosphofructokinase, 79
Phosphoinositidases, 170
Phosphoinositides, 172, 174–175
Phospholipase, 230
Phospholipids, 78, 170, 189
Phosphorylation, 73, 85, 98, 170, 267–268, 272, 310
  meiotic, 10, 16, 27, 40, 307
  mitotic, 4–5, 16, 301, 307
Pid gene, 133–135, 138
Pig, 231–235, 237, 306, 308, 319, 328
  embryo, 89–91, 116, 170–171, 206–207, 232–238, 308, 313, 334
  follicular fluid, 328, 338
  Meishan breed, 338
  oocyte, 11, 24–29, 310, 328, 338
Pinocytosis, 75
Placenta, 91, 192, 220, 230–233, 244, 247, 287
  early, 277, 284, 289–290
Plasma membrane, 192, 200–201, 204, 327
Plasma membrane transport, 66, 97–98, 108
Plasmin/plasminogen, 236, 247
Plasminogen activator, 236
Platelet activating factor, 277, 280–284, 292
Platelet-derived growth factor (PDGF), 83, 85, 87, 90, 122–124, 146, 149, 152, 176, 291
Platelet-derived growth factor receptor, 291, 324

Polar body, 3, 9–10, 51, 307, 309, 330
Polarity
  cell, 190–192, 202, 248–250, 314
  embryo, 3, 133, 200, 202, 204, 207, 248
  implantation, 255–256, 277, 335
  membrane, 200, 204, 322
Poly(A) binding protein (PABP), 40
Polyadenylic acid tail, 38–40
Polymerase chain reaction, 118, 120, 122–123, 216, 336
Polypeptides, 235–237
Polysomes, 39–40, 147, 326
Polyvinylalcohol (PVA), 59, 177
Position effect, 139
Posttranscription, 204
Posttranslation, 10–11, 15, 17, 19, 118, 204, 310
Potassium ion, 100, 107–108, 203
Prader-Willi syndrome (PWS), 153
Pregnancy
  biochemical, 185, 197
  ectopic, 185, 230, 245, 325
  establishment of, 264, 335
  rates in IVF, 185, 196, 276
  uterus in, 79, 88, 91, 229, 233–234, 276
Pregnant mare serum gonadotropin (PMSG), 12, 15–17, 29, 45–50, 250, 308, 311, 318, 326
Prehybridization, 119
Primate blastocysts, 90
Progesterone, 24, 27–33, 229–231, 233–234, 236–238, 245–246, 251, 264, 268–269, 273, 276, 302, 308, 318
Proline, 59, 62
Prometaphase, 8–9, 16
Promoter binding sites, 204
Pronucleus, 44, 120, 122, 131, 148, 184–185, 278, 301, 302, 338
Prophase, 4–5, 22
Prostaglandins, 233, 277
Protease, 247
Protease inhibitor, 230, 236–238
Protein
  content (of embryos), 169–171, 173, 179, 206–207
  kinase, 170, 267–269, 288, 301, 307, 312

Protein (cont.)
  loss, 74, 76–78
  secretion, 229, 233–234, 236, 238–239, 303, 318, 325
  synthesis. See Translation
Proteolysis, 236–237
Proton, 66, 97–99, 103–104, 106–108
  permeability, 107–108
  transport, 66–67
Pseudopregnancy, 150, 159, 234–235, 237, 302, 316
Purines, 22, 27–28, 32, 78
Puromycin, 302
Pyrimidines, 78
Pyruvate, 58–59, 61, 74–76, 78–79, 118, 178, 188–191, 308, 319, 320, 333, 334, 335

QO_2, 73, 76

Rabbit, 172, 229, 237, 246, 249
  embryo, 57–58, 66, 74, 76, 78, 89–91, 97, 116–117, 169–179, 246, 249, 306, 318
  oocyte, 7, 17, 23, 308
Radiation, 132
Radiography, 235
Radioimmunoassay, 50
Radioisotopes, 75–76
Rat
  embryo, 78, 116–117, 169, 171, 303, 304, 309
  implantation, 245, 248, 250, 268
  oocyte, 11, 24–27, 32, 44
  uterus, 230
Rate constant, 101–102
Recombination, homologous, 157
Reichert's membrane, 215–216
Relaxin, 318
Reptiles, 74, 212, 221
Restriction enzyme, 120, 136–138
Reticulocytes, 40
Retinoblastoma, 213
Retinol, 237–238
Retinol-binding proteins (RBP), 237–238
Retrovirus infection, 131, 140, 158–159
Reverse transcription, 120, 123, 216, 319, 336

RG 50862, 268–269, 271–272
RG 50864, 268–272
Rhabdomyosarcoma, 152
Rhesus, 7, 27, 116, 194, 330
Rhodamin phalloidin, 6–7
Riboprobe, 119, 136–137, 324
Ribose, 189
Ribose 5-phosphate, 78
Ribosomes, 189
RNA
  injection, 39
  messenger, 38–39, 43, 85, 87, 89, 115–118, 120, 123, 140, 176, 203–207, 216, 237–238, 287, 306, 327, 336
  polymerase II, 115
  ribosomal, 6, 117, 206
  small nuclear, 118, 120–122
  synthesis of, 78, 80, 86, 88, 303, 326
  total, 120, 123, 205–207, 336
  transfer, 38
RNAse protection assay, 147, 306

Sarcoma, 249
Sclerotome, 152
Sea urchin, 99, 120
Second messenger, 83, 170, 175–176
Seizures, 153
Selectin molecules, 251
Selenium, 285–286, 288–289, 330
Sephadex, 177
Serine, 62–63, 230, 236–237
Serpins, 230, 236
Shaker locus, 134–135, 138
Sheep
  embryo, 90–91, 116–117, 206, 306, 325, 334, 336
  oocyte, 27, 44
  uterus, 230
Sickle cell disease, 316
Signal transduction, 5, 18, 83, 85, 219, 287
Skeletal muscle, 147–149, 152, 213
Sodium, 67–68, 97–98, 100–103, 107–108, 173–174, 200–203, 206
  amino acid transport, 201–202
  channel, 201–202, 323
  cotransporters, 102–103, 174, 201–202, 304
  H exchanger, 201–202

Southern blot analysis, 123, 133, 319, 336
Spacer group, 253
Species differences, 10–11, 17, 19, 24–28, 74–75, 77, 79, 89–90, 92, 115–117
Specific-locus method, 131
Spermatocytes, 18, 39–40, 59, 63–64, 278, 327
Spindle
  meiotic, 8–17
  mitotic, 4–5, 10
Spleen, 135, 213, 234, 319
Splice-acceptor site, 132
Src gene, 132
Starfish oocytes, 24
Stem cells. *See* Embryonic stem cells
Steroids. *See* specific steroid
Stilbenes, 103–106
Stroma cells, 244–248, 251, 255, 259, 265, 312, 318
Strontium ion, 178
Success rate (IVF), 184, 196, 276
Superovulation, 61, 63, 119, 318, 322, 332
Superoxide dismutase-2 (Sod-2), 146
SV40 shuttle vector, 140, 158–159
Syncytiotrophoblast, 91, 288

T (brachyury) gene, 140
T maternal effect (Tme) gene, 146, 152, 222
T-complex, 131
T9H translocation, 146
Taurine, 62–63, 66–69
Taxol, 10–11
Tay-Sachs disease, 316
Telophase, 4–5
Testosterone, 28–32, 229
Tetraploid chimeras, 158–159
Threonine, 62, 323
Thymidine, 172–173, 178, 239, 301
Thymocytes, 78
Thyroid hormones, 204
Tight junction, 178, 202, 204, 249
Tissue plasminogen activator (tPA), 39
Tp gene, 134–135
Translocations, 221

Transcription, 6, 43, 51, 84–85, 115–118, 122, 124, 133, 139–140, 146, 153, 186, 188–189, 202, 204, 207, 216–217, 219–220, 303, 305, 307
Transcription factor binding site, 204
Transcripts, status, 38–40
Transcytose, 231
Transfection, 131, 140, 251
Transferrin, 285–289, 291–292, 330, 337
Transferrin receptor, 287
Transforming growth factor (TGF), 83, 85, 87–88, 90–91, 122–124, 264–265, 267, 271–273
Transgenic mice, 132, 229, 305
Transglutaminase, 230
Transition factors, 49, 51
Transition protein 1, 40
Translation, 10, 12, 15–17, 19, 38–41, 76–77, 80, 85–88, 92, 116–117, 202, 204, 229, 306, 310, 321, 327
Translocations, 139, 145–146, 150
Tricarboxylic acid (TCA) cycle, 66, 319, 334
Trisomies, 186
Trophectoderm, 120–121, 150, 176, 184, 190–196, 200–202, 205–207, 212–213, 215–219, 233, 244, 247, 252–253, 255–256, 266, 270, 285, 287–289, 291–292, 304, 305, 309, 311, 314, 315, 323, 334
Trophoblast (TR), 84–87, 92, 148–149, 206, 212–213, 215–216, 218–219, 231, 236–237, 244–248, 253–258, 264, 277, 280, 284, 289–291, 305, 306, 308, 314, 325, 333, 336
Trophoblast-endometrium interaction, 244–248, 252–259
Trypsin inhibitor, 236
Trypsinolysis, 203, 313
Tryptophan, 61–62
Tubulin, 5, 10–15
Tumors. *See* Carcinoma
Tyrode's solution, 58–59, 90, 314
Tyrosination, 10, 13
Tyrosine, 61–63, 268–269, 272
Tyrosine kinase, 85, 92, 324
Tyrphostin, 269, 271–272

Uridine, 25–26, 80, 117, 189, 308, 326
Uridine kinase, 80
Uteroferrin, 233–236, 238
Uteroglobin, 229–230, 302
Uterus
  cell co-culture, 331, 335
  flushing, 234, 323, 330
  lumen of, 75, 79, 244, 255–256, 276
  nutrients in, 73–74, 76, 88, 91, 172, 176
  pregnant, 79, 88, 91, 229, 233–234, 325
  secretions of, 176, 229–239
Uvomorulin, 202, 204, 207

Valine, 61–63
Vimentin, M-phase, 4
Vitamins, 169–170, 172, 175–176, 179, 237–238
V_m, 107–108

WEB 2086, 280–284, 292
Wilms' tumor, 153

Wnt-1, 132
Woolf plot analysis, 90

X-chromosome, 220–221, 316
Xanthine, 28, 300
Xenopus, 39–40, 175, 203, 324, 327

Y-chromosome, 302, 316
Yeast artificial chromosome (YAC), 138, 140
Yolk sac, 215, 219–220

ZO-1 protein, 202
Zona hatching. *See* Blastocyst hatching
Zona pellucida, 18, 79, 170, 172, 192, 200–201, 214, 244, 256, 278–279, 304, 311, 312, 313, 314, 322, 328, 333
Zonular junctions, 190
Zwitterions, 67–68
Zygote, 43, 51, 57, 60–65, 74–75, 115–117, 119–120, 123–124, 131, 146, 150, 169, 171, 173–174, 176
Zygotic message, 84